IMMUNOLOGICAL ASPECTS OF NEOPLASIA – THE ROLE OF THE THYMUS

Cancer Growth and Progression

Volume 17

Series Editor:

Hans E. Kaiser, D.Sc.
Professor,
Department of Pathology,
School of Medicine, University of Maryland, Baltimore, MD, U.S.A.
& Department of Clinical Pathology, University of Vienna, Austria

Associate Editor:

Aejaz Nasir, M.D., M.Phil.
Department of Interdisciplinary Oncology-Pathology,
H. Lee Moffit Cancer Centre & Research Institute,
University of South Florida, Tampa, FL, U.S.A.

Production Assistant:

Yasmin Qayumi, B.S.

Immunological Aspects of Neoplasia – The Role of the Thymus

by

Bela Bodey, M.D., D.Sc.

Professor
Department of Pathology and Laboratory Medicine,
Keck School of Medicine, University of Southern California, Los Angeles,
& Childrens Center for Cancer and Blood Diseases,
Childrens Hospital Los Angeles, Los Angeles, CA, U.S.A.

Stuart E. Siegel, M.D.

Professor
Department of Pediatrics,
Keck School of Medicine, University of Southern California, Los Angeles,
& Childrens Center for Cancer and Blood Diseases,
Childrens Hospital Los Angeles, Los Angeles, CA, U.S.A.

and

Hans E. Kaiser, D.Sc.

Professor
Department of Pathology,
School of Medicine, University of Maryland, Baltimore, MD, US.A.
& Department of Clinical Pathology, University of Vienna, Vienna, Austria

SPRINGER-SCIENCE+BUSINESS MEDIA, B.V.

A C.I.P. Catalogue record for this book is available from the Library of Congress.

ISBN 978-90-481-6608-4 ISBN 978-1-4020-2185-5 (eBook)
DOI 10.1007/978-1-4020-2185-5

Cover image:
Dog thymus (49 days of gestation; 118 mm long). The fully developed prenatal thymic cellular microenvironment. Methacrylate embedding. HE staining. Magnification: 1000x.

Printed on acid-free paper

Contents

SERIES PREFACE

The present multi-volume Book Series, CANCER GROWTH AND PROGRESSION, encompasses the widest possible framework of cutting edge research in the field of neoplastic pathology and other integrated fields. Normal and pathologic growth is one of the most intensively studied yet challenging areas in pathology. Thus the individual volumes in this series focus on the topics of highest scientific interest for basic and clinical researchers, pathologists, medical and surgical oncologists and allied multidisciplinary teams interested in the study of these aspects of neoplastic growth, progression and inhibition. The range of topics covered is extensive, including but not limited to autonomous growth characteristics of malignancy, phenomena of progression of malignant growth involving the various body systems, and recent advances being made in successful neoplastic inhibition and control.

Cell function may be described as producing progression or regression, often found as alternating features in tumors or as variations between normal tissues and tumors. The source of regression in normal melanin producing cells may not be the same as in melanomas. These functions of living matter persist in all phyla of eumetazoans vascular plants as well as in particular species of fungi. However, *homo sapiens* are the eumetazoan species, which interest us the most.

Normal growth processes cannot be entirely understood in all its diversity until we have a thorough knowledge of what constitutes normal growth in various organisms. Complex cellular metabolic pathways are the fundamental elements of growth processes with wide variation in different tissues and organs subjected to a host of carcinogenic influences. The etiology of neoplasms that includes inherent/acquired gene defects, chemical and physical carcinogens, radioactive emissions, viruses, bacteria and parasites are too numerous to catalogue. Therefore, it would be a challenging task to address every aspect of the diverse processes of neoplastic growth and progression. In order to accomplish this goal in the most practical manner, I have invited a highly select team of distinguished authors who are among the most knowledgeable authorities in their fields, to share their expertise in various areas of normal and neoplastic growth, progression, inhibition and control.

A specific purpose of this Book Series is to provide a broad yet comprehensive review of the topics covered that will enrich the reader with the latest and most authentic information at the cutting edge of the field from the world authorities that will be of practical utility to a wide range of professionals in the field of cancer. I have dedicated all my life to the study, teaching and research in cancer. It is my utmost desire to wish you all a great success in your fight against cancer. I hope the second edition of this series will serve as a landmark in our continued efforts to unravel the complexities of the neoplastic phenomena at the one end and to improve the quality of life and minimize the suffering of the patients with cancer on the other!

Hans E. Kaiser. D.Sc.
Series Editor

Contributors

Professor Bela Bodey, M.D., D.Sc.
Department of Pathology and Laboratory Medicine, Keck School of Medicine, University of Southern California, Los Angeles, & Childrens Center for Cancer and Blood Diseases, Childrens Hospital Los Angeles, Los Angeles, CA, USA

Professor Stuart E. Siegel, M.D.
Department of Pediatrics, Keck School of Medicine, University of Southern California, Los Angeles, & Childrens Center for Cancer and Blood Diseases, Childrens Hospital Los Angeles, Los Angeles, CA, USA.

Professor Hans E. Kaiser, D.Sc.
Department of Pathology, School of Medicine, University of Maryland, Baltimore, MD, USA & Department of Clinical Pathology, University of Vienna, Vienna, Austria.

Acknowledgements

We dedicate this monograph in loving memory of Dr. Bodey's beloved Mother, Rossitza Derebeeva.

Section 1

OVERVIEW OF MAMMALIAN THYMIC DEVELOPMENT BIOLOGICAL BASIS

Introduction

Until 1960, classical morphological observations were not able to discover an organ-specific function of the thymus, and no scientist was even interested in changing the situation. Indeed, until that time, clinicians, especially pediatricians, regarded a large thymus as detrimental to health. Children with minor respiratory distress were often observed to have a large thymic mass filling the supracardial mediastinum and wrapped closely around the trachea. We agree with Dr. Webb's and co-authors' opinion (1): "By a remarkable leap in logic, it was concluded that these children were suffering from chronic strangulation from an abnormally large thymus. The treatment offered for this condition of 'status thymolymphaticus' was that of local irradiation of the thymus! Only in later years was it appreciated that the thymus is naturally a very large organ in childhood."

It is well established today that at least four major directions of thymic function exist:

1. the thymus is a member of endocrine regulation, with production of own hormones (thymosins, thymulin, and thymopoietin) in addition to other *in situ* acting active substances (lymphokines, cytokines);
2. an important member of the neuroendocrine-immune regulation axis, with cell surface receptors for oxytocin and vasopressin, allowing for neuroendocrine immunomodulation;
3. an old, but true, view that the thymus is simply a "factory for lymphocytes" and
4. starting from the early intrauteral life, the thymus is a permanently involuting and reconstructing cellular and humoral microenvironment.

Descriptions of the lymphopoietic function can be traced back to the works of Beard (2, 3) and even earlier, as early as the times of Henle (4) and Koelliker (5). Beard suggested, after systematic morphological study on the lymphatic tissue of a stingray:

> "... Just as the Anglo-Saxon stock has made its way from its original home into all parts of the world, and has there set up colonies for itself and for its increase, so the original leukocytes, starting from their birthplace and home in the thymus, have penetrated into almost every part of the body, and have there created new centers for growth, for increase, and for useful work for themselves and for the body." (2)

The lack of interest in the thymus over the following 50 years can be attributed to the lowly scientific reputation of lymphocytes. Lymphocytes were pictured, as cells without a known function, except to act in the thymus as *in situ* 'trephocytes', eaten by cells that are more important. The recovery of the interest to the "small lymphocyte" during the 1960s was driven by three significant studies:

1. Gowans' (6) critical discovery that small lymphocytes are capable of transforming into lymphoblastic cells in response to histocompatibility antigens;
2. Miller (7) and Good and co-workers (8, 9), at about the same time, reported their classic discoveries on the effects of neonatal (newborn) thymectomy.

Unfortunately, the last two simply led to the now firmly established idea that the main and perhaps sole function of the mammalian thymus is the production of immunocompetent T lymphocytes, which control cellular immune responses. In the last two decades, experimental thymic research has increasingly focused on the identification of the potent immunomodulatory effects of several hormones and neuropeptides. Consequently, numerous research groups have observed the actions of steroid hormones, and their effects on the thymus have been reviewed (10-21).

Our purpose in writing this monograph is to share our thirty years of experience and knowledge.

This book is intended to share our research material and methods with thymologists, research investigators, medical doctors and medical students, discussing thymic embryology, morphology, histology, histochemistry, FACS analyses, immunocytochemistry and *in vitro* tissue culture investigations.

REFERENCES

1. Webb S, Morris C, Sprent J: Extrathymic tolerance of mature T cells: clonal elimination as a consequence of immunity. Cell *63:* 1249-1256, 1990.
2. Beard J: The source of leukocytes and the true function of the thymus. Anat Anz *18:* 550-573, 1900.
3. Beard J: The origin and histogenesis of the thymus in *Raja batis*. Zool Jahrbücher, Abt Ontogenie Tiere *17:* 403-480, 1902.
4. Henle J: Ueber Hassall's konzentrische Körperchen des Blutes. Z Radiation Med *7*, 1849.
5. Kölliker A von: Entwicklungsgeschichte des *Menschen* und der höheren *Tiere*. 2nd Aufl, Leipzig, W Engleman Publ, 1879.
6. Gowans JL: Life-span, recirculation and transformation of lymphocytes. Int Rev Exp Pathol *5:* 1-24, 1966.
7. Miller JFAP: Effect of thymectomy in adult mice on immunological responsiveness. Nature *208:* 1337-1338, 1965.
8. Stutman O, Yunis EJ, Martinez C, Good RA: Reversal of post-thymectomy wasting disease in mice by multiple thymus grafts. J. Immunol *98:* 79-87, 1967.
9. Cooper MD, Peterson RD, Good RA: Development and physiology of the lymphoid system. Rheumatology, *3:* 61-102, 1970.
10. Sunder-Plassmann P: Problem of thymus and Basedow's disease: further studies on biology, pathology and clinical aspects of neurohormonal cell system. Deutsche Zschr Chir *255:* 523-615, 1942.
11. Constant GA, Porter EL, Andronis A, Rider JA: Effect of thymic extracts on neuromuscular response. Texas Rep Biol Med *7:* 350-354, 1949.
12. Selye H, Marion D: Direct thymolytic effect of hexestrol (estrogen). Acta Anat *23:* 180-184, 1955.
13. Savino W, Cirne-Lima EO, Soares JF, Leite-de-Moraes M do C, Ono IP, Dardenne M: Hydrocortisone increases the numbers of KL1+ cells, a discrete thymic epithelial cell subset characterized by high molecular weight cytokeratin expression. Endocrinology *123:* 2557-2564, 1988.
14. Goya RG, Sosa YE, Quigley KL, Reichhart R, Meites J: Homeostatic thymus hormone stimulates corticosterone secretion in a dose- and age-dependent manner in *rats*. Neuroendocrinology *51:* 59-63, 1990.
15. Timsit J, Savino W, Safieh B, Chanson P, Gagnerault MC, Bach J-F, Dardenne M: Growth hormone and insuline-like growth factor-I stimulate hormonal function and proliferation of thymic epithelial cells. J Clin Endocrinol Metab *75:* 183-188, 1992.
16. Lannes-Vieira J, Dardenne M, Savino W: Extracellular matrix components of the *mouse* thymus microenvironment: ontogenetic studies and modulation by glucocorticoid hormones. J Histochem Cytochem *39:* 1539-1546, 1991.
17. Villa-Verde DM, Defresne MP, Vannier-dos-Santos MA, Dussault JH, Boniver J, Savino W: Identification of nuclear triiodothyronine receptors in the thymic epithelium. Endocrinology *131:* 1313-1320, 1992.
18. Wick G, Hu Y, Schwarz S, Kroemer G: Immunoendocrine communication via the hypothalamo-pituitary-adrenal axis in autoimmune diseases. Endocr Rev *14:* 539-563, 1993.
19. Anselmino KJ: Thymotropic hormone of anterior pituitary and thymus hormone. Klin Wchnschr *21:* 611-612, 1942.
20. Sacks J: Thytuitary (thymus-posterior pituitary preparation). JAMA *122:* 1083-1084, 1943.
21. Geenen V, Legros JJ, Franchimont P, Baudrihaye M, Defresne MP, Boniver J: The neuroendocrine thymus: coexistence of oxytocin and neurophysin in the *human* thymus. Science *232:* 508-511, 1986

Chapter 1

Embryology of the Mammalian Thymus

1. INTRODUCTION

The unique macroscopical appearance of the thymus gland, already noticed in the days of Herofilos and Erasistratos, was first described by Galen (lived 130 to 200 AD), and could be characterized as a complex cellular and humoral microenvironment (1-9). Since the 18th century, thymic research has emerged from descriptive studies concerning normal embryology (i.e. the pre- and postnatal histogenesis of the thymic cellular microenvironment) in a variety of vertebrates by a great number of famous morphologists (10-143).

Developmental variations in the thymus have attracted the attention of authors during the first one hundred years of thymology, if one can judge by the great number of embryological papers. The rapid ontogenetic cell and tissue changes required complex and extremely well coordinated histogenetic events, as well as interactions between stem cells of various origins (cells of the embryonal tissues). Developmental arrest may occur at any stage and at times half or the entire organ can be absent.

The primary thymic anlagen, containing only pure, omnipotent, endodermal epithelial cells, appear in the beginning of the sixth embryonal week or 10-somite embryo of the prenatal ontogenesis (11, 14, 16, 20, 26, 27). The third and fourth pharyngeal pouches in the mammalian and human embryos are characterized at their distal extremity by the dorsal and ventral wings (36, 40, 42, 45, 46, 53, 56, 57, 61, 66, 67, 77, 84, 100, 105, 131). The thymic anlagen develop mainly from the ventral wing (out-pocketing) of the endodermal part of third pharyngeal pouch on each side. The dorsal wings of the same pouch give rise to the inferior parathyroid glands. By that time there is a contact between the pharyngeal endoderm and the ectoderm just posterior to the first aortic arch. In the proliferation of the pharyngeal endoderm, just as the median thyroid component, the cells probably divide at a more rapid rate than those which do not take part in the process (in a 4 mm long human embryo, the thymic primordium is increasing in size, and the third aortic arch have already reached its full development). The formation of a network of vessels around the rapidly growing thymic anlagen is of major ontogenetic importance. The original thymic anlagen does not contain blood vessels, therefore the primitive feeding is *via* pure diffusion of oxygen, making possible the needed rapid cell proliferation and size expansion.

The development of the epithelial thymus, representing the "ideal" microenvironment for the immigration of pluripotent hematopoietic stem cells depends upon cellular interactions between the thymic epithelium (endodermal in origin) and the ectomesenchyme developed from the neural crest. One of the leading principles of prenatal thymic histogenesis is that cells of the thymic epithelial anlagen do not proliferate in the absence of an inductive ectomesenchymal interaction. Auerbach (141, 142) was able to demonstrate this "inductive" interaction of mesenchyme in thymic tissue culture, resulting in rapid epithelial proliferation. If the mesenchyme added to the thymic cultures originated from lung or submandibular gland, only delayed epithelial proliferation was detected.

The ectomesenchyme that surrounds the primary thymic anlagen and the connective tissue cells that form the primary capsule and septae were demonstrated to originate from the neural crest. The neural crest, through the ectomesenchyme, monitors early ontogenetic thymic events. The origin of the "embryonal thymus mesenchyme" was defined by Le Douarin and Jotereau (143-145) employing a transplantation experiment (quail into chicken), using the recognized "biological marker" of the Japanese quail's nuclear morphology (DNA rich nucleolus; characteristic accumulation of nuclear heterochromatin in large clumps, and small amounts attached to the nuclear membrane as a marker for lymphocytes, and one large central mass of heterochromatin as a marker for reticular and connective tissue cells).

Part of the mesenchymal cells around the primary thymic anlagen and the connective tissue cells of the primitive capsule and the intrathymic septae were demonstrated to originate from the neural crest (141-145). Since this special, embryonal type mesenchyme is derived from ectoderm, it is referred to as ectomesenchyme in the International Histologic Nomenclature (IHN). The neural crest, through formation of the thymic ectomesenchyme, and through production of local, *in situ* growth factor(s) (probably before the secretion of an 11 kD chemotactic factor, thymotaxin), monitors the differentiation of epithelial and connective tissue components of the early thymic microenvironment (146-169). Interestingly, later the *in vivo* route of entry for committed hematopoietic stem cells is via the highly vascularized cortico-medullary junction, using special homing receptors (adhesion molecules) for contact with endothelial cells, and integrin molecules like addressin or ICAM-1 (LFA-1), L-Selectin, as well as s.E- and s.P-Selectin and VCAM-1 (VLA-4) related structures. In the second step, the stem cells migrate to the so-called subcapsular thymic, mostly reticulo-epithelial (RE) cell microenvironment, where the integrins $\alpha4\beta1$ and $\alpha5\beta1$ induce their proliferation. A major portion of "double positive" ($CD4^+CD8^+$ immunophenotype-IP) thymocytes also express $\alpha4\beta1$ integrin in its active form. These integrin molecules are also involved in cellular traffic and positive intrathymic selection of immunocompetent, non self-reactive T lymphocytes. The tissue composition of the initial, large blood vessels of the human thymus, which appear at the embryonic length of 20-30 mm except for the endothelial cell layer, is also of ectomesenchymal (i.e. neural crest) origin.

Experimental or spontaneous *in vivo* early neural crest ablation during ontogenesis always resulted in the abnormal development of the thymus, but also of the heart, great vessels, thyroid, and parathyroid glands (170). This defective thymic onto- and histogenesis can be used in mouse experiments because this interesting immune-neuroendocrine association was successfully developed in a mouse mutation called "staggerer", with defective developments in the cerebellum and thymus.

The thymic anlagen that develop from the fourth pharyngeal pouch from the beginning appear to be quite small and rudimentary. They may give rise only to vestigial tissue masses. The superior parathyroid glands develop from the dorsal wings (sacculation) of the fourth pharyngeal pouches.

By the end of the sixth embryonal week, the connections of the four thymic anlagen with the pharyngeal pouches are severed. The next developmental stage represents significant changes in the external form of anlagen. Each anlage elongates caudally and medially, forming a typical tubular structure (thymopharyngeal tract).

During the seventh week, this tract soon begins to obliterate by rapid proliferation of the epithelial cells. Weller (134) reported a formation of parathymus glands by some differentiated cells of the thymic primordia at 10 mm embryonal length. The lumen of the thymus is still in contact with that of pharynx. The thymic components become detached from the pharynx at an embryonal length of 14.5 mm. The thymus in this age appears intimately adjacent to the ductus cervicalis of either side (43, 52). It is believed that the ductus cervicalis contact exerts no influence upon the development of the thymus gland (134).

The eighth week of ontogenesis is crucial, probably the most important in thymic ontogenesis, not only because of the non-complete fusion of anlagen in the midline, but also because this is the real time for beginning of the thymic journey in a caudal direction (descensus thymi or thymic descent) under the sternum to the supracardiac mediastinum. The fusion of the right and left thymic anlagen is never really complete, so that the organ never loses entirely its paired character. At the end of the thymic journey (to the parietal pericardial tissue), the upper end of the thymus becomes drawn in and generally vanishes.

At embryonal length 16.8 mm, the thymus increases significantly its size, the compact tissue mass is about six times the size found at 14.5 mm embryonal length, the contact with the thyroids is established and is moved below the lower border of the junction of the subclavian and carotid arteries. Most of the caudal portion of the thymus is surrounded by loose mesenchyme, so it will be not difficult for the compact thymic tissue to change its place. At the embryonal length of 23 mm, the thymic gland occupies a position little different from that in newborns. Its caudal part has descended to the upper margin of the arch of the aorta. The subsequent histological and cytological changes that occur following this stage of ontogenesis are described in detail in the chapter about thymic histogenesis.

2. WEEKLY CHRONOLOGY OF IMMUNO-MORPHOLOGIC CHANGES DURING THYMIC ORGANOGENESIS

The following immunobiologic chronology of human immunomorphologic ontogenesis represents an attempt to summarize cellular, microenvironmental, intrathymic changes during the thymic organogenesis. The systematic explanation of the maturation and differentiation events of T lymphocyte subpopulations and non-lymphatic cell types (RE and stromal or accessory) will, I hope, complete an overview of the functional unity of the thymic microenvironment:

4-6 Weeks

The primary thymic epithelial anlagen (two) of endodermal origin are formed from the third pharyngeal pouch. This is a typically cuboidal epithelium in appearance, but in Hammar's observation it is described as prismatic (43, 51) and rapidly proliferating under the induction stimulus of ectomesenchymal tissue. The ectomesenchyma around the anlagen, from neural crest origin, contains primitive blood vessels and TdT-hemopoietic stem cells from the yolk sac or, later, from the fetal liver (171-177). Cells of typical macrophage appearance are detected in the yolk sac and ectomesenchyme during the 4th week of prenatal development. During the 5th week, the fetal liver is in pre-hemopoietic condition. A chemotactic factor, responsible for hemopoietic stem cell immigration, is secreted in the ectomesenchymal (neural crest) cells or in the pure epithelial cells as a response to the induction chemistry of ectomesenchymal tissue (178, 179).

7 Weeks

The left and right thymic anlage has already moved in a caudal direction (descensus thymi or thymic descent), to lie on each side of the cervical region. No thymocytes are present within the thymic epithelium (which already reacts positively with MoABs A2B5 and TE4), but around the thymus in the perithymically located ectomesenchyme and around the embryonal liver, TdT^+ (terminal transferase), BrdU (bromo-2-deoxyuridine incorporated into cells during the S phase) positive, cytoplasmic $CD3^+$ or $CD3^-$, $CD7^+$, $CD34^+$, $CD38^+$, $CD44^+$, $T200^+$, and $Ki67^+$ hemopoietic stem cells, committed to the T lymphocyte cell lineage are present (180). The embryonal ectomesenchyme contains a number of small blood vessels (endothelial organ and non organ receptors) located near the surface of the thymic anlagen. The caudally located epithelial cells have already commenced their transformation into cells later forming a reticulo-epithelial (RE) network. $CD7^+$, triple negative stem cells are present in the fetal yolk sac, fetal liver and thorax. Rare lysozyme positive cells are present in the fetal liver at approximately 7-8 weeks of prenatal development. Precursors of interdigitating cells (ID) are detectable during the 13th week of ontogenesis and share three features with the Langerhans' cells:

1. Birbeck granules;
2. membrane antigen CD1; and
3. S100 cytoplasmic antigen. Early cortical RE cells and ID cells express HLA-DR, -DQ and -DP MHC Class II antigen molecules (180-184).

The triple positive cells give rise, *in vitro*, to CFU-T in the presence of TCM and IL-2 (185, 186). Phenotypic analysis of cells of CFU-T has demonstrated that they reciprocally express CD4 or CD8 molecules and surface CD3 (sCD3) and TCR α, β ($WT31^+$). CD7 and cytoplasmic CD3 (cCD3) expression occurs early, extrathymically, prior to T cell precursor entry into the pure epithelial thymic rudiment.

The early reticulo-epithelial (RE) cell surface antigenic distribution profile is as follows: $TE4^+$, $A_2B_5^+$, Thy-1^+, MHC Class I antigens positive, and $vimentin^+/cytokeratin^+$ (detected with MoAB AE2-ICN). Neuroectodermal, opioid and neuroendocrine receptors are also present. Secretion of multiple active hormone-like substances, among them a humoral chemotactic factor by the cells of ectomesenchymal connective tissue allows immigration of pluripotent, hemopoietic stem cells, which are, however, already committed to lymphocytic cell lineage, which, following functional cell to cell contact with the RE cells, express surface antigens typical of early, immature T lymphocytes. Expression of *c-myc* has been detected all over the chest, in lymph nodes, and spleen.

8-9 Weeks

Pre-T, pluripotent, hemopoietic stem cells have already colonized the primarily pure epithelial cell thymus anlagen. As a result of the intrathymic maturation process, there is the first expression of

CD2, an early stage dependent, highly specific, differentiation antigen (other names: 3A1, p40 molecule, T11, 50kD sheep E-rosette receptor, 9.6, 35.1, LFA-2 adherence molecule) (187). This antigen is also a surface component of the alternative or antigen independent pathway of T cell activation. CD2 is a 50-55 kD, single-chain glycoprotein expressed on the majority of peripheral T cells and thymocytes (188). CD2 contains three epitopes:

1. T11.1, or the sheep erythrocyte-binding epitope;
2. T11.2, expressed on "resting" T cells; and
3. T11.3 alternatively, CD2R is a conformational epitope, which only appears after thymocyte activation.

Membrane expression of TCR α, β-CD3 complex (182, 189, 190). The close cell to cell contact with the RE cells (process of self recognition "teaching") results in expression of MHC class I receptor on thymocytes (181, 191-194).

Induction of the first proliferative wave in immature thymocytes by the specialized and subcapsullarly seated RE cells ($A_2B_5^+$, $TE4^+$, Thy-1^+ subpopulation). They are operated through the LFA-3 (CD58; gp 50) adherence molecules of RE cells, serving as receptors for the *de novo* expressed CD2 antigens.

The immature embryonal thymocytes in all investigated species lack nuclear expression of TdT. The thymic RE cells control TdT expression. The functional changes within the RE cells are so fast and powerful that 1-2 days difference in age represent various inductive power. The presence or absence of TdT in thymocytes roughly corresponds to steroid sensitivity or steroid resistance of thymocyte subsets.

The commencement of the main histogenetic changes is marked by organization of the loose structure of the RE network from the initial, pure prismatic epithelial thymic anlagen. By 8.5 weeks, the right and left thymic anlagen fuse. The $A_2B_5^+$, $TE4^+$ and Thy-1^+ RE cells form the central portion of the thymus. The cell surface antigenic characteristics of the remaining RE cell subpopulation: $TE3^+$, LFA-3^+ and MHC class I antigens positive (restricted!). In the 24 mm long human embryo, the first blood vessels have already migrated from the ectomesenchyma, but mesodermal cellular elements are also present.

10 Weeks

The T lymphocyte antigen CD1 (also termed T6, NA1/34, VIT 6, Leu 6, triad of 43, 46, and 49 kD cell surface antigens) is expressed on cortical thymocytes and peripheral T cells.

CD4 and CD8 first appeared, to define a basic dichotomy, characteristic for the development of T lymphocytes. CD4 (OKT4, Leu3) represents a 62 kD cell surface antigen expressed on helper/inducer peripheral T cells, most thymocytes, and, in low density on monocytes, tissue macrophages, and Langerhans cells. The CD4 molecule is physically linked to the CD3/TCR complex, stabilizing Ti antigen interactions by binding to MHC class II molecules. CD8 (OKT8, Leu2) is a cell surface, 76 kD mol weight antigen, expressed on most thymocytes and peripheral blood T cells involved in cytotoxic/suppressor cellular immunological interactions. The CD8 antigen participates in the alloantigen recognition by stabilizing the Ti antigen during the interactions between T cells and MHC class I antigen bearing thymic macrophages or dendritic cells. This differentiation pathway is referred to as the double negative - double positive - single positive (DN-DP-SP) maturation. DP populations are generated through an immature $CD4^-8^+$ "transit" cell population. These membrane associated ligands stabilize MHC restriction, the interactions of TCR with "other self cells" carrying the MHC class I and II antigens.

As a result of the extremely high proliferation rate, the thymic mass markedly expands in all dimensions and its cellular organization changes: numerous microlobules are formed. No medullary area is distinguishable yet. A $TE3^+$ cortical RE zone is formed surrounding the $TE4^+$, $A_2B_5^+$ and Thy-1^+ subcapsular, already endocrine, RE part.

In morphologic terms, the thymic cell types, already immuno-characterized above with their cell surface antigenic profile are classified as:

1. LARGE THYMOBLASTS ($CD7^+$, $CD2^+$, $CD3^-$, $CD4^-$, $CD8^-$, the "triple negative immunophenotype", a great number of them are located subcapsullarly);
2. CORTICAL (immature, non-immunocompetent) THYMOCYTES ($CD7^+$, $CD2^+$, $CD3^+$, $CD4^+$, $CD8^+$);
3. MEDULLAR (mature, immunocompetent) THYMOCYTES a mixture of $CD4^+8^-$ (majority, helper/inducer) and $CD4^-8^+$ (minority, cytotoxic/suppressor) T lymphocytes; and
4. THYMIC NURSE CELLS (TNC) widely accepted function derived name of a specialized form of $A_2B_5^+$, $TE4^+$, and Thy-1^+ subcapsullarly located, large RE cells.

11-12 Weeks

Mature T cell antigens, the receptor associated CD3 molecules, first appear. Mature T lymphocytes are identified by the expression of the following, surface differentiation markers: CD7, CD2, CD3, CD5, CD6, CD4 or CD8 positivity is required and T cell receptor antigens as well as the CD3 segment and MHC class I antigens.

Co-expression of the mature thymocyte expresses antigen CD45 (also named T200, because it has a molecular weight of 200-220 kD), leukocyte common antigen (LCA) is also present on "virgin" or unprimed T cells. This antigen is an enzyme-CD45 phosphotyrosine phosphatase and regulates the biologic action of p59 (fyn) protein tyrosine kinase (195).

The mature T lymphocytes (TCR α, β and WT31$^+$ or TCR δ, γ) in the human thymus do not remain in the proliferative cycle. It has recently been reported that only 60-70% of DP cortical thymocytes is also Ki67$^+$ and 20-30% demonstrates BrdU positivity. This suggests that some thymocytes in this maturation stage leave the cell cycle to become a "resting" population.

Finally, recent studies have detected that only the TCR α, β cell lineage is subjected to MHC guided "education," while the TCR γ, δ lineage does not appear to be MHC restricted. Such "uneducated" cells were detected within the youngest (investigated) embryonal avian thymus, but their number remains rare (<1%) and does not change during the stages of fetal and postnatal development.

By the 12th week of prenatal development, the TE3$^+$ RE cells have proliferated and assumed their position in the inner cortical zone.

13-16 Weeks

The morphologically well-known thymic tissue is already formed. At the end of this stage, the first Hassall's bodies (HBs) appear. The HBs have very special antigen expression, unique only for their antigen characteristics: the outer layer of the bodies demonstrate the presence of antigens typical for the cells of the RE network: TE8$^+$, TE16$^+$, and TE19$^+$ (antigen present only in HB). The bodies are organized in a concentric manner in which the inner cellular layer is also present. These cells are different from the cells forming the outer layer, and demonstrate expression of TE15$^+$ and TE19$^+$ antigens. These four mouse-anti human MoABs were raised against thymic epithelial cells. They reacted positively not only with RE cells and the HB of human thymic medulla, but also stained cells of the terminally differentiated layers of the epidermis and dermis.

The medullary RE cells also contain high and low molecular weight cytokeratins and intermediate filaments (IF) also typical for the HBs.

Employing a library of mouse anti-human thymic RE cells specific MoABs [developed by Haynes and co-workers (185, 186, 196, 197) and from the laboratory of van Vliet (198, 199), the RE cellular network was subdivided by cell surface IP into three zones. Migrating to the medulla, the thymocyte cell surface differentiation antigenic profile (CD antigens) has assumed the mature pattern seen in peripheral, immunocompetent T lymphocytes.

From 16 Weeks to Birth

Recent knowledge of thymus histogenesis indicates that this organ is fully developed during this late prenatal period. The cytogenetical changes are still significant, such as continuous, extremely high rate of cell proliferation, resulting in expansion in size of both the cortex and medulla, and formation of immunophenotypically very similar (TE4$^+$, Thy-1$^+$ and $A_2B_5^+$) subcapsular cortical and medullary RE cell zones (representing the endocrine portion of the thymus). These cells produce the so-called thymic hormones such as thymosins, thymulin (FST), and thymopoietin, whereas the TE3$^+$ RE cells (inner cortical zone) do not secrete them. The thymic tissue, after a wave of rapid expansion of both cortex and medulla, undergoes a *de novo* lobulation. After the 16th week, the cell surface IP changes are only minimal in the observed thymic cell populations (thymocytes, RE cells and thymic stromal dendritic elements).

At birth, the thymus is large relative to total body weight (average weight, 13 g). Its absolute weight increases in the first two years after birth and at approximately 11-12 years of age reaches its maximum weight of 30-40 g. After this age the thymic weight begins to decrease to an adult weight of approximately 15 g or less.

REFERENCES

1. Andreasen E: Function of the thymus. I. Critical remarks on the theory and experiments of Bomskov. Acta Anat *2:* 275-286, 1947.
2. Miller JFAP: Immunological function of the thymus. Lancet *II:* 748-749, 1961.
3. Trainin N: Thymic hormones and the immune response. Physiol Rev *54:* 272-315, 1974.
4. Rask-Nielsen R, Andreasen E: Growth of normal thymic tissue. Acta Pathol Microbiol Scand *36:* 1-9, 1955.
5. Kendall MD: *Avian* thymus glands: a review. Dev Comp Immunol *4:* 191-209, 1980.
6. Ritter MA, Rozing J, Schuurman HJ: The true function of the thymus? Immunol Today *9:* 189-193, 1988.
7. Kendall MD: Functional anatomy of the thymic microenvironment. J Anat *177:* 1-29, 1991.
8. Izard J: Introduction to the thymus. Microsc Res Tech *38:* 207-208, 1997.
9. Schuurman HJ, Kuper CF, Kendall MD: Thymic microenvironment at the light microscopic level. Microsc Res Tech *38:* 216-226, 1997.
10. His W: Zur Anatomie der *menschlichen* Thymusdrüse. Z Zool *11:* 164, 1862.
11. His W: Anatomie *menschlicher* Embryonen. Leipzig, 1880.
12. Kölliker A von: Entwicklungsgeschichte des *Menschen* und der höheren *Tiere*. Leipzig, W Engleman Publ, 1861.
13. Paulitzky A: Disquisitiones de stratis glandulae thymi corpusculis. Halis, 1863.
14. Toldt: Über lymphoide Organe der Amphibien. Sitzungsber Akad Wissensch *58 (2)*, 1868.
15. Krause C: Handbuch des *Menschen*. Hannover, 1876, p 359.
16. Kölliker A von: Entwicklungsgeschichte des *Menschen* und der höheren *Tiere*. 2nd Aufl, Leipzig, W Engleman Publ, 1879.
17. Watney H: Note on the minute anatomy of the thymus. Proc Roy Soc *27*, 1878.
18. Watney H: The minute anatomy of the thymus. Phil Trans Roy Soc *3:* 1063-1123, 1882.
19. Stieda L: Untersuchungen ueber die Entwicklung der Glandula thymus, Glandula thyreoidea und Glandula carotica. Leipzig, *4:* 1-38, 1881.
20. Born G: Über die Derivate der embryonalen Schlundbogen und Schlundspalten bei *Säugetieren*. Arch Mikr Anat *22:* 271-318, 1883.
21. Fischels P: Beiträge zur Kenntnis der Entwickelungsgeschichte der Gl. Thyreoidea und Gl. Thymus. Arch Mikr Anat *25:* 405-440, 1885.
22. Schedel J: Zellvermehrung in der Thymusdrüse. Arch Mikr Anat *24:* 352-354, 1885.
23. Kastschenko N: Das Schicksal der embryonalen Schlundspalten bei *Säugetieren*. Arch Mikr Anat *30:* 1-26, 1887.
24. Pierson G: Ueber die Entwicklung der embryonalen Schlundspalten und ihre Derivate bei *Säugetieren*. Zeitschr Zool *47:* 155-189, 1888.
25. Maurer F: Schilddrüse und Thymus der *Teleostier*. Morph Jahrb *11:* 129-175, 1886.
26. Maurer F: Schilddrüse, Thymus und Kiemenreste der *Amphibien*. Morph Jahrb *13:* 296-382, 1888.
27. Maurer F: Die Schilddrüse, Thymus und andere Schlundspaltenderivate bei der *Eidechse*. Morph Jahrb *27:* 119-172, 1899.
28. Capobianco F: Della natura dei corpuscoli di Hassall, contribuzione alle conoscenze morphologiche del timo. Boll Soc Natur Napoli *1(4)*, 1890.
29. Antipa G: Ueber die Beziehungen der Thymus zu den sog. Kiemenspaltenorganen bei *Selachiern*. Anat Anz *7:* 690-692, 1892.
30. Prenant A: Contribution a l'étude du développement organique et histologique du thymus, de la glande thyroide et de la glande carotidienne. La Cellule *10:* 85-184, 1894.
31. Bemmelen van JF: Ueber vermuthliche rudimentäre Kiemenspalten bei *Elasmobranchiern*. Mitt Zool Stat Neapel *6:* 165-184, 1895.
32. Beard J: The development and probable function of the thymus. Anat Anz *9:* 476-486, 1894.
33. Beard J: The source of leukocytes and the true function of the thymus. Anat Anz *18:* 550-573, 1900.
34. Beard J: The origin and histogenesis of the thymus in *Raja batis*. Zool Jahrbücher, Abt Ontogenie Tiere *17:* 403-480, 1902.
35. Tourneux F, Verdun P: Sur les premiers developments de la thyroide, du thymus et des glandules parathyroidennes chez *l'homme*. J de l'Anat *33:* 305-325, 1897.
36. Ebner V von: Der Thymus. *In:* Handbuch der Gewebelehre des *Menschen*. (Koelliker A, hg) Leipzig, 1899, p 328.
37. Groschuff K: Uber das Vorkommen eines Thymussegmentes der vierten Kiementasche beim *Menschen*. Anat Anz *17:* 161-170, 1900.
38. Prymak T: Beiträge zur Kenntniss des feinern Baues und der Involution der Thymusdrüse bei den *Teleostieren*. Anat Anz *21:* 164-177, 1902.
39. Wallish M: Zur Bedeutung der Hassallschen Körperchen. Arch Mikr Anat *63:* 274-282, 1903.
40. Zuckerkandl E: Die Entwicklung der Schilddrüse und der Thymus bei der *Ratte*. Anat Hefte *21:* 1-28, 1903.
41. Goodall A: The postnatal changes in the thymus of *guinea-pigs* and the effect of castration on thymus structure. J Physiol *32:* 191-198, 1905.
42. Lewis T: The *avian* thymus. J Physiol *22:* 40, 1905.
43. Hammar JA: Zur Histogenese und Involution der Thymusdrüse. Anat Anz *27:* 23-30 & 41-89, 1905.
44. Hammar JA: Zur Kenntnis der *Teleostier*thymus. Arch Mikr Anat *73:* 1-68, 1908.
45. Hammar JA: Zur gröberen Morphologie und Morphogenie der *Menschen*thymus. Anat Hefte *43:* 203-242, 1911.
46. Hammar JA: Fünfzig Jahre Thymusforschung. Ergebn Anat Entwicklunggesch *19:* 1-274, 1909.
47. Hammar JA: Der gegenwärtige Stand der Morphologie und Physiologie der Thymusdrüse. Wien Med Wschr *2:* 2746-2750, 2795-2801, & 2910-2916, 1909.
48. Hammar JA: Zur Kenntnis der *Elasmobranchier*thymus. Zool Jahrb Anat Ontog *32:* 135-180, 1911.
49. Hammar JA: Methode, die Menge der Rinde und des Marks der Thymus, sowie die Anzahl und die Größe der Hassallschen Körper zahlenmäßig festzustellen. Z Angew Anat *1:* 311-396, 1914.
50. Hammar JA: The new views as to the morphology of the thymus gland and their bearing on the problem of the function of the thymus. Endocrinology *5:* 543-573 & 731-760, 1921.
51. Hammar JA: Zur Frage der Histogenese der Thymusdrüse. Ztrblt Allg Pathol *33:* 505-513, 1923.
52. Hammar JA: Die *Menschen*thymus in Gesundheit und Krankheit: Ergebnisse der numerischen Analyse von mehr als tausend *menschlichen* Thymusdrüsen. Teil I. Das normale Organ - zugleich eine kritische Beleuchtung der Lehre des "Status thymicus". Z eitschr mikrosk-anat Forsch Erg.-Bd *6:* 1-570, 1926.
53. Hammar JA: Thymus. Encycl Med Technik *3:* 2161-2168, 1927.
54. Hammar JA: Die *Menschen*thymus in Gesundheit und Krankheit. Teil II. Das Organ unter anormalen Körperverhältnissen. Zugleich Grundlagen der Theorie der

Thymusfunktion. Zeitschr mikrosk-anat Forsch Erg.-Bd *16*, 1929.
55. Maximow AA: Untersuchungen über Blut und Bindegewebe. II. Über die Histogenese der Thymus bei *Saugetieren*. Arch mikrosk Anat *74:* 525-621, 1909.
56. Maximow AA: Untersuchungen über Blut und Bindegewebe. IV. Über die Histogenese der Thymus bei *Amphibien*. Arch mikrosk Anat *79:* 560-611, 1912.
57. Maximow AA: Untersuchungen über Blut und Bindegewebe. V. Über die embryonale Entwicklung der Thymus bei *Selachiern*. Arch mikrosk Anat *80:* 39-88, 1912.
58. Maximow AA: Development of Non-granular leucocytes (Lymphocytes and Monocytes) into Polyblasts (Macrophages) and Fibroblasts *in vitro*. Proc Soc Exp Biol *24:* 570-572, 1927.
59. Bell ET: The development of the thymus. Am J Anat *5:* 29-62, 1905.
60. Dantschakoff V: Untersuchungen von Blut und Bindegewebe bei *Vögeln*. Arch Mikr Anat *73:* 117-181, 1908.
61. Fox H: The pharyngeal pouches and their derivates in the *Mammalia*. Am J Anat *8:* 187-250, 1908.
62. Mietens H: Zur Kenntnis des Thymusretikulum und seiner Beziehungen zu dem der Lymphdrüsen nebst einigen Bemerkungen über Winterschlafdrüse. Jenaische Z *44:* 149-192, 1908.
63. Klose H, Vogt H: Klinik und Biologie der Thymusdrüse. Beitr Klin. Chir *69:* 1-200, 1910.
64. Zotterman A: Die *Schweine*thymus als eine Thymus ecto-entodermalis. Anat Anz *38:* 514-530, 1911.
65. Fiore G: Fisiopatologia del timo. Morgagni *53:* 577-591, 625-640 & 680-686, 1911.
66. Sokolow AJ: Zur Entwicklungsgeschichte der Thymusdruse bei den *Sterletten*. Mem Sci Univ Kazan *79:* 1-33, 1912.
67. Hamilton B: Zur Embryologie der *Vogel*thymus. II. Die Thymusentwicklung bei der *Ente*, neben einigen Beobachtungen über die Kiemenspaltorgane dieses Tieres. Anat Anz *44:* 417-439, 1913.
68. Laguesse E: Structure et évolution du thymus. Echo Méd Nord *17:* 461-469, 1913.
69. Pappenheimer AM: A contribution to the normal and pathological histology of the thymus gland. J Med Res *22:* 1-73, 1910.
70. Pappenheimer AM: Further studies of the histology of the thymus. Am J Anat *14:* 299-326, 1913.
71. Pappenheimer AM: The effects of early extirpations of the thymus in albino *rats*. J Exp Med *19:* 319-338, 1914.
72. Hart C: Thymusstudien. II. Die Thymuselemente. Virchows Arch *207:* 255-277, 1912.
73. Hart C: Thymusstudien. III. Die Pathologie der Thymus. Vichows Arch *214:* 1-83, 1913.
74. Hart C: Thymusstudien IV. Die Hassallschen Körperchen. Virch Arch Pathol Anat *217:* 239-255, 1914.
75. Jolly J, Levin S: Sur les modifications du thymus ŕ la suite du jeûne. C R Soc Biol *71:* 374-377, 1911.
76. Jolly J, Levin S: Evolution des corpuscles de Hassall dans le thymus de l'animal jeűneur. C R Soc Biol *72:* 642-644, 1912.
77. Rabl H: Die Entwicklung der Derivate des Kiemendarmes beim *Meerschweinchen*. Arch mikr Anat *82:* 79-147, 1913.
78. Lerch O: A contribution on the thymus. Med Rec *85:* 648-655, 1914.
79. Marine D: The frequency of duct-like spaces in the thymus gland, with remarks on the formation and fate of Hassall's corpuscles. Cleveland Med J *14:* 186-195, 1915.
80. Dustin A-P: Contribution ŕ l' étude du thymus des *Reptiles*. C R Ass Anat *11:* 66-71, 1909.
81. Dustin A-P: Contribution ŕ l' étude du thymus des *Reptiles*. Cellules épithéloďdes, cellular myoďdes et corps de Hassal. Arch Zool Exp *42:* 43-227, 1909.
82. Dustin A-P: Le thymus de l' *Axolotl*. Arch Biol *26:* 557-616, 1911.
83. Dustin A-P: Influence du jeűne sur le développement du thymus de *Rana fusca*. Bull Soc Sci Med Natur *70:* 174-177, 1912.
84. Dustin A-P: Recherches d' histologie normale et expérimentale sur le thymus des *Amphibiens anoures*. P.1. Structure normale, variations saisonničres, variations expérimentales. Arch Biol *28:* 1-110, 1913.
85. Dustin A-P: Quelques mots ŕ propos du thymus *humain*. Bull Soc Sci Med Natur *72:* 30-31, 1914.
86. Dustin A-P: Les réversions épithéliales dans le thymus *humain*. Arch Zool Exp *56:* 73-87, 1916/17.
87. Dustin A-P: Recherches d' histologie normale et expérimentale sur le thymus des *Amphibiens anoures*. Arch Biol *30:* 601-690, 1920.
88. Dustin A-P: A propos d' une nouvelle explicative de l' histogęnese du thymus. Arch Biol *45:* 339-352, 1934.
89. Dustin A-P, Baillez G: Recherches sur les cultures de thymus *"in vitro"*. Bull Soc Sci Med Natur *71:* 1-6, 1913.
90. Dustin A-P, Baillez G: Sur la lobation et la disposition des zones médullaires dans le thymus du *Chat*. C R Soc Biol *84:* 1237-1238, 1920.
91. Dustin A-P, Grégoire Ch: Greffes de thymus de mammifčres aux différents stades de l' histogęnese. C R Ass Anat *28:* 269-274, 1933.
92. Ankarsvärd G, Hammar JA: Zur Kenntnis der *Ganoiden*thymus (*Amia calva, Lepidosteus osseus*). Zool Jahrb Abt Anat *36:* 293-306, 1913.
93. Coplin WML: Morphology of the *human* thymus. Publ Jefferson Med Coll *6:* 116-126, 1915.
94. Dock G: The thymus gland. Ohio Med J *11:* 555-562, 1915.
95. Fenger F: On the size and composition of the thymus gland. J Biol Chem *20:* 115-118, 1915.
96. Flesch M Jr: Experimentelle Thymusstudien. I. Thymus and Milz bei der *Ratte*. Beitr Klin Chir *95:* 376-402, 1915.
97. Badertscher JA: The development of the thymus in the *pig*. Am J Anat *17:* 317-337 & 437-493, 1915
98. Kingsbury BF: The development of the human pharynx. Am J Anat *18:* 329-386, 1915.
99. Kingsbury BF: On the nature and significance of the thymic corpuscles (of Hassall). Anat Rec *39:* 141-159, 1928.
100. Kingsbury BF: On the *mammalian* thymus, particularly thymus IV: the development in the *calf*. Am J Anat *60:* 149-183, 1936.
101. Badertscher JA: The fate of the ultimobranchial body in *pig* (*Sus scrofa*). Am J Anat *23:* 89-131, 1918.
102. Baldwin FM: Pharyngeal derivatives of *Amblystoma*. J Morphol *30:* 605-680, 1918.
103. Stewart FW: On the (so called) thymus IV and the ultimobranchial body of the cat (*Felis domestica*). Am J Anat *24:* 191-223, 1918.
104. Hagstrom M: Die Entwicklung der Thymus beim *Rind*. Anat Anz *53:* 545-566, 1921.
105. Rabl H: Weitere Beiträge zur Entwicklungsgeschichte der Derivate des Kiemendarmes beim *Meerscheinchen*. Arch mikr Anat *96:* 210-339, 1922.
106. Anderson EL: The development of the pharyngeal derivatives in the *calf* (*Bos taurus*). Anat Rec *24:* 25-36, 1922.
107. Johnson CE: Branchial derivative in *turtles*. J Morphol *36:* 299-319, 1922.
108. Goldner J: Histogénčse du corpuscule de Hassall; unité cytogénique des cellules de charpenta, des placards épithéaux, des corpuscules unicellaires et des corpuscules

hassalliens. C R Soc Biol *88:* 947-950, 1923.
109. Birk H: Thymusdrüse. Monat Kinderhlk *27:* 321-342, 1923/24.
110. Jaffe HL, Plavska A: Experimental studies on the formation of Hassall's corpuscles. Proc Soc Exp Biol *23:* 91-93, 1925.
111. Bronnikova M: Zur Frage des Gewichtes der Glandula Thymus. Moskovskij Med Zurnal 6*:* 1-6, 1926.
112. Popoff NW: The histogenesis of the thymus as shown by tissue cultures, transplantation and regeneration. Proc Soc Exper Biol Med *24:* 148-151, 1926.
113. Popoff NW: The histogenesis of the thymus as shown by tissue cultures. Arch Exp Zellforsch *4:* 395-418, 1927.
114. Popoff M: Über die Thymus in vergleichend-anatomischer und physiologischer Beziehung. Biol Zbl *47:* 55-57, 1927.
115. Scammon RE: The prenatal growth of the *human* thymus. Proc Soc Exp Biol *24:* 906-909, 1927.
116. Firket J, Linhoff C: Pycnoses thymique et régénération. C R Soc Biol *97:* 1818-1820, 1927.
117. Deanesly R: The structure and development of the thymus in *fish*, with special reference to *Salmo fario*. J Micr Sci *71:* 113-145, 1927.
118. Deanesly R: Experimental studies on the histology of the *mammalian* thymus. J Micr Sci *72:* 245-275, 1928.
119. Dearth OA: Late development of the thymus in the *cat*: nature and significance of the corpuscles of Hassall and cystic formations. Am J Anat *41:* 321-345, 1928.
120. Jordan HE, Horsley GW: The distribution of reticulum in the thymus. Anat Rec *38:* 50, 1928.
121. Dynn HL, Scammon RE: Growth and variability of the thymus in the *human* fetus and newborn. Anat Rec *38:* 44-45, 1928.
122. Hanson AM: The *bovine* thymus. Minnesota Med *13:* 17-21, 1930.
123. Greenwood AW: Some observations on the thymus gland in the *fowl*. Proc R Soc Edinburgh *50:* 26-37, 1929/30.
124. Mason VA: The fate of the ultimobranchial body in the *cat* (*Felis domestica*), with special reference to cyst formation within the thyroid. Am J Anat *49:* 43-63, 1931.
125. Choi MH: Experimental study of migration of the thymus gland in *amphibian* embryos. Fol Anat Jpn *9:* 487-494, 1931.
126. Choi MH: Experimental study about histogenesis of the *amphibian* thymus gland. Fol Anat Jpn *9:* 495-503, 1931.
127. Berblinger W: Thymus. Physiologie und Pathologie dieses Organs. N D Klinik *10:* 457-477, 1932.
128. Florentin P: Constitution et rôle du thymus. Rev Méd de l' Est *60:* 581-586, 1932.
129. Florentin P, Weis M: Structures hassalliennes et pseudohassalliennes. C R Ass Anat *27:* 273-283, 1932.
130. Harrison BM, Mohn LA: Some stages in the development of the pharynx of the embryo *horse*. Am J Anat *50:* 233-250, 1932.
131. Woollard HH: The potency of the pharyngeal endoderm. J Anat *66:* 242-260, 1932.
132. Clark GA: Analysis of forty thymus glands from infants. With special reference to Hassall's corpuscles. Pennsylvania Med J *35:* 449-461, 1932.
133. Fedolfi N: Sullo sviluppo del timo. Monit Zool *43 (Suppl):* 330-333, 1933.
134. Weller GL: Development of the thyroid, parathyroid and thymus glands in *man*. Publ Carnegie Inst Washington, Contrib to Embryology No 443, *24:* 93-142, 1933.
135. Webster WD: The development of the thymus bodies in *Necturus maculosus*. J Morphol *56:* 295-323, 1934.
136. Selle RML: The embryology of the thyroid, parathyroid and the thymus of the *pacific pallid bat* (*Antrozous pacificus Merriam*). Am J Anat *56:* 161-191, 1935.
137. Peter H: Das histologische Bild des *Igel*thymus im jahreszeitlichen Zyklus. Zeitschr Anat *104:* 294-326, 1935.
138. Crisan C: Die Entwicklung des thyreo-parathyreothymischen Systems der weiβen *Maus*. Z Anat Entwicklungsgeschichte *104:* 327-358, 1935.
139. Hill BH: The early development of the thymus glands in *Amia calva*. J Morphol *57:* 61-89, 1935.
140. Norris EH: The morphogenesis and histogenesis of the thymus gland in man: in which the origin of the Hassall's corpuscles of the *human* thymus is discovered. Publ Carnegie Inst Washington, Contrib to Embryology No 166, *27:* 193-207, 1938.
141. Auerbach R: Morphogenetic interactions in the development of the *mouse* thymus gland. Dev Biol *2:* 271-284, 1960.
142. Auerbach R: Experimental analysis of the origin of cell types in the development of the *mouse* thymus. Dev Biol *3:* 336-354, 1961.
143. Le Douarin N, Jotereau FV: Tracing of cells of the avian thymus through embryonic life in interspecific chimeras. J Exp Med *142:* 17-40, 1975.
144. Le Lievre CS, Le Douarin NM: Mesenchymal derivatives of the neural crest: Analysis of chimeric quail and chick embryos. J Embryol Exp Morphol *34:* 125-154, 1975.
145. Le Douarin NM: The neural crest. Cambridge Univ. Press, London, pp. 1-259, 1982.
146. Burnet FM: The immunological significance of the thymus: an extension of the clonal selection theory of immunity. Austr Ann Med *11:* 79-91, 1962.
147. Waksal SD, Cohen IR, Waksal HW, Wekerle H, St Pierre RL, Feldman M: Induction of T-cells differentiation *in vitro* by thymus epithelial cells. Ann N Y Acad Sci *249:* 492-498, 1975.
148. Sato VL, Waksal SD, Herzenberg LA: Identification and separation of pre T cells from *nu/nu mice* by preculture with thymic reticuloepithelial cells. Cell Immunol *24:* 173-185, 1976.
149. Stutman O: Intrathymic and extrathymic T cell maturation. Immunol Rev *42:*138-184, 1978.
150. Kruisbeek AM: Thymic factors and T cell maturation in vitro: a comparison of the effects of thymic epithelial cultures with thymic extracts and thymus dependent serum factors. Thymus *1:* 163-185, 1979.
151. Kyewski BA, Rouse RV, Kaplan HS: Thymocyte rosettes: multicellular complexes of lymphocytes and bone marrow-derived stromal cells in the *mouse* thymus. Proc Natl Acad Sci USA *79:* 5646-5650, 1982.
152. Jason JM, Janeway CA: *In vitro* cultivation of nonlymphoid thymic cells: morphological and immunological characterization. Clin Immunol Immunopathol *30:* 214-226, 1984.
153. Vakharia DD, Mitchison NA: Helper T cell activity demonstrated by thymic nurse cells (TNC-T). Immunol *51:* 269-273, 1984.
154. Schuurman HJ, van de Wijngaert FP, Huber J, Schuurman RK, Zegers BJ, Roord JJ, Kater L: The thymus in "bare lymphocyte" syndrome: significance of expression of major histocompatibility complex antigens on thymic epithelial cells in intrathymic T-cell maturation. Hum Immunol *13:* 69-82, 1985.
155. Singer KH, Wolf LS, Lobach DF, Denning SM, Tuck DT, Robertson AL, Haynes BF: *Human* thymocytes bind to autologous and allogeneic thymic epithelial cells *in vitro*. Proc Natl Acad Sci USA *83:* 6588-6592, 1986
156. Singer KH, Haynes BF: Epithelial-thymocyte interactions in *human* thymus. Hum Immunol *20:* 127-144, 1987.
157. Kurtzberg J, Denning SM, Nycum LM, Singer KH, Haynes BF: Immature *human* thymocytes can be driven to differentiate into nonlymphoid lineages by cytokines from thymic epithelial cells. Proc Natl Acad Sci USA *86:* 7575-

7579, 1989.

158. Zuniga-Pflucker JC, Jones LA, Longo DL, Kruisbeek AM: Both TCR/MHC and accessory molecule/MHC interactions are required for positive and negative selection of mature T cells in the thymus. Cold Spring Harb Symp Quant Biol *54 Pt 1:* 153-158, 1989.
159. Singer KH, Denning SM, Whichard LP, Haynes BF: Thymocyte LFA-1 and thymic epithelial cell ICAM-1 molecules mediate binding of activated *human* thymocytes to thymic epithelial cells. J Immunol *144:* 2931-2939, 1990.
160. Haynes BF, Denning SM, Le PT, Singer KH: *Human* intrathymic T cell differentiation. Semin Immunol *2:* 67-77, 1990.
161. Ritter MA, Boyd RL: Development in the thymus: it takes two to tango. Immunol Today *14:* 462-469, 1993.
162. Martin-Fontecha A, Schuurman HJ, Zapata A: Role of thymic stromal cells in thymocyte education: a comparitive analysis of different models. Thymus *22:* 201-213, 1994.
163. Defresne MP, Nabarra B, van Vliet E, Willemsen R, van Dongen H, van Ewijk W: The ER-TR4 monoclonal antibody recognizes *murine* thymic epithelial cells (type 1) and inhibits their capacity to interact with immature thymocytes: immuno-electron microscopic and functional studies. Histochemistry *101:* 355-363, 1994.
164. Deschaux P, Khan NA: Immunophysiology: the immune system as a multifunctional physiological unit. Cell Mol Biol Res *41:* 1-10, 1995.
165. Anderson G, Anderson KL, Tchilian EZ, Owen JJ, Jenkinson EJ: Fibroblast dependency during early thymocyte development maps to the $CD25^+$ $CD44^+$ stage and involves interactions with fibroblast matrix molecules. Eur J Immunol *27:* 1200-1206, 1997.
166. de Mello-Coelho V, Villa-Verde DM, Dardenne M, Savino W: Pituitary hormones modulate cell-cell interactions between thymocytes and thymic epithelial cells. J Neuroimmunol *76:* 39-49, 1997.
167. Mihovilovic M, Denning S, Mai Y, Whichard LP, Patel DD, Roses AD: Thymocytes and cultured thymic epithelial cells express transcripts encoding α-3, α-5 and β-4 subunits of neuronal nicotinic acetylcholine receptors: preferential transcription of the α-3 and β-4 genes by immature $CD4^+ 8^+$ thymocytes. J Neuroimmunol *79:* 176-184, 1997.
168. Mihovilovic M, Denning S, Mai Y, Fisher CM, Whichard LP, Patel DD, Roses AD: Thymocytes and cultured thymic epithelial cells express transcripts encoding α-3, α-5, and β-4 subunits of neuronal nicotinic acetylcholine receptors. Preferential transcription of the α-3 and β-4 genes by immature $CD4^+8^+$ thymocytes and evidence for response to nicotine in thymocytes. Ann NY Acad Sci *841:* 388-392, 1998.
169. Miralles GD, Smith CA, Whichard LP, Morse MA, Haynes BF, Patel DD: $CD34^+CD38^-$lin- cord blood cells develop into dendritic cells in *human* thymic stromal monolayers and thymic nodules. J Immunol *160:* 3290-3298, 1998.
170. Trenkner E, Hoffmann MK: Defective development of the thymus and immunological abnormalities in the neurological mouse mutation "straggerer". J Neurosci *6:* 1733-1737, 1986.
171. Restifo NP, Kawakami Y, Marincola F, Shamamian P, Taggarse A, Esquivel F, Rosenberg SA: Molecular mechanisms used by tumors to escape immune recognition: immunogenetherapy and the cell biology of major histocompatibility complex class I. J Immunother *14:* 182-190, 1993.
172. Rooney CM, Rowe M, Wallace LE, Rickinson AB: Epstein-Barr virus positive Burkitt's lymphoma cells not recognized by virus specific T-cell surveillance. Nature (London) *317:* 629-631, 1985.
173. Rosenberg SA, Terry W: Passive immunotherapy of cancer in animals and man. Adv Cancer Res *25:* 323-388, 1977.
174. Sinkovics JG, Horvath J: New developments in the virus therapy of cancer: a historical review. Intervirology *36:* 193-214, 1993.
175. Somers SS, Guillou PJ: Isolation and expansion of lymphocytes from gastrointestinal tumour tissue. Surg Oncol *2:* 283-291, 1993.
176. Tahara H, Shiozaki H, Kobayashi K, Yano T, Yano H, Tumnia S, Oku K, Miyata M, Wakasa K, Sakurai M: Phenotypic characteristics of tumour-infiltrating-lymphocytes in human oesophageal cancer tissues defined by quantitative two colour analysis with flow-cytometry. Virchows Arch *416:* 329-334, 1990.
177. Uchiyama A, Hoon DS, Morisaki T, Kaneda Y, Yuzuki DH, Morton DL: Transfection of interleukin gene into human melanoma cellsaugments cellular immune response. Cancer Res *53:* 949-952, 1993.
178. Potworowski EF, Thibodeau L, Zelechowska MG: Specific adherence of thymocytes to a thymic medullary epithelial cell line. Immunol Lett *13:* 89-94, 1986.
179. Potworowski EF, Hugo P, Couture C: Binding of cortical thymocytes to a medullary epithelial cell line: a brief review. Thymus *13:* 237-243, 1989.
180. Dargemont C, Dunon D, Deugnier MA, Denoyelle M, Girault JM, Lederer F, Ho Diep Le K, Godeau F, Thiery JP, Imhof BA: Thymotaxin, a chemotactic protein, is identical to β2-microglobulin. Science *246:* 803-806, 1989.
181. Brown JH, Jardetzky T, Saper MA, Samraoui B, Bjorkman PJ, Wiley DC: A hypothetical model of the foreign antigen binding site of class II histocompatibility molecules. Nature (London) *332:* 845-850, 1988.
182. Clevers H, Alarcon B, Willeman T, Terhorst C: The T cell receptor/ CD3 complex: a dynamic protein ensemble. Annu Rev Immunol *6:* 629-662, 1988.
183. Doyle C, Strominger JC: Interaction between CD4 and class II MHC molecules mediates cell adhesion. Nature (London) *330:* 256-259, 1987.
184. Engel I, Ottenhoff THM, Klausner RD: High-efficiency expression and solubilization of functional T cell antigen receptor heterodimers. Science *256:* 1318-1321, 1992.
185. Haynes BF: Human T lymphocyte antigens as defined by monoclonal antibodies. Immunol Rev *57:* 127-161, 1981.
186. Haynes BF: Human thymic epithelium and T cell development: Current issues and future directions. Thymus *16:* 143-157, 1990.
187. Cavender DE: Organ-specific and non-organ-specific lymphocyte receptors for vascular endothelium. J Invest Dermatol *94:* 41S-48S, 1990.
188. Dunon D, Kaufman J, Salomonsen J, Skjoedt K, Vainio O, Thiery JP, Imhof BA: T cell precursor migration towards β 2-microglobulin is involved in thymus colonization of chicken embryos. EMBO J 9: 3315-3322, 1990.
189. de la Hera A, Muller U, Olsson C, Isaaz S, Tunnacliffe A: Structure of the T cell antigen receptor (TCR): two CD3 β subunits in a functional TCR/CD3 complex. J Exp Med *173:* 7-17, 1991.
190. Gold DP, Clevers H, Alarcon B, Dunlap S, Novotny J, Williams AF, Terhorst C: Evolutionary relationship between the T3 chains of the T-cell receptor complex and the immunoglobulin supergene family. Proc Natl Acad Sci USA *84:* 7649-7653, 1987.
191. Benoist C, Mathis D: Positive and negative selection of the T cell repertoire in MHC class II transgenic mice. Semin Immunol *1:* 117-124, 1989.
192. Bjorkmann P, Saper MA, Samaoui B, Bennett WS, Strominger JL, Wiley DC: Structure of the human class I histocompatibility antigen, HLA-A2. Nature (London) *329:*

506-512, 1987.

193. Borgulya P, Kishi H, Muller U, Kirberg J. von Boehmer H: Development of the CD4 and CD8 lineage of T cells: instruction versus selection. EMBO J *10:* 913-918, 1991.

194. Claverie J-M, Kourilsky P: The peptidic self model: A reassessment of the role of the major histocompatibility complex molecules in the restriction of the T-cell response. Annu Immunol 137: 425-442, 1987.

195. Mustelin T, Pessa-Morikawa T, Autero M, Gassmann M, Andersson LC, Gahmberg CG, Burn P: Regulation of the p59(fyn) protein tyrosine kinase by the CD45 phosphotyrosine phosphatase. Eur J Immunol *22:* 1173-1178, 1992.

196. Haynes BF: The human thymic microenvironment. Adv Immunol *36:* 81-142, 1984.

197. Haynes BF: Use of monoclonal antibodies to identify antigens of human endocrine thymic epithelium. In: Thymic hormones and lymfokines: biochemical chemistry and clinical applications. AL, Goldstein (ed), Plenum Press, pp. 43-56, 1984.

198. van Vliet E, Melis M, van Ewijk W: Monoclonal antibodies to stromal cell types of the mouse thymus. Eur J Immunol *14:* 524-529, 1984.

199. van Vliet E, Melis M, van Ewijk W: Immunohistology of thymic nurse cells. Cell Immunol *87:* 101-109, 1984.

Section 2

HISTOGENESIS OF THE MAMMALIAN THYMUS

Chapter 2

The Reticulo-Epithelial (Re) Cellular Network of the Mammalian Thymus

Abstract: The thymus provides an optimal humoral and cellular microenvironment for the development of immunocompetent T lymphocytes. Although yolk sac-derived pre-T, committed hematopoietic stem cells enter the thymus using a homing receptor, the immigration process also requires secretion of a peptide called thymotaxin by the cells of the reticulo-epithelial (RE) network of the thymic cellular microenvironment. The majority of RE cells have a round or irregular pale nucleus, which contains few, scattered, chromatin granules with a defined, spherical nucleolus, rich in basic histones. Their cytoplasm occasionally displays RNP granules, and is rich in non-histone proteins, fine phospholipid, lipid or cholesterin granules, and vacuoles filled with secreted substances. The cells of the subcapsular, "endocrine" type RE cell layer (giant or nurse cells), characterized by PAS positive granules, express A_2B_5/TE4 cell surface antigens and major histocompatibility complex (MHC) Class I (HLA A, B, C) molecules. In contrast to medullar RE cells, these subcapsular nurse cells also produce thymosins β3 and β4. Thymic nurse cells (TNCs) display a neuroendocrine cell specific immunophenotype (IP): Thy-1$^+$, $A_2B_5^+$, TT$^+$, TE4$^+$, UJ13/A$^+$, UJ127.11$^+$, UJ167.11$^+$, UJ181.4$^+$, and presence of common leukocyte antigen (CLA$^+$). Medullar RE cells display MHC Class II (HLA-DP, HLA-DQ, HLA-DR) molecule restriction. These cells also contain transforming growth factor-β (TGF-β) type II receptors and participate in the positive selection of T lymphocytes. Transmission electron microscopic (TEM) observations have defined four functional subtypes of medullar RE cells: undifferentiated, squamous, villous, and cystic. All subtypes are connected by desmosomes. Immunocytochemical observations have shown that the secreted thymic hormones, thymosin α1 and thymopoietin (and its short form, thymopentin or TP5), are produced by the same RE cells. Thymic RE cells also produce numerous cytokines including IL-1, IL-6, G-CSF, M-CSF, and GM-CSF that likely are important in various stages of thymocyte activation and differentiation. The co-existence of pituitary hormone and neuropeptide secretion, such as growth hormone, prolactin, adrenocorticotropic hormone, thyroid stimulating hormone, triiodothyronine, somatostatin, oxytocin, follicle stimulating hormone, luteinizing hormone, arginine vasopressin, growth hormone releasing hormone, corticotropin releasing hormone, nerve growth factor, vasoactive intestinal peptide, (pro) enkephalin, and β-endorphin, production of a number of interleukins and growth factors, as well as the expression of receptors for all, by the same RE cell is an unique molecular biological phenomenon. These data illustrate the immensely important and diverse immuno-neuroendocrine functions of the thymic RE cellular network. Based on our systematic observations of the thymus in humans and other mammalian species, we suggest that the thymic RE cell network represents an extremely important cellular and humoral microenvironment in homeopathic regulatory mechanisms of the multicellular organism. Intrathymic T lymphocyte selection is a complex, multi-step process, influenced by several functionally specialized RE cell subtypes and under constant immuno-neuroendocrine regulation, reflecting the dynamic changes of the organism.

Key words: Thymic histogenesis; Thymic cellular and humoral microenvironment; Thymic reticulo-epithelial (RE) cells; Transmission electron microscopy (TEM); Scanning electron microscopy (SEM); Cell surface antigens; Cellular immunophenotype (IP); Major histocompatibility complex (MHC); Monoclonal antibodies (MoABs); Immunocytochemistry; Thymic hormones; Intrathymic pituitary hormones; Intrathymic neuropeptides; Immuno-neuroendocrine regulation.

1. INTRODUCTION

After detailed descriptions of histogenesis, thymic research has increasingly aimed at a better understanding of intrathymic T lymphocyte maturation and differentiation, one of the main functions of the thymic stromal microenvironment (1-11). We agree with Trainin (12) that Maximow (13, 14) was the first thymologist to mention a close cellular interaction between thymic RE cells and the maturating thymocytes. He first suggested the production of a chemotactic (humoral) substance by the RE cells of the thymic microenvironment involved in the immigration of lymphopoietic stem cells into the thymus. As demonstrated in the experiments of Jolly and Grégoire, the regulation of thymic lymphopoiesis is critically related to the function of stromal accessory cells and the RE cell network of the thymus, organized in so-called "lympho-epithelial symbiosis" (15-23).

The thymic RE cells provide the optimal cellular and humoral microenvironment for intrathymic T lymphocyte differentiation, which manifests itself in a number of regulatory pathways:

1. secretion of a great number of polypeptides, including the thymic hormones and cytokines;
2. cell-cell interactions through adhesion molecules and those between major histocompatibility complex (MHC) class I and class II proteins and the TCR within the context of CD4 and CD8 molecules, respectively; and
3. RE cells can bind and interact with differentiating T lymphocytes using extracellular matrix (ECM) ligands and their respective receptors.

In all mammalian species, the prenatal thymic histogenesis can be immunomorphologically divided into three stages:

1) Epithelial Thymus

The embryonic, epithelial pharynx serves as the origin for different organs. Originally, it is lined with an epithelial cell layer of endodermal origin that expands into the pharyngeal pouches. In human embryos of 4-9 mm length, thymic, pure epithelial anlagen are formed from the dorsolateral portions of the 3rd pharyngeal pouch, as a tube-like expansion (5, 24-34).

In the absence of humoral and direct cellular interactions with the ectomesenchyme, the primary epithelial tissue is unable to proliferate (the humoral interaction probably represents the earliest production of epithelial growth factor (EGF) and expression of its receptor on the surface of thymic epithelial cells, earliest expression of erb-B2 oncogene intramembrane protein and the cell-cell contact relationship) (35-39). The earliest expression of the oncogene *c-myc* was detected between 7-10 weeks of human embryonic differentiation (19, 40-42). To date, such an early human thymus was not directly investigated, but around the thymic anlagen, over 20% of cells in other tissues expressed *c-myc*: brain, intestines, kidneys, lung, and skin (43). *C-myc* expression is not a simple marker of proliferation, but rather reflects tissue specific gene regulation (43). In general, the nuclei of these epithelial cells are large and more euchromatic than nuclei of reticular cells. The vacuolated appearance of their cytoplasm, as seen in the light microscope (LM), is characteristic for the primary epithelial cells in more advanced stages of development and differentiation. The epithelial cell cytoplasm contains large accumulations of β-glycogen granules. In this stage of histogenesis the mitotic activity of epithelial cells is extremely high

2) Lymphopoietic or Lympho-Epithelial Thymus

Large magnitude of lymphopoiesis is initiated by the immigration of pluripotent, but already lymphocyte cell lineage committed hemopoietic stem cells during the 6-7th week of intrauterine life (44-73). The "early stem cell chemotactic factor", required for immigration is produced by the cells of the ectomesenchyme or by the surface molecules of epithelial cells of thymic anlagen, developed under the ectomesenchymal regulation influence (74-83). In this early embryonal period, three mechanisms of regulation are present in the lymphopoietic thymus:

1. intracellular regulatory interactions between cortical and specialized RE cells, endodermal in origin, and immature thymocytes (thymic nurse cell - TNC);
2. cell to cell type interactions between "embryonal thymic stromal cells" and thymocytes (formation of different forms of T cell rosettes) (84-95); and
3. secretion of humoral type active substances that play a regulatory or *in situ* role.

Since the 10th week of gestation, the rapid daily cellular proliferation has enabled the thymus to produce 40-50 X 10^6 thymocytes within the thymic cortex (48, 77, 96-108), but only 1-3 x 10^3 leave the organ as "mature", immunocompetent T lymphocytes (48, 99). The constant proliferation of thymocyte subpopulations results from the permanent presence of a high percentage of "activated thymoblasts", an autocrine stimulus for T cell growth and proliferation (109). The humoral regulatory interactions result in permanent combinant production of lymphokines, interleukins 1, 2, 4 and 7 (IL-1, IL-2, IL-4, IL-7) (53, 56, 58, 60, 79, 91, 96, 97, 100, 101, 110-125).

3) Differentiation of the Cellular Thymic Microenvironment

There is rapid proliferation and differentiation of all the various cell types in the thymus, which results in the expansion of the thymic RE network, and thymic stroma and their reorganization as an unique thymic cellular microenvironment. The thymic cellular and humoral microenvironment (49, 77, 84-91, 93, 110, 126-141) also contains non-lymphopoietic, accessory (antigen presenting) cells (dendritic cells, macrophages, interdigitating cells, Langerhans cells) (47, 83, 106, 123, 142-157). Their influential action on the intrathymic T lymphocyte differentiation pathway is still not clear. The rapid proliferation rate of thymocytes also predicts a dynamic cell surface antigenic redistribution, rearrangement of T cell receptor (TCR) genes and heterodimer surface molecules. The extremely high mitotic activity of the thymocytes causes them to accumulate at the periphery of the newly formed thymic lobules, thus an early cortex and medulla are distinguishable. Within small medullary areas, few hypertrophic RE cells are already present, recognizable by their large eosinophilic cytoplasmic granules. Occasionally, some erythropoietic and granulopoietic stem cells are also located in the perilobular loose connective tissue. The differentiation of ectomesenchymal cells is marked by the first appearance of mast cells within the developing thymic parenchyma and in the loose connective tissue of the primary capsule.

The first appearance of the Hassall's bodies (HBs) occurs between 50-60 mm fetal length in the third lunar month of human prenatal development and 38-39th day of gestation in dogs (86, 110, 126, 158-170). The thymic organogenesis is complete at this stage, also marked by clear separation of the cortex from the medulla.

2. CYTOMORPHOLOGY OF THE THYMIC RE CELLS

Cytomorphological observations by several research groups (171-174) in human and other mammalian thymus glands enumerated the following characteristics of the thymic, endodermal in origin RE cells: round, oval, irregular or grooved nucleus, with a sharply defined spherical nucleolus, a thin, fragile nuclear membrane, and a few scattered, chromatin granules. In contrast, the always present reticular cells of mesenchymal origin had a nucleus richer in chromatin, more clumping of chromatin granules, and a heavier nuclear membrane than the thymic RE cells (175). The cytoplasm demonstrated little affinity for any stain, sometimes contained numerous vacuoles full of secreted mucosal substance or simple lipid granules. The cytoplasmic processes were either vacuolated or contained already phagocytized cellular debris. Cortical RE cells have a pale nucleus which only contains fine chromatin dots and is between 6.8 and 10.7 µm in diameter (174). The cortex contains numerous proliferating RE cells. The largest mitotic figures are scarce and show a colorless and poorly limited cytoplasm. These, therefore, are considered to be RE cell mitoses, and are quite numerous in the prenatal thymic cortex. In contrast the reticular cells of mesenchymal origin had a more chromatic nucleus, with a heavier nuclear membrane than the thymic RE cells, and more clumping of chromatin granules (176).

Some cortical RE cells are differentiated to save a minimal number of thymocytes from *in situ* (within the thymus) occurring death, forming lymphoepithelial complexes. These cells were named Thymic Nurse Cells (TNC) by Wekerle and co-workers (177, 178), and represent a "one cell thymic microenvironment" for T lymphocyte maturation, education and MHC restriction.

In the medulla, the RE cells are numerous, presenting the most important cell type (RE cells are three times as abundant in the medulla as in the cortex). Numerous RE cells of the medulla degenerate through the process described by Jolly in 1923 (179), consisting of the clumping of chromatin into dots, each one appearing as it enclosed in a vacuole. These dots then progressively disappear so that, at the end, the nuclei look like empty spaces. Such degenerating nuclei are seen either in isolated RE cells or in those making up the structures of the Hassall's bodies (HB). In some cases, RE cells could be found with two nuclei (180-182) and in one cell layer, under the capsule, the so-called "giant cells" (162, 181, 183-185). These "giant cells" were examined by Metcalf and Ishidate (185) and their conclusions were that these cells are RE cells in hypertrophy and contain a strong PAS positive granular substance in their cytoplasm (these giant RE cells are in the same place than the TNCs, they are HLA, Class I restricted and Thy-1 antigen is present on their cell surface). The nucleolus of RE cells is very rich in basic histones (186) and the cytoplasm always contains RNP granules rich from non-histone proteins (42, 187, 188). Concerning the histochemistry of the RE cells, we found a large collection of publications in the literature. PAS positive granules are always present in the cytoplasm, of course the different physiological condition might be important for the different means (162, 184, 185, 189-191). The histochemical nature of these granules is glycogen. The presence and appearance of the different lipids and phospholipids in the thymic tissue, especially in the RE meshwork, is very important during the organogenesis as a reserve and during the process of involution after different agent as remarkable degeneration. There are numerous publications about the lipids in the thymic literature, fine granules were described in all RE cells, but phospholipids and cholesterol consists were also found (37, 162, 181, 192-200). With the aging, the mass of the lipid granules in the cytoplasm of the RE cells is increasing rapidly (foamy cells), as an important characteristic of the normal age-involution of the organ (37). The process begins in the thymic cortex, with a big depletion of the thymocyte number and these lipid containing cells are very easy to recognize. Naturally, with aging, the mesenchymal lymphocytes are also present in the interlobular septa (110, 201).

3. HISTOCHEMISTRY

The histochemistry of thymic RE cells has been extensively described in the literature. Granules with strong PAS positivity are always present all over the cytoplasm, but different physiological conditions result in various cytoplasmic distributions of these glycogen-containing granules (162, 202-207). The nucleolus of RE cells is very rich in basic histones (186) and the cytoplasm always contains RNP granules rich in non-histone proteins (42, 208).

The appearance of lipid and phospholipid droplets within the cytoplasm of RE cells is biologically important during organogenesis (as a nutritional reserve) and during the process of acute involution caused by various agents (as cell degeneration products). Fine lipid granules have been described in all RE cells, but phospholipids and cholesterol have also been observed (162, 181, 194-196, 198, 209-212). With aging, the total mass of the lipid granules in the RE cells increases rapidly (transformation to foamy cells), as an important characteristic of the normal age-related involution of the organ (37, 195, 213, 214). This process begins in the thymic cortex, with a great depletion of thymocytes, which makes this lipid containing RE cells very easy to recognize. Naturally, with the progression of aging-related tissue changes, the mesenchymally located lymphocytes also move to the interlobular septa (110).

The histochemistry and histoenzymology of RE cells and their significance in metabolistic relations to other cells of the microenvironment have not been extensively observed (110, 176, 215-240). The RE cells are richer in mitochondrial enzymes than the lymphatic elements of thymic tissue. The presence of these enzymes, the oxydoreductases (LDH, SDH, MDH, and G-6 PDH), is a morphological characteristic of very active cellular metabolism. The level of lysosomal enzymes (hydrolases) in RE cells is smaller, but they are always present (acid and alkaline phosphatase, β-glucoronidase, G-6-phosphatase, and specific and non-specific esterases). The appearance of alkaline-phosphatase (AP) is directly related to the functional activity of RE cells (236-239). Active medullar, AP rich RE cells have been detected and, later, named alkaline phosphatase rich or APR cells. The presence of lipase was described by Day (226-229) and its appearance plays a role in the active lipid metabolism within RE cells. Szigeti (234, 235) also detected acid paranitro-phenol-phosphatase activity in these cells, and reported that the level of this enzyme is age-dependent, being higher in older thymuses and thus its presence might be an important indicator for the new formation of HBs.

In terms of research on enzyme production and the role of enzymes in the metabolism of RE cells, the thymic literature is poor (110, 176, 224, 226, 232, 236-239, 241). The RE cells are richer in so-called mitochondrial enzymes than the lymphatic elements of the thymic tissue. These oxyreductive enzymes (LDH, SDH, MDH, and G-6 PDH) are morphological characteristics of the very active cellular metabolism. The level of the lysosomal enzymes (hydrolases) are smaller, but they are always present (acid and alkaline phosphatase, β-glucoronidase, G-6-phosphatase, and specific and non-specific esterases). The appearance of the alkaline-phosphatase is always a characteristic of the functional activity of the RE cells, in some cases, by Timperley and co-workers (236-239) RE cells were described with a rich contain and later named APR cells. The presence of lipase was described by Day (226) and his appearance plays a role in the active lipid metabolism of the RE cells. Szigeti (234) had found in these cells acid paranitro-phenol-phosphatase activity, and described that the level of this enzyme is age-dependent higher by older thymuses and its presence and appearance might be an important indicator for the formation of the HBs.

Our systematic histochemical and histoenzymatic observations of the human fetal thymus revealed that the level of several dehydrogenates was higher during prenatal ontogenesis. The enzyme levels showed an age-dependency during the maturation of the thymic cellular microenvironment. The enzymatic levels were extremely high in the so-called alkaline phosphatase rich (APR) mostly medullar RE cells

(110, 242, 243), and in RE cells included in the periphery of HBs (243). The presence of δ5-3β-steroid-dehydrogenase in the medullar RE cell subpopulation was detected from the beginning of the third lunar month. The most interesting result in our material, was the detection of δ 5-3 β-steroid-dehydrogenase in the cells of the RE meshwork from the third lunar month. The level of this enzyme was higher in the hypertrophied RE cells, which represent the functionally most active variation of the RE cells.

4. ELECTRON MICROSCOPY

In the last 50 years, several works employing transmission electron microscopy (TEM) were published focusing on the mammalian thymus in general; for example in mice (244-246), in rats (247, 248), in hamsters (249, 250), in monkeys (251), and in guinea pigs (252). Very few TEM studies were performed on the thymus of birds. Clawson, Cooper and Good (34) studied the compared characteristics of lymphocytes in the bursa of Fabricius, the thymus and peripheral lymphoid germinal centers in chickens and Gilmore and co-authors (253) described the presence of intercellular cysts and myoid cells in the thymus of the cockerel. Systematic observations for TEM of the human fetal thymus had been described by Kameya and Watanabe (254), Pinkel (255), Hirokawa (256), Haar (257), and Bartel (258).

Hoshino (246) described two categories of cells in the thymic reticulum: The reticulo-epithelial cells and the macrophages or mesenchymal cells. The presence of desmosomes at the cell surface and bands of cytoplasmic tonofibrils have allowed a clear distinction between the RE cells, which possess these features and the mesenchymal "reticular" cells. Izard (259) also described two cell categories in guinea pigs: desmosomal reticular cells and phagocytic reticular cells. The desmosomal type cells were the most numerous and were always situated within the groups of lymphocytes. Their nucleus was large and stellate. Its chromatin was not as dense as the surrounding thymocytes'. The nucleolus was spherical and central with a prominent nucleolonema. The cytoplasm was extended in thin processes between the thymocytes. These RE cells and their processes were connected by typical desmosomes. The cytoplasmic content differed, according to the age of the guinea pigs. In fetus and newborn, the principal feature consisted of abundant granules of glycogen, clear vacuoles and lipid droplets. Moreover, there were mitochondria, microbodies, a Golgi apparatus and an endoplasmic reticulum. In some cells, many intracytoplasmic cysts, lined by microvilli were found. No morphological evidence of a granular secretion resembling that of exocrine or endocrine gland was ever seen. In the adult thymus, the cytoplasmic content was poorer. Both the cortical and the medullar RE cells seemed to be morphologically identical. Near the Hassall's bodies some RE cells contain numerous peculiar granules of different size (1.300 A to 1.5 μm) and very dense, in close association with the tonofilament bundles. Around these granules, numerous ribosomes were scattered. The RE cells, the only dense granules usually observed were keratohyalin granules linked to keratinization (259). The mucous droplets were very rare and resemble more a degenerating form, than an active secretory process (260). The reticular cells from the so-called deep cortex, close to the cortico-medullary junction, resembled "undifferentiated medullary epithelial cells" (252). Mandel (252) classified the medullary RE cells of the thymus in guinea pigs as being undifferentiated, squamous, villous, and cystic. The RE cells present in the outer cortex were characterized by the presence of a pale nucleus (usually stellate), which contained a prominent central nucleolus. Their perinuclear cytoplasm was sparse, but long cytoplasmic processes extended from the cell body passing in between the closely packed lymphoid cells. Well-developed desmosomes joined these processes. The cytoplasm contained bands of tonofibrils, scattered polyribosomes, a few cisternae of endoplasmic reticulum, always lipid droplets and occasional mitochondria. A feature of these cortical RE cells was the presence of numerous membrane-bound bodies. They usually consisted of a large membrane-bound space, which was partially filled with smaller vesicles. The number and size of these vesicles varied, ranging from numerous to a relatively small number aggregated near the bounding membrane of the parent body. These structures did not stain for acid phosphatase. The majority of cells in the cortex were separated by an intercellular space of 200A. It is possible, described Mandel (252), that the medullary "undifferentiated RE cells may act as stem cells for the structurally specialized cortical epithelium." We never recognized various morphological or functional forms of the thymic RE cells in dogs. Our view, according to Lundin and Schelin (248), is that the medullary and cortical RE cells are members of a united thymic cellular meshwork. It has been suggested that the subcapsular RE cells (thymic nurse cells), play a role in the initiation or control of thymocyte maturation. The close membrane contact and possible cytoplasmic continuity may facilitate an easy, humoral transfer between RE cells and thymocytes.

Ito and Hoshino (261) investigated the ultrastructure of the thymus in mice (5-8 weeks old Japanese dd-strain mice). In the medulla of the thymus, the RE cells form a network and they are interconnected with desmosomes. These cells often contain large vesicles of round or somewhat round shape, which are surrounded by a limiting membrane with a number of microvilli projecting into the lumen. The ciliated vesicles containing RE cells are usually polygonal in shape. Generally, they contain a single ciliated vesicle located in the cytoplasm adjacent to nucleus. Such vesicles are spherical, measuring 5 to 9 µm in diameter. The vesicles contain amorphous substance of low density, and small dense granules measuring 20 to 30 µm in diameter. With light microscope examination, the PAS staining reveals that these contents in the lumen are strongly PAS positive. In the cytoplasm, smooth surfaced vesicles, which were smaller than 200 µm, were present in large numbers. A small number of oval or rod-shaped mitochondria and Golgi apparatus were also seen in the cytoplasm. The presence of the intracytoplasmic vesicles might suggest a secretory activity of these cells. These ciliated vesicle-containing RE cells can be distinguished from the cells in the wall of the cysts, because the vesicles are intracytoplasmic.

Frazier (262) observed the ultrastructure of the thymus in chicken (Rhode Island Red strain, ranging in age from 1 day to 12 weeks). He reported that cortical RE cells have long cytoplasmic processes and display two varieties (pale and dark) of the morphological form. Most of these cells contain a pale nucleus, one or two nucleoli and the nuclear membrane is often indented. The cytoplasm is pale, and cytoplasmic organelles include ribosomes, a few small mitochondria, a Golgi apparatus, vacuoles with amorphous contents and dense granules of various sizes. Intracytoplasmic fibrils are always present and desmosomes connect the long cytoplasmic processes with those of adjacent RE cells. Other cortical RE cells have a much darker nucleus with an irregular outline, and its form is often elongated. Their cytoplasm is dark, but vacuoles, similar to those seen in the pale RE cells, are present, and sometimes the dense granules are evident in the cytoplasm. Desmosomes connect the plasma membranes to adjacent RE cells. Morphological forms intermediate between these two are also present. The pale RE cells are typical for the cortex, the dark RE cells are a common feature of the medulla. They often appear in small groups where desmosomes connecting adjacent cells are sometimes seen. Another type of RE cell is found in the medulla and the cortico-medullary junction of the thymus, the so-called "undifferentiated epithelial (UE) cell". This cell is characterized by a pale, spherical or oval nucleus containing one or two nucleoli. The nuclear membrane is not indented. The cytoplasm contains a Golgi apparatus and centrioles are often seen in the Golgi region. Mitochondria, polyribosomes, endoplasmic reticulum and a few small dark granules are also present, but intracytoplasmic fibrils are rare and desmosomes are infrequent. In the medulla, these cells are often in small groups. Numerous medullary RE cells are involved in the formation of intracellular and intercellular cysts. These cystic RE cells vary considerably in their morphological features. The morphologic characteristics are similar to other RE cells, but very few cytoplasmic fibrils were detected, while other subtypes of RE cells contained numerous cytoplasmic fibrils which were often present around the borders of the cyst. The cells usually contain only one large cyst which was lined by microvilli, but rarely, two or three smaller cysts were seen. The cysts contain dense granules, droplets and amorphous material. Associated with both intercellular and intracellular cyst-forming cells, and present throughout the medulla, were other cells which, characterized by the presence of numerous dense, spherical granules or droplets in their cytoplasm. Some of the granules have an outer limiting membrane but often this is absent. The nuclei of these cells were spherical or oval, with one or two prominent nucleoli. The cytoplasm contains some endoplasmic reticulum and intracytoplasmic fibrils were also present in some cases. These cells have epithelial characteristics, since occasional desmosomes can be seen. Intercellular cysts vary considerably in complexity. Some of the simpler cysts were formed from two or three cells which vary in morphology from very pale to very dark. The nuclear membrane of these cells was often highly indented and the cytoplasm contained a variety of organelles including mitochondria, endoplasmic reticulum, spherical inclusions resembling lipid droplets and very small, dark granules, which were possibly glycogen. Intercellular cysts were in some cases seen to be formed by junctions between the dark and pale cells. Microvilli project into the lumina of the cysts and some cysts appear to be empty, while others contain dense granules and amorphous material. Cysts that are more complex were formed from three or more RE cells or from a variety of cells. Very rarely, cilia were present in lumen wall. Another type of RE cell in the medulla may be named as a squamous epithelial (SE) cell (252), since the prominent features of this cell type was an abundance of cytoplasmic fibrils and desmosomes, and they resemble keratinizing SE cells. The nucleus was oval or round and slightly indented. A nucleolus was usually present and moderate numbers of mitochondria and ribosomes were present in the cytoplasm. These cells may be found in groups in the medulla and they are also concerned with the formation of the HBs. Although many of the RE cells in chick thymus can be classified as "undifferentiated", "squamous", and "cystic". Villous cells were not particularly prominent. Some authors (261, 262) are of the opinion that these cystic epithelial (CE) cells show cytoplasmic features suggesting a secretory function. Gad and Clark (263) provided some evidence to suggest that cystic RE cells in mouse thymus secrete a sulfated mucoid lymphopoietic hormone and Henry (264) reported that a mucoid substance is regularly secreted by the RE cells (squamous?) forming the HBs and by isolated RE cells in the human thymus. Pfoch (265, 266) suggests that "clear vesicles" present in the RE cells of the rat thymus, are the morphological equivalent of the production of a thymic humoral factor (THF). Other authors consider that granular inclusions in cystic RE cells (249-251) may be related to a lymphocyte stimulating hormone (LSH). In the chick thymus (262) there are several structures suggesting secretory function. Many of the smaller intracellular and intercellular cysts were composed of morphologically diverse cells, containing a variety of droplets, inclusions and granules in their cytoplasm.

Hirokawa (256) in his systematic observations on six late fetal human thymus (four during 4-6 lunar months and one in the eighth and another one in the tenth lunar month) with TEM, described as basic structure of the thymic tissue, a cellular meshwork, loose in cortex and dense in medulla, composed of two types of RE cells. In cortex (at the 22nd to 24th week the cellular network was composed of one type of RE cell, having slender, long cytoplasmic processes adhering one to another by desmosomes. The nucleus was larger and paler than that of thymocytes. The cytoplasm was narrow and contains a small amount of tonofibrils and smooth endoplasmic reticulum. In the medulla, he recognized two types of RE cells: designated as "slender" and "hypertrophic". The basic cellular elements of the medullary meshwork are the cytoplasmic processes of RE cells, like in the cortex, but smaller in size and darker in the density of the nucleus. The narrow cytoplasm was poor in organelles except for the numerous bundles of tonofibrils. "Slender" RE cell: long cytoplasmic processes adhere one to another by frequent desmosome formation. The "hypertrophic" type of RE cells, arranged in groups, have a large and pale nucleus, with one or two nucleoli, and wide, irregular cytoplasm. The cytoplasm is characterized by the presence of numerous small smooth surfaced vesicles (0.1-1.0 μm in diameter) and occasionally several large vesicles (1.0-3.0 μm in diameter), the wall of which is provided with microvillous processes into the lumen. These small vesicles contain a large amount of fine granular substance, which corresponds to PAS positive granules on light microscopy. In addition, there were tonofibrils of various thicknesses, saccular rough surfaced endoplasmic reticulum, microtubules and Golgi apparatus, variable in amount. Microbodies with homogenous interior of low density were also found in some cases. The small vesicles and the large ones with microvillous processes in the hypertrophic type were most prominent at the 22nd and 24th fetal week. In the thymus at the 16th fetal week, these cells were poor in organelles and contained no vesicles. In the neonatal period, the hypertrophic type cells with small vesicles were rather frequently observed, but the large ones with microvillous processes were rarely seen. As an evidence of secretory function of the thymus, PAS positive granules of colloid-like inclusions of Metcalf (267, 268) were present in the RE cells of hypertrophic type localized in the medulla and identified electron microscopically as large vesicles with microvillous cytoplasmic processes, containing fine granular substance of moderate density. The vesicles were most frequently observed in the second half of the intrauterine life, when the thymic endocrine activity is considered to be the most active. These RE cells of hypertrophic type also composed the HBs.

Bartel (258) provided observations on the thymic primordium in the initial period of gestation (4th to 14th week of gestation). In the youngest primordium (4th week of intrauterine life), the detailed TEM analysis shows that the differences between the two types of cells described as dark, basophilic and scarce light cells in light microscope, are not generally significant. They are only noticeable regarding the aggregates of glycogen, which are present in "light" cells. Tonofilaments and desmosomes can be observed in some cells. The nuclei were big and occupied a large part of the cell. They contained a small quantity of granular chromatin. In the "light" cells, the nucleoli were more distinct, whereas granular chromatin was scarce and evenly scattered. On the surface of both cell types, some cytoplasmic processes can be seen. The oval thymus anlage is surrounded by a delicate outer basal lamina, outside which there were numerous typical in appearance mesenchymal cells. The way, as described, in early thymic primordium (4th week of gestation) was not alymphopoietic, but instead, we believe that dark lymphatic stem cells were already present. The vacuoles and granules start to appear in the endodermal RE cells, together with the growth of the Golgi complex since the 10th intrauterine week.

During our electron microscopical (TEM, SEM) observations, we were able to distinguish various ultrastructural forms of thymic RE cells in dogs, but we support a view, similar to that of Lundin and Schelin (248) and Ropke and co-workers (269), that the medullary and cortical RE cells are members of an united thymic RE cellular meshwork, within the complex cell composition of the thymic stromal microenvironment (Fig. 1; Fig. 2; Fig. 3; Fig. 4; Fig. 5; Fig. 6; Fig. 7). Functional investigation may reveal a RE cell specialization, resulting in specialized ultrastructural features: for example, it is clear that the subcapsular RE cells (giant or thymic nurse cells), play a role in the initiation or control of thymocyte maturation. The close intracellular cytoplasm-cell membrane contact between RE cells and the vast number of thymocytes may facilitate an easy cell surface molecule mediated and also a humoral type transfer between these cells. They often degenerate with more clumping of chromatin granules. Observations with TEM distinguished four different, probably functionally specialized types of medullary RE cells: undifferentiated, squamous, villous, and cystic, but all of them were connected with desmosomes. Scanning electron microscopic (SEM) observations revealed a close contact between the cytoplasmic processes of RE cells with thymocytes (Fig. 7; Fig. 8).

A number of authors consider the cytoplasmic vacuoles to be morphological evidence of hormonal secretion in RE cells. In the literature, the function "humoral secretion" of these primitive, undifferentiated epithelial cells is accepted, recently. A human 9 weeks old fetus had epithelial cells on the outer surface of the thymus, columnar in shape (257). These epithelial cells were joined to each other with desmosomes and contained aggregates of glycogen, numerous mitochondria and rather short profiles of rough endoplasmic reticulum (RER).

Hammar (161) conducted his studies in the frog, Rana temporaria: "Einige Male habe ich aber unentwickelte Schleimzellen mit erhaltenen charakteren von Retikulumzellen gesehen. Sie gehen also offenbar aus solchen Zellen hervor....Die Entleerung des Sekrets muss notwendigerweise in die Maschen des Markretikulums hinein erfolgen". Maximow (13, 14), Hartmann (270), and Harland (271) reported in other mammalian species similar formations of vacuoles in the early RE cells. Maximow (13, 14) in his studies on the thymic ontogenesis observed the supporting effect of the endodermal epithelial anlage, seeming to attract the immigrating lymphatic stem cells (from the yolk sac) and to stimulate their rapid proliferation and maturation within the thymic humoral microenvironment. A similar thymocyte differentiation stimulating influence of the RE cells has been reported in regenerating thymus implants (19, 272, 273). In the serial investigations with injection or implantation of tissue composed predominantly of thymic RE cells, in the regional lymph nodes were demonstrating hyperplastic changes of lymphopoietic nature (or lymphoblastic reaction by some authors) (274). Various effects of stimulation of hematopoiesis by extracts or suspension of the whole thymus have been reported by Hammar (189). These stimulations of lymphopoiesis by thymic RE cells provides additional support to the concept of the lympho-epithelial synthesis. Attraction of donor lymphocytes into the epithelial remnant of the thymus after total body irradiation (TBI), subsequent stimulation of the proliferation of these lymphocytes within the RE meshwork, were also recorded as a continuation of the early histogenetical process, in the reconstitution of the thymus by immigrating lymphocytes (19, 272, 273). These results suggest that an active interaction occurs in the thymic cortex between the RE cells and the immigrated lymphatic stem cells.

A pure thymic cell line was produced, the so-called IT-45R1 cell line, and was tested in vitro the effect of direct cell-cell contact with the rat bone marrow cells by Itoh, *et al.* (275-277). Among the five fractions isolated from the bone marrow by ovine serum albumin (BSA) density gradient, two types of T lymphocyte progenitors were separated, fraction 5 the so-called prethymic T lymphocyte progenitors and fraction 4, cells which are considered as postthymic. Stutman (2) pointed out the existence of two types of T cell progenitors, pre and postthymic. Postthymic cells, the population that has once contacted with the thymic RE meshwork, are capable of differentiating into mature T cells under the humoral activity of the thymus, while the prethymic stem cells require contact with the thymic RE cells for their further maturation. Pahwa and co-workers (278) considered the precursor cells in MAB Thy 1.1 assay as prethymic and the ones in RFC assay as postthymic. The results of experimental observations by Itoh and co-authors (276, 277) determined that T lymphocyte precursor cells can be separated by BSA density gradient, and that the isolated fraction 5 is considered to be formed by prethymic stem cells and that these prethymic progenitor cells differentiate into postthymic T cells through direct cell-cell contact with the pure IT-45R1 cell line RE cells. These results suggested that Touraine *et al.* (279-281), who pointed out the sequence of human T cell maturation, was right. They were able to demonstrate that T cell progenitors first expressed the cell surface differentiation antigens and only then acquired the capacity to form rosettes. A number of laboratories have described production of partially or fully purified thymic factors which can replace in part the function of thymus in inducing T lymphocyte maturation (282). These factors are secreted by the thymic RE cells, and they could be produced in the first two weeks in vitro by culturing thymic RE cells.

Kater and co-workers (283) examined the effects of human thymus epithelial conditioned medium (HTECM) on in vitro mitogen responses of human and mouse lymphocytes. The anti-thymosin fraction 5 and 6 had a clear inhibitory effect on the HTECM and in the PHA and Con A responses of mouse lymphocytes. Surprisingly the MAB, antithymosin α-1 did not inhibit.

5. IMMUNOMORPHOLOGICAL AND FUNCTIONAL MATURATION OF THE RE CELLULAR NETWORK OF THE THYMIC MICROENVIRONMENT

The discovery of monoclonal antibody (MoAB) production technology and the employment of thousands of MoABs in immunomorphological studies of thymic ontogenesis also reopened the histogenetic and functional observations of the human thymic microenvironmental cellular networks (284-313). The immunocytochemistry of the RE cell histogenesis and functional differentiation has been observed in details (314-386).

Eisenbarth and co-workers (298) developed and characterized the A_2B_5 MoAB that reacts with a complex of neuronal ganglioside expressed on neural crest derived cells and on peptide secreting endocrine elements. Tetanus toxin (TT) also binds to GD and GT gangloisides present on thymic neuroendocrine cells. $A_2B_5^+$ and TT^+ RE cells were detected in two thymic zones, in the subcapsular cortex and in the medulla. During an International Conference ("The Thymus Histophysiology and Dynamics in the Immune System", in Rolduc, April 1989), a special workshop was held to characterize MoABs to thymic RE cells in the thymus of man, mouse, and rat. The staining patterns of twenty-five RE cell specific MoABs were evaluated immunohistologically on reference thymuses, thymuses during ontogeny, various other organs (all tissues obtained from the species against which the particular MoAB was raised), and thymuses of other species. In the reports of von Gaudecker (357), Kampinga (358) and Colic (359) and their co-investigators, only immunohistological results of reference thymuses and thymuses of other species were included. Based on the staining patterns on reference thymuses, the observed MoAB library was subdivided into five main groups. The use of these clusters of thymic RE cell staining patterns (CTES) was also recommended, to designate any further anti-RE MoAB, awaiting the possible incorporation in existing CD nomenclature for leukocyte differentiation antigens. The present immunohistological approach will be extended by additional analysis for which a protocol was designed. The ultimate goal of this important workshop on MoABs against thymic RE cells was to get well-characterized reagents in the analysis of RE cell associated molecules with putative function in intrathymic T lymphocyte processing and immuno-neuroendocrine regulation.

The immunohistochemical characteristics of the human thymic cellular microenvironment have also been observed with extensive panels of poly- and monoclonal antibodies (314-385). Obviously, cell lineage specific immunoreactivity of various antibodies has served well in assessing cell differentiation (maturation) pathways and/or the origin of cell types in the thymic stromal microenvironment. The so-called endodermal region of the third pharyngeal pouch, the widely proposed and accepted by classical embryology as the "birthplace" of medullary RE cells, is frequently arranged in a duct-like manner and undergoes keratinization and transformation into squamous epithelia (357-359, 368-371, 376, 377). Thymic RE cells, both cortical and medullary, are epithelial in origin and nature (e.g. they contain tonofilaments and desmosomes), and the reticular appearance of these cells have not been sufficient to confirm their mesenchymal derivation (320, 337, 338, 349, 376, 377).

In a number of immunocytochemical studies concerning prenatal human thymic histogenesis, Haynes and co-workers (133, 296, 297, 301, 309, 325, 328, 329, 331, 332) repeated the widely accepted statement of classical thymic histogenetic descriptions: "Endodermal epithelial tissue from the third pharyngeal pouch gives rise to $TE3^+$ cortical thymic epithelium while ectodermal epithelial tissue from the third pharyngeal cleft invaginates and splits during development to give rise to $A_2B_5^+/TE4^+$ medullary and subcapsular cortical thymic epithelium". Earlier experiments defined that normal thymic RE cells express antigens of early epithelial cell maturation (A_2B_5, TE4) throughout prenatal thymic ontogeny and acquire antigens 12/1-2, TE8, and TE15 at the fetal age of 14 to 16 weeks. Haynes and co-workers (133, 329, 331, 332) also demonstrated that the $A_2B_5^+$ and TT^+ subcapsullarly located "giant" or hypertrophied RE cells and medullary ones contained (secreted) the thymic hormones thymopoietin and α1-thymosin, while in the $TE3^+$ cortical RE cells these hormones were absent. In addition, it was detected that only the RE cells of the subcapsular cortex produced thymosin-β3 and thymosin-β4 (325, 328). Employment of anti-keratin MoAB AE-1 revealed presence of keratin in all $A_2B_5^+$ thymic endocrine RE cells (133). The use of the anti-TE4 MoAB developed by Haynes (133, 329, 331, 349) in an indirect fluorescent assay also defined the functionally specialized Thy-1^+, $A_2B_5^+$, α1-thymosin containing subcapsular RE cell subpopulation. During immunocytochemical screening, the anti-TE4 MoAB did not react with any other endocrine tissues, but did, however, give positive reaction with basal keratinocytes of squamous epithelium in skin, tonsil, upper esophagus and conjuctiva (similar reactivity pattern was also detected with MoAB anti-A_2B_5) (298). The initial immunocytochemical characterization suggested that MoAB anti-TE4 does not bind to characterized keratin-like molecules. All $TE4^+$ RE cells contain keratin as defined by MoAB AE-1, but not all keratinized thymic cortical RE cells exhibited a $TE4^+$ IP. Moreover, thymic RE cells grown in culture always demonstrated presence of keratin intermediate filaments, and absence of TE4 antigen (AE-1^+, $TE4^-$ IP). In later stages of prenatal thymic histogenesis, the medullary HBs always remained TE4 negative (337, 341).

The great majority of cortical RE cells expressed the cytoplasmic TE3 antigen, detected by immunocytochemistry employing anti-TE3 MoAB, also developed in Dr. Haynes' laboratory. This TE3 antigen defines an intracellular regulatory molecule that reflects a specialized function of the RE cells located in the thymic inner cortex. Extrathymically, TE3 antigen expression was detected in a great variety of tissues: thyroid, pancreas, anterior pituitary cells, and brain neurons. This antigen was absent in epidermal keratinocytes and from the cells of the adrenal gland. In addition, macrophages and large dendritic cells located in the thymic medulla also showed TE3 positivity (349, 379). It is significant that RE cells close to the TNCs formed a specialized subset of cortical RE cells, which also displayed a mixed, probably transitory $TE3^+$, $TE4^+$, $A_2B_5^+$ IP. They also demonstrated presence of keratin, as well as production of thymosin α1.

The Thy-1 antigen (expressed on approximately 1% of pediatric human thymocytes) has a differential expression amongst thymic epithelial cells, being confined to those in the subcapsular cortex and to TNCs (320). The former represent the site to which thymocyte precursors first migrate upon entering the thymus (387-394, 308, 327). The latter are large, functionally specialized RE cells, located within the cortex, whose plasma membranes totally enclose a number of thymic lymphocytes; these cells are therefore good candidates for the mediators of direct contact (stromal) induced thymocyte maturation.

Another MoAB, anti-p19 (developed against a 19 kD structural protein of anti-human T cell leukemia/lymphoma virus (HTLV)) reacted with the majority of human thymic RE cells (325, 328, 329, 331, 332). Thus, the p19 antigen is specific for the neuroendocrine thymic RE subpopulation.

Drenckhahn and co-workers (395) employed anti-smooth-muscle myosin and anti-actin antibodies for immunofluorescent microscopic observation of the human thymus. The concentrically arranged cells comprising HBs demonstrated strong immunoreactivity with both antibodies, but the intense immunofluorescence also defined antigen expression in some RE cells. Epidermis specific antigens have also been identified in thymic RE cells and in the peripheral layers of HBs (323, 330, 337, 341, 351). Laster and co-workers (345, 346) detected specific keratins involved in early and advanced stages of epidermal keratinocyte maturation in thymic RE cells.

In the human thymus, MoAB anti-TE7 reacted with mesodermal in origin stromal cells (133, 329, 332). It is specific for fibroblasts, reacted with cartilage, vessel and fetal interstitial mesenchymal cells. Normal tissue screening showed that MoAB anti-TE7 reacts with fibrous connective tissue in practically all tissues investigated. In malignant tissues, the antibody only reacts with neoplasms of fibrous connective tissue (e.g. fibrosarcoma). All cells located within the thymic capsule, interlobular and intralobular septas, and some individual cells dispersed in the thymic microenvironment demonstrated presence of TE7 antigen. TE7 positive cells in the thymic cortex and medulla uniformly displayed a thymosin $\alpha 1^-$, $A_2B_5^+$, keratin$^-$, TE3$^-$, and TE4$^-$ IP.

Lobach and co-investigators (in the laboratory of Dr. Haynes) (337, 441, 349) developed four additional MoABs (TE8, TE15, TE16, and TE19). These MoABs selectively demonstrated immunoreactivity with human HBs. TE8 and TE16 reacted with the hypertrophied cells of the outer cell layer of HBs. MoAB TE19 stained the entire cellular content of HBs, as well as the medullary RE cells adjacent to HBs. More than 90% of HBs displayed immunoreactivity with MoAB TE15 (345, 346, 351, 354, 355). Three of these antibodies, TE8, TE16, and TE19 also reacted positively with the stratum granulosum while TE15 reacted with the stratum corneum of human skin. Employing these same antibodies, we also defined HBs as antigenically distinct parts of the thymic secretory cellular microenvironment (243, 375)

It seems that immuno-electron microscopy on thymic thin sections may be the best method to identify stromal cell IPs (354). The use of MoABs against Mac-1 and Mac-2, which were developed against macrophage antigens, as well as a MoAB against MHC class II antigens determined the IP differences between macrophages and RE cells, as well as bone marrow derived thymic stromal cells (354). The macrophages were Mac-1$^+$, Mac-2$^+$, and only 50% of them showed Ia positivity. The cortico-medullary interdigitating cells (IDC) were characterized by a Ia$^+$, Mac-1$^+$ and Mac-2$^+$ IP (354), while RE cells displayed cell surface immunoreactivity with Ia, but neither Mac-1 nor Mac-2. These results strongly suggest that macrophages and IDCs are derived from a common precursor stem cell (354, 379). A recent study has described the use of a novel panel of four MoABs to classify subsets of cells within the thymic microenvironment, two for medullary RE cells and HBs (His-39 and RMC-20) and two for cortical RE cells (His-37 and RMC-17). On the 15th day of gestation, MHC class I and class II molecules can also be identified on thymic RE cells, although they reach their highest density around birth (316, 319, 336).

The three well defined and biochemically characterized thymic hormones, Facteur Serique Thymique [FTS/thymulin (321, 322)], thymosin-α and thymopoietin (317, 318) have also been detected in the cells of the RE network and in HBs. These results allowed us to suggest that HBs possess an active secretory role and thereby may participate in intrathymic T lymphocyte differentiation (243). Furthermore, it is significant that immunocytological evidence for the simultaneous presence of all three thymic hormones in the same RE cells of the human thymic microenvironment has also been published (334).

In a recent study, Gaudecker and co-investigators (377) used a library of MoABs together with gold-coupled reagents in immuno-electron microscopy to study the fine cellular distribution of the molecules to which the antibodies bind. Anti-cytokeratin antibodies were employed to identify the epithelial nature of all thymic RE cells, while the distribution of MHC class II molecules was revealed with MoABs to shared nonpolymorphic determinants. MoAB MR6 showed strong surface labeling of cortical RE cells and subcapsular TNCs and very weak surface staining of thymocytes, medullary macrophages, and interdigitating cells. MoAB 8.18 also labeled another cell surface molecule present on HBs and associated hypertrophied medullary RE cells. The two molecules detected by MoABs MR6 and 8.18 are therefore located in a position where they are available to interact with external cellular and soluble signals within the thymus. In contrast, MoABs MR10 and MR19 recognized intracellular molecules within subcapsular, perivascular, and medullary RE cells. A similar distribution pattern was seen with MoAB 4β, developed against the thymic hormone thymulin, although, in addition to the expected intracellular RE cell staining, large lymphoblasts in the subcapsular zone showed surface positivity, indicating the presence of thymulin bound to surface receptors on these early thymoblasts. As expected, MHC class II molecules were expressed on some medullary and essentially cortical RE cells. However, although most subcapsular epithelium was MHC class II-negative, some cells did express these MHC molecules on their apical surface and on the surface of their cytoplasmic extensions into the cortex. Interestingly, some cortical RE cells surrounding capillaries expressed MR6 and MR10, MR19 antigens, and presence of thymulin was revealed. Double-labeling experiments, using both MR6 and MR19 MoABs simultaneously, revealed the existence of a double-positive perivascularly located RE cell subpopulation in the thymic cortex. It remains a possibility that these "double positive" RE cells represent a common thymic RE cell progenitor.

Similar results were published after immunocytochemistry on RE cells derived from thymic tissue cultures (269). By the use of rapidly expanding cultures of thymic RE cells from 14 to 16 day-old murine fetuses and by specific antibodies against cortical and medullary epithelium, respectively, the authors were able to demonstrate a small subpopulation of double-labeled RE cells in the cultures. These cells were not present in RE cell cultures initiated from thymuses of neonatal mice. Double-labeled RE cells were also detected in prenatal thymic tissue sections. These findings may indicate that the "double positive" RE cell subpopulation represents a common epithelial stem cell for the thymic RE networks in the cortex and the medulla.

Amongst the most important signal transduction molecules involved in regulating growth and differentiation are the protein tyrosine kinases (PTK). Since intrathymic T lymphocyte maturation and differentiation is a consequence of cellular and humoral interactions between thymic stromal cells (TSC) and thymocytes, identification of the PTK in both compartments is required to dissect this process. PTK was observed in mouse TSC, using RT-PCR to survey the repertoire of PTK mRNAs expressed in a freshly isolated TSC preparation (375, 396). Ten different PTK cDNAs were identified among the 216 cDNAs sequenced. Transcripts of three of those (*ufo*, *fyn* and *fer*) were widely expressed among a large panel of immortalized thymic RE cell lines and in primary cultures of TSC. Of the other seven, none were expressed in established RE cell lines but, instead, displayed distinct expression patterns in cell types likely to have contaminated the fresh TSC preparation (i.e., macrophages, B cells, T cells and fibroblasts). Among the three PTK expressed in RE cell lines, only one, *ufo*, exhibited expression exclusively in cells of non-hemopoietic origin. Although expression of *ufo* (also named *tyro 7*, *axl* or *ark*) is not thymus specific, it is also expressed in cell types of mesodermal origin, placed in other tissues, its presence in thymic RE cells suggests a role for *ufo* in differentiation of the TSC microenvironment. Consistent with this notion, high-level expression of this receptor PTK at the protein level could be documented in every thymic RE cell line observed, as well as in fresh, frozen thymus tissue sections. These results provided the initial example of expression of a receptor PTK in TSC and opened a new direction in the study of the molecular regulation of thymic stromal microenvironment differentiation (375).

Our immunomorphological studies revealed that the so-called giant, subcapsular RE cells are TNCs, and are MHC (HLA-A,B,C) class I molecule restricted and express a neuroendocrine cell specific IP: Thy-1$^+$, $A_2B_5^+$, TT$^+$, TE4$^+$, UJ13/A$^+$, UJ127.11$^+$, UJ167.11$^+$, UJ181.4$^+$, and presence of common leukocyte antigen (CLA$^+$). In contrast, medullar RE cells showed MHC (HLA-DP, HLA-DQ, HLA-DR) class II molecule restriction. Essentially every CD2$^+$ thymocyte binds to RE cells using the CD2/CD58 pathway, therefore RE cells express the CD54 antigen. Activated thymocytes display the high affinity form of CD11a/CD18 (LFA-1) and bind to CD54$^+$ RE cells. In tissue culture conditions, RE cells lose their MHC restriction and also become CD54- but both antigens can be re-induced by γ-interferon (IFN-γ) treatment. Glucocorticoid hormone, prolactin and thyroid hormone receptors were mostly detected on cortical RE cells (intensity of staining: A to C and number of stained cells: ++, between 10% and 50%).

MoAB UJ 308 has been defined to react positively with cell surface receptors on immature myeloid cells and mature leukocytes. A strong immunoreactivity employing this MoAB was observed in the HBs in frozen tissue sections. Therefore, it is possible that certain leukocytes may contribute to the formation of HBs. The central core of the HBs also demonstrated strong expression of antigens detectable with MoABs UJ 308 and UJ 223.8, and this reactivity was independent of the presence of necrotic cellular changes. The hypertrophied, physiologically active RE cells of the peripheral cell layer of the HBs reacted positively with medium to strong intensity when stained with MoABs UJ127.11, J1153, A_2B_5, 215.D11, and 275.G7. These results further suggest that HBs are not exclusively degenerative structures (243). It is certain that the medullar RE cells stained by MoAB anti- A_2B_5 represent an active endocrine, peptide secreting RE cell functional subpopulation, associated with or already involved in the structure of HBs.

In human fetal thymic tissue, over 90% of the hypertrophied RE cells that were located in the outer core of HBs expressed TGF-βIIRs (372). A markedly decreased proportion of RE cells within the HBs of postnatal thymuses demonstrated presence of TGF-βIIR: only 10 to 50% during the early postnatal period which decreased further to between 1 and 10% during the first year of age (372). High density of this important growth factor receptor was also established in medullar RE cells adjacent to HBs.

Several groups of scientists isolated partially or fully purified thymic hormones, which can partially replace the function of the thymus in inducing T lymphocyte maturation (397-399). These hormones are secreted by thymic RE cells, and they could be produced during the first two weeks *in vitro* by culturing thymic RE cells. Two decades ago, Kater and co-workers (3, 283, 400) examined the effects of human thymus epithelial conditioned medium (HTECM) on *in vitro* mitogen responses of human and mouse lymphocytes. The anti-thymosin fraction 5 and 6 had a clear inhibitory effect on the HTECM and in the PHA and Con A proliferation responses of mouse lymphocytes. Surprisingly, the treatment with anti-thymosin α-1 MoAB did not inhibit these responses. The presence of thymic hormones in RE cells was also detected *in vivo* in connection with intrathymic T lymphocyte differentiation. Thymosin α-1, thymosin β-4, and thymopoietin (also its short form, called TP5), all biochemically well characterized thymic hormones, were detected immunocytochemically to be produced by the same RE cells (334). In the last four decades, experimental thymic research has increasingly focused on the identification of the potent immunomodulatory effects of these thymic hormones and neuropeptides. Consequently, the actions of steroid hormones have been observed by numerous research groups, and their effects on the thymus have been reviewed (401-408). It is amazing that the presence of intrathymic receptors for and actual production of the following collection of pituitary hormones and neuropeptides has been established: growth hormone, prolactin, adrenocorticotropic hormone, thyroid stimulating hormone, triiodothyronine, somatostatin, oxytocin, follicle stimulating hormone, luteinizing hormone, arginine vasopressin, growth hormone releasing hormone, corticotropin releasing hormone, nerve growth factor, vasoactive intestinal peptide, (pro) enkephalin, and β-endorphin (409-449). All of the experimental data define a regulatory, immuno-neuroendocrine function of the thymic RE cellular network.

It has long been known that malnutrition often results in zinc deficiency and that the biologically active form of FTS or thymulin requires presence of zinc (450). Zinc was identified in secretory granules associated with cytoplasmic vacuoles in thymic RE cells. Zinc is an essential component of over 200 enzymes, it is required for all DNA- and RNA-polymerases and Zn-metalloenzymes, all being participants in replication or repair of DNA. Mitotic activity of thymocytes increases when either calcium or magnesium is elevated in the cellular microenvironment. Parathyroidectomy results in the abolition of fluctuations in plasma level of calcium and cell mitosis, causing severe thymic atrophy. Enhanced epithelial growth and an increase in the number of macrophages were detected in thymic cultures with the addition of glutarine (a parathyroid hormone) to the culture medium. Recently calcitrol (biologically the most active metabolite of vitamin D) has been shown to inhibit IL-1 and IL-2 dependent thymocyte proliferation.

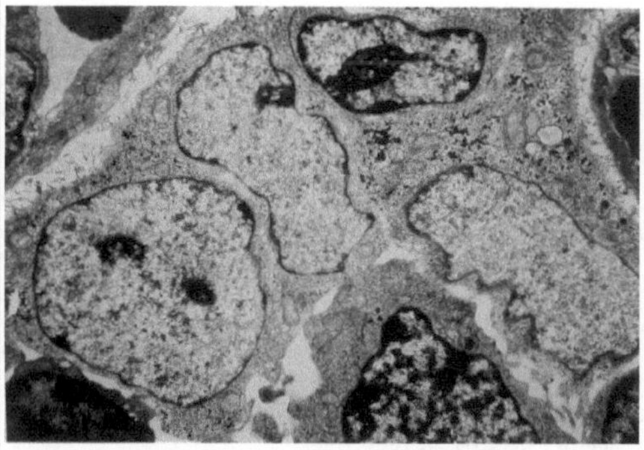

Figure 5. Dog fetus (91mm body length, 46th day of gestation). Thymic cortex. RE cell and differentiating thymocytes Characteristic for the thymus lympho-epithelial intercellular cooperation via cell surface molecules and humoral. Epon embedding. TEM picture. Magnification 18500x.

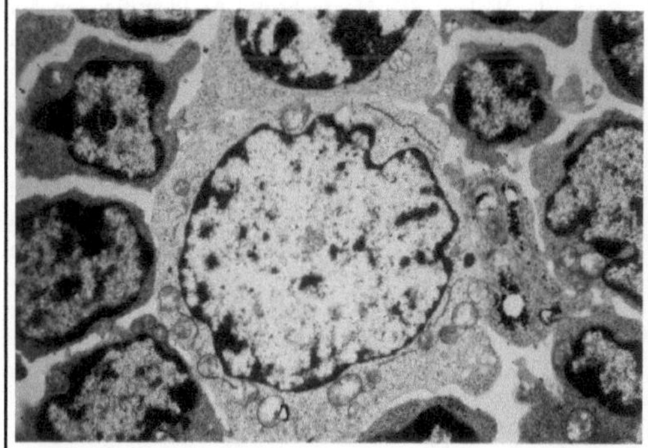

Figure 7. Thymic cortex. Dog fetus of 67 mm body length (early 5th lunar month). RE cell surrounded with thymocytes in maturation, and showing several nuclear envelope related cell to cell contacts. The type of cellular contact is not desmosomal. Epon embedding. TEM picture. Magnification 17500x.

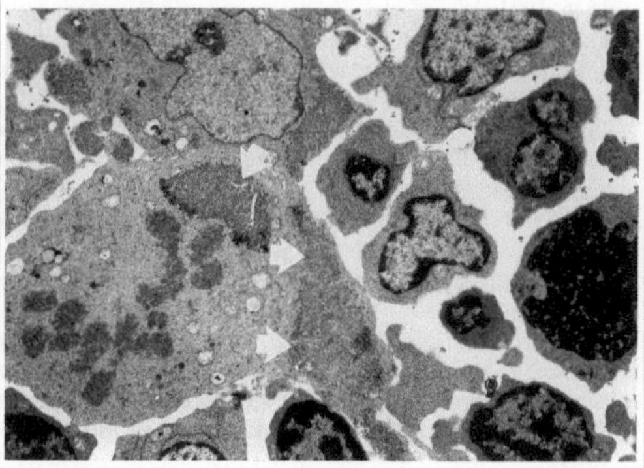

Figure 6. Thymic cortex. Dog fetus (43 days of intrauterine gestation). Several developing cortical thymocytes around a typical RE cell. The main function of the thymic RE network is to educate the immunocompetent T lymphocytes. Epon embedding. TEM picture. Magnification 18.500x.

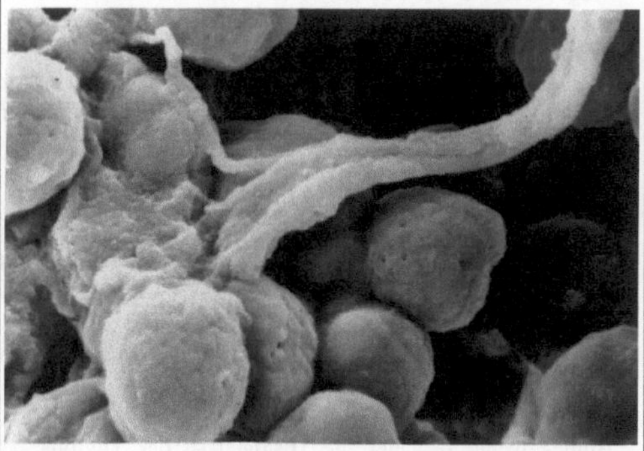

Figure 8.. Thymic cortex. Three months old mouse. Cytoplasmic process of a RE cell in intimate contact with a number of cortical immature thymocytes. SEM picture. Magnification 34000x.

REFERENCES

1. Sato VL, Waksal SD, Herzenberg LA: Identification and separation of pre T cells from *nu/nu mice* by preculture with thymic reticuloepithelial cells. Cell Immunol *24:* 173-185, 1976.
2. Stutman O: Intrathymic and extrathymic T cell maturation. Immunol Rev *42:*138-184, 1978.
3. Kruisbeek AM: Thymic factors and T cell maturation in vitro: a comparison of the effects of thymic epithelial cultures with thymic extracts and thymus dependent serum factors. Thymus *1:* 163-185, 1979.
4. Kyewski BA, Rouse RV, Kaplan HS: Thymocyte rosettes: multicellular complexes of lymphocytes and bone marrow-derived stromal cells in the *mouse* thymus. Proc Natl Acad Sci USA *79:* 5646-5650, 1982.
5. Jason JM, Janeway CA: *In vitro* cultivation of nonlymphoid thymic cells: morphological and immunological characterization. Clin Immunol Immunopathol *30:* 214-226, 1984.
6. Vakharia DD, Mitchison NA: Helper T cell activity demonstrated by thymic nurse cells (TNC-T). Immunol *51:* 269-273, 1984.
7. Schuurman HJ, van de Wijngaert FP, Huber J, Schuurman RK, Zegers BJ, Roord JJ, Kater L: The thymus in "bare lymphocyte" syndrome: significance of expression of major histocompatibility complex antigens on thymic epithelial cells in intrathymic T-cell maturation. Hum Immunol *13:* 69-82, 1985.
8. Singer KH, Wolf LS, Lobach DF, Denning SM, Tuck DT, Robertson AL, Haynes BF: *Human* thymocytes bind to autologous and allogeneic thymic epithelial cells *in vitro*. Proc Natl Acad Sci USA *83:* 6588-6592, 1986.
9. Singer KH, Haynes BF: Epithelial-thymocyte interactions in *human* thymus. Hum Immunol *20:* 127-144, 1987.
10. Kurtzberg J, Denning SM, Nycum LM, Singer KH, Haynes BF: Immature *human* thymocytes can be driven to differentiate into nonlymphoid lineages by cytokines from thymic epithelial cells. Proc Natl Acad Sci USA *86:* 7575-7579, 1989.
11. Zuniga-Pflucker JC, Jones LA, Longo DL, Kruisbeek AM: Both TCR/MHC and accessory molecule/MHC interactions are required for positive and negative selection of mature T cells in the thymus. Cold Spring Harb Symp Quant Biol *54 Pt 1:* 153-158, 1989.
12. Trainin N: Thymic hormones and the immune response. Physiol Rev *54:* 272-315, 1974.
13. Maximow AA: Untersuchungen über Blut und Bindegewebe. II. Über die Histogenese der Thymus bei *Saugetieren.* Arch mikrosk Anat *74:* 525-621, 1909.
14. Maximow AA: Untersuchungen über Blut und Bindegewebe. IV. Über die Histogenese der Thymus bei *Amphibien.* Arch mikrosk Anat *79:* 560-611, 1912.
15. Jolly J: Sur les organes lympho-épithéliaux. C R Soc Biol *74:* 540-543, 1913.
16. Jolly J: Le tissue lymphod'de, considéré comme un tissu de resérve. Arch Anat Micr *83:* 1209-1212, 1920.
17. Jolly J: Le thymus et les organes lymphod'des. C R Ass Anat. *19:* 1-10, 1924.
18. Jolly J: Le thymus est-il un organe lymphod'de. Bull d'Histol *1:* 176-179, 1924.
19. Jolly J and Lieure C: Recherches sur la greffe du thymus. Phénomčnes histologiques de la reconstitution du thymus greffé. Arch Anat Micr *28:* 159-221, 1932.
20. Jolly J, Tannenberg C: Sur l' existence de centres germinatifs dans le substance médullaire d' un thymus de *chat.* C R Soc Biol *90:* 405-407, 1924.
21. Jolly J, Tannenberg C: Centres germinatifs dans le thymus. Arch Anat Micr *27:* 127-147, 1931.
22. Grégoire C: Recherches sur le symbiose lymphoépithéliale au niveau du thymus de *Mammifere*. Arch Biol *46:* 717-820, 1935.
23. Grégoire C, Duchateau G: A study on lympho-epithelial symbiosis in thymus. Reactions of the lymphatic tissue to extracts and to implants of epithelial components of thymus. Arch Biol *68:* 269-296, 1956.
24. Krause C: Handbuch des Menschen. Hannover, 1876, p 359.
25. Kölliker A von: Entwicklungsgeschichte des Menschen und der höheren Tiere. 2nd Aufl, Leipzig, W Engleman Publ, 1879.
26. Pierson G: Ueber die Entwicklung der embryonalen Schlundspalten undihre Derivate bei Säugetieren. Zeitschr Zool *47:* 155-189, 1888.
27. Maurer F: Schilddrüse und Thymus der Teleostier. Morph Jahrb *11:* 129-175, 1886.
28. Popoff NW: The histogenesis of the thymus as shown by tissue cultures, transplantation and regeneration. Proc Soc Exper Biol Med *24:* 148-151, 1926.
29. Popoff NW: The histogenesis of the thymus as shown by tissue cultures. Arch Exp Zellforsch *4:* 395-418, 1927.
30. Deanesly R: The structure and development of the thymus in fish, with special reference to Salmo fario. J Micr Sci *71:* 113-145, 1927.
31. de Mello-Coelho V, Villa-Verde DM, Dardenne M, Savino W: Pituitary hormones modulate cell-cell interactions between thymocytes and thymic epithelial cells. J Neuroimmunol *76:* 39-49, 1997.
32. Lannes-Vieira J, Dardenne M, Savino W: Extracellular matrix components of the mouse thymus microenvironment: ontogenetic studies andmodulation by glucocorticoid hormones. J Histochem Cytochem *39:* 1539-1546, 1991.
33. Takigawa M, Imamura S, Ofuji Sh: Demonstration of epidermis-specific heteroantigens in thymic epithelial cells. Int Arch Allergy Appl Immunology *55:* 58-60, 1977.
34. Clawson CC, Cooper MD, Good RA: Lymphocyte fine structure in the bursa of fabricius, the thymus, and the germinal centers. Lab Invest *16:* 407-421, 1967.
35. Beard J: The source of leukocytes and the true function of the thymus. Anat Anz *18:* 550-573, 1900.
36. Zuckerkandl E: Die Entwicklung der Schilddrüse und der Thymus beider Ratte. Anat Hefte *21:* 1-28, 1903.
37. Hammar JA: Die Menschenthymus in Gesundheit und Krankheit. Teil II. Das Organ unter anormalen Körperverhältnissen. Zugleich Grundlagen der Theorie der Thymusfunktion. Zeitschr mikrosk-anat Forsch Erg.-Bd 16, 1929.
38. Zotterman A: Die Schweinethymus als eine Thymus ecto-entodermalis. Anat Anz *38:* 514-530, 1911.
39. Dustin A-P, Baillez G: Recherches sur les cultures de thymus "in vitro". Bull Soc Sci Med Natur *71:* 1-6, 1913.
40. Pappenheimer AM: Further studies of the histology of the thymus. Am J Anat *14:* 299-326, 1913.
41. Auerbach R: Experimental analysis of the origin of cell types in the development of the mouse thymus. Dev Biol *3:* 336-354, 1961.
42. Mysliwska J, Mysliwski A: Studies on binding of Fast-green and eosin Y with histones of rat lymph node cells. Folia Morphol *32:* 263-269, 1973.
43. Geenen V, Cormann-Goffin N, Martens H, Vandersmissen E, Robert F, Benhida A, Legros JJ, Martial J, Franchimont P: Thymic neurohypophysial-related peptides and T cell selection. Regul Pept *45:* 273-278, 1993.
44. Agus DB, Surh CD, Sprent J: Reentry of T cells to the adult thymus is restricted to activated T cells. J Exp Med *173:*

1039-1046, 1991.
45. Burgess A, Nicola N: Growth factors and stem cells. Academic Press, New York, pp. 1-38, 1983.
46. Clark MJ, Gagnon J, Williams AF, Barclay AN: MRC OX-2 antigen: A lymphoid/neuronal membrane glycoprotein with a structure like a single immunoglobulin light chain. EMBO J *4:* 113-118, 1985.
47. Dunon D, Kaufman J, Salomonsen J, Skjoedt K, Vainio O, Thiery JP, Imhof BA: T cell precursor migration towards β 2-microglobulin is involved in thymus colonization of chicken embryos. EMBO J 9: 3315-3322, 1990.
48. Fowlkes BJ, Pardoll DM: Molecular and cellular events of T-cell development. Adv Immunol *44:* 207-264, 1989.
49. Haynes BF, Denning SM, Singer KH, Kurtzberg J: Ontogeny of T-cell precursors: a model for the initial stages of human T-cell development. Immunol Today *2:* 87-89, 1989.
50. Janossy G, Pizzolo G: Lymphocytes. 2. Differentiation. Differentiation of lymphoid precursor cells. J Clin Pathol (suppl.) *13:* 48-58, 1979.
51. Jotereau FV, Le Douarin NM: Demonstration of a cyclic renewal of the lymphocyte precursor cells in the quail thymus during embryonic and perinatal life. J Immunol *129:* 1869-1877, 1982.
52. Le Douarin N, Jotereau FV: Tracing of cells of the avian thymus through embryonic life in interspecific chimeras. J Exp Med 142: 17-40, 1975.
53. Lesley J, Hyman R, Schulte R: Evidence that Pgp-1 glycoprotein is expressed on thymus-homing progenitor cells of the thymus. Cellular Immunol *91:* 397-403, 1985.
54. Lesley J, Trotter J, Hyman R: The Pgp-1 antigen is expressed on early fetal thymocytes. Immunogenetics *22:* 149-157, 1985.
55. Mathieson BJ, Fowlkes BJ: Cell surface antigen expression on thymocytes: Development and phenotypic differentiation of intrathymic subsets. Immunol Rev *82:* 141-173, 1984.
56. Matsumoto K, Yoshikai Y, Matsuzaki G, Asano T, Nomoto K: A novel CD3-J11d$^+$ subset of CD4$^+$8$^+$ cells repopulating thymus in radiation bone marrow chimeras. Eur J Immunol *19:* 1203-1208, 1989.
57. Nakano N, Hardy R, Kishimoto T: Identification of intrathymic T progenitor cells by expression of Thy-1, IL2 receptor and CD3. Eur J Immunol *17:* 1567-1571, 1987.
58. Palacios R, Sideras P, von Boehmer H: Recombinant mouse interleukin 4/BSF-1 promotes growth and differentiation of intrathymic T cell precursors from fetal mice in vitro. EMBO J *6:* 91-95, 1987.
59. Pardoll DM, Fowlkes BJ, Lechler RI, Germain RN, Schwartz RH: Early genetic events in T cell development analyzed by in situ hybridization. J Exp Med *165:* 1624-1638, 1987.
60. Penit C: In vivo proliferation and differentiation of prothymocytes in the thymus. Immunol Res *6:* 271-278, 1987.
61. Reinherz EL, Schlossman SF: The differentiation and function of human T lymphocytes: a review. Cell *19:* 821-827, 1980.
62. Rothenberg EV: Death and transfiguration of cortical thymocytes: a reconsideration. Immunol Today *11:* 116-119, 1990.
63. Scollay R: T-cell subset relationships in thymocyte development. Curr Opinion Immunol *3:* 204-209, 1991.
64. Scollay R, Andrews P, Boyd R, Shortman K: The role of the thymic cortex and medulla in T cell differentiation. Adv Exp Med Biol *186:* 229-234, 1985.
65. Scollay R, Chen W-F, Shortman K: The functional capabilities of cells leaving the thymus. J Immunol *132:* 25-30, 1984.
66. Scollay R, Shortman K: Thymocyte subpopulations: an experimental review, including flow cytometric cross-correlations between the major murine thymocyte markers. Thymus *5:* 245-295, 1983.
67. Scollay R, Smith J, Stauffer V: Dynamics of early T cells: Prothymocyte migration and proliferation in the adult mouse thymus. Immunol Rev *91:* 129-157, 1986.
68. Scollay R, Weissman I: T cell maturation: thymocyte and thymus migrant subpopulations defined with monoclonal antibodies to the antigens Lyt-1, Lyt-2 and ThB. J Immunol 124: 2841-2844, 1980.
69. Scollay R, Wilson A, D'Amico A, Kelly K, Egerton M, Pearse M, Wu L, Shortman K: Developmental status and reconstitution potential of subpopulations of murine thymocytes. Immunol Rev *104:* 81-120, 1988.
70. Spooncer E, Boettiger D, Dexter TM: Continuous in vitro generation of multipotential stem cell clones from src-infected cultures. Nature (London) *310:* 228-230, 1984.
71. Toribio ML, Martinez AC, Marcos MAR, Marquez C, Cabrero E, de la Hera A: A role for T3$^+$4$^-$6$^-$8$^-$ transitional thymocytes in the differentiation of mature and functional T-cells from human prothymocytes. Proc Natl Acad Sci USA *83:* 6985-6988, 1986.
72. Weiss A, Imboden JB: Cell surface molecules and early events involved in human T lymphocyte activation. Adv Immunol *41:* 1-38, 1987.
73. Weissman IL: Thymus cell maturation. Studies on the origin of cortisone resistant thymic lymphocytes. J Exp Med *137:* 504-510, 1973.
74. Dargemont C, Dunon D, Deugnier MA, Denoyelle M, Girault JM, Lederer F, Ho Diep Le K, Godeau F, Thiery JP, Imhof BA: Thymotaxin, a chemotactic protein, is identical to β2-microglobulin. Science *246:* 803-806, 1989.
75. Giunta M, Favre A, Ramarlio D, Grossi CE, Corte G: A novel integrin involved in thymocyte-thymic epithelial cell interactions. J Exp Med *173:* 1537-1548, 1991.
76. Ignotz RA, Heino J, Massague J: Regulation of cell adhesion receptors by transforming growth factor β. J Biol Chem *264:* 389-392, 1989.
77. Kina T, Majumdar AS, Heimfeld S, Kaneshima H, Holzmann B, Katsura Y, Weissman IL: Identification of a 107-kD glycoprotein that mediates adhesion between stromal cells and hematolymphoid cells. J Exp Med *173:* 373-381, 1991.
78. Le PT, Kurtzberg J, Brandt SJ, Niedel JE, Haynes BF, Singer KH: Human thymic epithelial cells produce granulocyte and macrophage colony stimulating factors. J Immunol *141:* 1211-1217, 1988.
79. Potworowski EF, Hugo P, Couture C: Binding of cortical thymocytes to a medullary epithelial cell line: a brief review. Thymus *13:* 237-243, 1989.
80. Potworowski EF, Thibodeau L, Zelechowska MG: Specific adherence of thymocytes to a thymic medullary epithelial cell line. Immunol Lett *13:* 89-94, 1986.
81. Roson E, Garcia-Caballero T, Heimer EP, Felix AM, Dominguez F: Cellular distribution of prothymosin α and parathymosin in rat thymus and spleen. J Histochem Cytochem *38:* 1889-1894, 1990.
82. Springer TA, Dustin ML, Kishimoto TK, Marlin SD: The lymphocyte function-associated LFA-1, CD2, and LFA-3 molecules: cell adhesion receptors of the immune system. Annu Rev Immunol *5:* 223-252, 1987.
83. Zelechowska MG, Lecomte J, Potworowski EF: Immunocytochemical localization of a thymic medullary epithelial glycoprotein. Scand J Immunol *23:* 561-565, 1986.
84. Brelinska R: Thymic nurse cells: division of thymocytes within complexes. Cell Tissue Res *258:* 637-643, 1989.

85. Brelinska R, Warchol JB, Boniver J, Houben-Defresne MP: Distribution of Lyt antigens on the surface of thymocytes associated with thymic macrophages and dendritic cells. Histochemistry *88:* 553-556, 1988.
86. Colic M, Jovanovic S, Mitrovic S, Dujic A: Immunohistochemical identification of six cytokeratin-defined subsets of the rat thymic epithelial-cells. Thymus *13:* 175-185, 1989.
87. Colic M, Matanovic D, Hegedis L, Dujic A: Heterogeneity of rat thymic epithelium defined by monoclonal anti-keratin antibodies. Thymus *12:* 123-130, 1988/89.
88. Colic M, Matanovic D, Hegedis L, Dujic A: Immunohistochemical characterization of rat thymic non-lymphoid cells. I. Epithelial and mesenchymal components defined by monoclonal antibodies. Immunology *65:* 277-284, 1988.
89. Colic M, Popovic LJ, Gasic S, Dragojevic-Simic V, Milicevic NM, Matanovic D, Dujic A: Immunohistochemical characterization of rat thymic non-lymphoid cells. II. Macrophages and granulocytes defined by monoclonal antibodies. Immunology *69:* 416-422, 1990.
90. Eshel I, Savion N, Shoham J: Analysis of thymic stromal cell subpopulations grown in vitro on extracellular matrix in defined medium. I. Growth conditions and morphology of murine thymic epithelial and mesenchymal cells. J Immunol *144:* 1554-1562, 1990.
91. 91. Eshel I, Savion N, Shoham J: Analysis of thymic stromal cell subpopulations grown in vitro on extracellular matrix in defined medium. II. Cytokine activities in murine thymic epithelial and mesenchymal cell culture supernatants. J Immunol *144:* 1563-1570, 1990.
92. Fink PJ, Weissman IL, Kaplan HS, Kyewski BA: The immunocompetence of murine stromal cell associated thymocytes. J Immunol *132:* 2266- 2272, 1984.
93. Galy AH, Dinarello CA, Kupper TS, Kameda A, Hadden JW: Effects of cytokines on human thymic epithelial cells in culture. II. Recombinant IL1 stimulates thymic epithelial cells to produce IL6 and GM-CSF. Cellular Immunol *129:* 161-175, 1990.
94. Hiramine C, Hojo K, Koseto M, Nakagawa A, Mukasa A: Establishment of a murine thymic epithelial cell line capable of inducing both thymic nurse cell formation and thymocyte apoptosis. Lab. Invest. *62:* 41-54, 1990.
95. Kosaka H, Ogata M, Sano H, Sakuma S, Mizushima Y, Nakamura E, Teraoka H, Hamaoka T, Fujiwara H: Thymic stroma-derived T cell growth factor (TSTGF): III. Its ability to promote T cell proliferation without stimulating interleukin 2 or 4-dependent autocrine mechanism. J Leukocyte Biol *47:* 121-128, 1990.
96. Dalloul AH, Mossalayi MD, Dellagi K, Bertho JM, Debre P: Factor requirements for activation and proliferation steps of human $CD2^+CD3^-CD4^-CD8^-$ early thymocytes. Eur J Immunol *19:* 1985-1990, 1989.
97. Fabien N, Auger C, Monier JC: Immunolocalization of thymosin α-1, thymopoietin and thymulin in mouse thymic epithelial cells at different stages of culture: a light and electron microscopic study. Immunology *63:* 721-727, 1988.
98. Fabris N, Mocchegiani E, Mariotti S, Pacini F, Pinchera A: Thyroid function modulates thymic endocrine activity. J Clin Endocrinol Metab *62:* 474-478, 1986.
99. Fujiwara H: Thymic stroma-derived T cell growth factor: its role in the growth of immature thymocytes and T cell repertoire selection. Int Symp Princess Takamatsu Cancer Res Fund *19:* 115-126, 1988.
100. Fujiwara H, Ogata M, Mizushima Y, Tatsumi Y, Takai Y, Hamaoka T: Proliferation and differentiation of immature thymocytes induced by a thymic stromal cell clone. Thymus *16:* 159-172, 1990.
101. Fujiwara S, Kobayashi H, Awaya A: The distribution and structure of FTS immunoreactive cells in the thymus of the mouse. Arch Histol Jpn *51:* 467-472, 1988.
102. Fujiwara S, Fisher RJ, Bhat NK, Diaz de la Espina SM, Papas TS: A short-lived nuclear phosphoprotein encoded by the human ets-2-protooncogene is stabilized by activation of protein kinase "C". Mol Cell Biol *8:* 4700-4706, 1988.
103. Geenen V, Defresne M-P, Robert F, Legros J-J, Franchimont P, Boniver J: The neurohormonal thymic microenvironment: immunocytochemical evidence that thymic nurse cells are neuroendocrine cells. Neuroendocrinology *47:* 365-368, 1988.
104. Inaba K, Steinman RM: Accessory cell - T lymphocyte interactions. Antigen-dependent and -independent clustering. J Exp Med *163:* 247-261, 1986.
105. Kinashi T, Tashiro K, Lee KH, Inaba K, Toyama K, Palacios R, Honjo T: Growth factors involved in lymphocyte differentiation. Phil Trans R Soc (London) *B327:* 117-125, 1990.
106. Kurtzberg J, Denning SM, Nycum LM, Singer KH, Haynes BF: Human immature thymocytes can be driven to differentiate into non-lymphoid lineages by thymic epithelial-derived cytokines. Proc Natl Acad Sci USA *86:* 7575-7579, 1989.
107. Lafontaine M, Landry D, Blanc-Brunat N, Pelletier M, Montplaisir S: IL-1 production by human thymic dendritic cells: studies on the interrelation with DC accessory function. Cellular Immunol *135:* 431-444, 1991.
108. Pfeifer-Ohlsson S, Rydnert J, Scott Goustin A, Larsson E, Betsholtz C, Ohlsson R: Cell-type-specific pattern of c-myc protooncogene expression in developing human embryos. Proc Natl Acad Sci USA *82:* 5050-5054, 1985.
109. Greenbaum LA, Horowitz JB, Woods A, Pasqualini T, Reich E-P, Bottomly K: Autocrine growth of $CD4^+$ T cells. J Immunol 140: 1555-1560, 1988.
110. Bodey B (1977) Histomorphology and histochemistry of the human thymus during the prenatal development. pp. 1-360, Ph. D. Thesis Sofia (590 references).
111. Brune B, Hartzell P, Nicotera P, Orrenius S: Spermine prevents endonuclease activation and apoptosis in thymocytes. Exp Cell Res *195:* 323-329, 1991.
112. Defresne MP, Humblet Ch, Rongy AM, Greimers R, Boniver J: Effect of interferon-γ and tumor necrosis factor-α on lymphoepithelial interactions within thymic nurse cells. Eur J Immunol *20:* 429-432, 1990.
113. Denning SM, Tuck DT, Singer KH, Haynes BF: Human thymic epithelial cells function as accessory cells for autologous mature thymocyte activation. J Immunol *138:* 680-686, 1987.
114. Grabstein KH, Namen AE, Shanebeck K, Voice RF, Reed SG, Widmer MB: Regulation of T cell proliferation by IL-7. J Immunol *144:* 3015-3020, 1990.
115. Granstein RD, Staszewski R, Knisely TL, Zelinsky DM: Aqueos humor contains transforming growth factor-β and a small (<3500 daltons) inhibitor of thymocyte proliferation. J Immunol *144:* 3021-3027, 1990.
116. Ingold AL, Landel C, Knall C, Evans GA, Potter TA: Co-engagement of CD8 with the T cell receptor is required for negative selection. Nature (London) *352:* 721-723, 1991.
117. Jung LKL, Haynes BF, Nakamura S, Pahwa S, Fu SM: Expression of early activation antigen (CD69) during human thymic development. Clin exp Immunol *81:* 466-474, 1990.
118. Kyewski BA, Jenkinson EJ, Kingston R, Altevogt P, Owen MJ, Owen JJT: The effects of anti-CD2 antibodies on the differentiation of mouse thymocytes. Eur J Immunol *19:* 951-954, 1989.
119. Mackay CR: T-cell memory: the connection between

function, phenotype and migration pathways. Immunol Today *12:* 189-192, 1991.
120. Nakashima M, Mori K, Maeda K, Kishi H, Hirata K, Kawabuchi M, Watanabe T: Selective elimination of double-positive immature thymocytes by a thymic epithelial cell line. Eur J Immunol *20:* 47-53, 1990.
121. Nonoyama S, Nakayama M, Shiohara T, Yata J: Only dull $CD3^+$ thymocytes bind to thymic epithelial cells. The binding is elicited by both CD2/LFA-3 and LFA-1/ICAM-1 interactions. Eur J Immunol *19:* 1631-1635, 1989.
122. Saalmuller A, Hirt W, Reddehase MJ: Phenotypic discrimination between thymic and extrathymic CD4-CD8- and $CD4^+CD8^+$ porcine T lymphocytes. Eur J Immunol *19:* 2011-2016, 1989.
123. Spangelo BL, Ross PC, Judd AM, McLeod RM: Thymic stromal elements contain an anterior pituitary hormone-stimulating activity. J Neuroimmunology *25:* 37-46, 1989.
124. Weiss A, Manger B, Imboden J: Synergy between the T3/antigen receptor complex and T44 in the activation of human T cells. J Immunol *137:* 819-825, 1986.
125. Zlotnik A, Ranson J, Frank G, Fischer M, Howard M: Interleukin 4 is a growth factor for activated thymocytes: possible role in T cell ontogeny. Proc Natl Acad Sci U.S.A. *84:* 3856-3860, 1987.
126. Bodey B, Calvo W, Prummer O, Fliedner TM, Borysenko M: Development and histogenesis of the thymus in dog. A light and electron microscopical study. Developmental Compar Immunol *11:* 227-238, 1987.
127. Couture C, Patel PC, Potworowski EF: A novel thymic epithelial adhesion molecule. Eur J Immunol *20:* 2769-2773, 1990.
128. Farr A: Epithelial heterogeneity in the murine thymus: a cell surface glycoprotein expressed by subcapsular and medullary epithelium. J Histochem Cytochem *39:* 645-653, 1991.
129. Godfrey DI, Izon DJ, Tucek CL, Wilson TJ, Boyd RL: The phenotypic heterogeneity of mouse thymic stromal cells. Immunology *70:* 66-74, 1990.
130. Gutierrez JC, Palacios R: Heterogeneity of thymic epithelial cells in promoting T-lymphocyte differentiation in vivo. Proc Natl Acad Sci USA 88*:* 5685-5689, 1991.
131. Hadden JW, Galy A, Chen H, Hadden EM: A pituitary factor induces thymic epithelial cell proliferation in vitro. Brain Behav Immunol *3:* 149-159, 1989.
132. Haynes BF: Human T lymphocyte antigens as defined by monoclonal antibodies. Immunol Rev 57: 127-161, 1981.
133. Haynes BF: The human thymic microenvironment. Adv Immunol *36:* 81-142, 1984.
134. Haynes BF: Use of monoclonal antibodies to identify antigens of human endocrine thymic epithelium. In: Thymic hormones and lymfokines: biochemical chemistry and clinical applications. AL, Goldstein (ed), Plenum Press, pp. 43-56, 1984.
135. Haynes BF: Human thymic epithelium and T cell development: Current issues and future directions. Thymus *16:* 143-157, 1990.
136. Haynes BF, Crummett B, Scearce RM, Wolf LS, Tuck D, Singer KG: Summary of reactivity patterns of the thymic epithelial panel of antibodies in human skin and thymus. In: Leucocyte Typing III., AJ, McMichael (ed), Oxford University Press, Oxford, pp. 262-269, 1987.
137. Haynes BF, Denning SM, Le PT, Singer KH: Human intrathymic T cell differentiation. Semin Immunol *2:* 67-77, 1990
138. Haynes BF, Singer KH, Denning SM, Martin ME: Analysis of expression of CD2, CD3, and T cell antigen receptor molecule during early human fetal thymic development. J Immunol *141:* 3776-3784, 1988.
139. Heldin C-H, Westermark B: Growth factors as transforming proteins. Eur J Biochem *184:* 487-496, 1989.
140. Janossy G, Bofill M, Trejdosiewicz LK, Willcox HN, Chilosi M: Cellular differentiation of lymphoid subpopulations and their microenvironments in the human thymus. Curr Top Pathol *75:* 89-125, 1986.
141. Janossy G, Thomas JA, Bollum FJ, Granger S, Pizzolo G, Bradstock KF, Wong L, McMichael A, Ganeshaguru K, Hoffbrand AV: The human thymic microenvironment: an immunohistologic study. J Immunol *125:* 202-212, 1980.
142. Ardavin C, Wu L, Ferrero I, Shortman K: Mouse thymic dendritic cell subpopulations. Immunol Letters 38: 19-25, 1993.
143. Bjorkmann P, Saper MA, Samaoui B, Bennett WS, Strominger JL, Wiley DC: Structure of the human class I histocompatibility antigen, HLA-A2. Nature (London) *329:* 506-512, 1987.
144. Brown JH, Jardetzky T, Saper MA, Samraoui B, Bjorkman PJ, Wiley DC: A hypothetical model of the foreign antigen binding site of class II histocompatibility molecules. Nature (London) *332:* 845-850, 1988.
145. Claverie J-M, Kourilsky P: The peptidic self model: A reassessment of the role of the major histocompatibility complex molecules in the restriction of the T-cell response. Annu Immunol *137:* 425-442, 1987.
146. Dellabona P, Peccoud J, Kappler J, Marrack P, Benoist C, Mathis D: Superantigens interact with MHC Class II molecules outside of the antigen groove. Cell *62:* 1115-1121, 1990.
147. Goldberg AL, Rock KL: Proteolysis, proteasomes and antigen presentation. Nature (London) *357:* 375-379, 1992.
148. Landry D, Lafontaine M, Barthelemy H, Paquette N, Chartrand C, Pelletier M, Montplaisir S: Human thymic dendritic cell-thymocyte association: Ultrastructural cell phenotype analysis. Eur J Immunol *19:* 1855-1860, 1989.
149. Morale MC, Farinella Z, Marchetti B: Central nervous system (CNS) modulation of immune system development: role of the thymic β2-adrenergic receptor. Pharmacol Res *22:* 47-48, 1990.
150. Ostergaard HL, Trowbridge IS: Coclustering CD45 with CD4 or CD8 alters the phosphorylation and kinase activity of p56 lck. J Exp Med *172:* 347-350, 1990.
151. Peault B, Chen C-LH, Cooper MD, Barbu M, Lipinski M, Le Douarin NM: Phylogenetically conserved antigen on nerve cells and lymphocytes resembles myelin-associated glycoprotein. Proc Natl Acad Sci USA, *84:* 814-818, 1987.
152. Tucek CL, Boyd RL: Surface expression of CD4 and Thy-1 on mouse thymic stromal cells. Int Immunol *2:* 593-601, 1990.
153. van Vliet E, Jenkinson EJ, Kingston R, Owen JJT, van Ewijk W: Stromal cell types in the developing thymus of the normal and nude mouse embryo. Eur J Immunol *15:* 675-681, 1985.
154. von Boehmer H: Self recognition by the immune system. Eur J Biochem *194:* 693-698, 1990.
155. von Boehmer H, Schubiger K: Thymocytes appear to ignore class I major histocompatibility complex antigens expressed on thymus epithelial cells. Eur J Immunol *14:* 1048-1052, 1984.
156. von Boehmer H, Teh HS, Kisielow P: The thymus selects the useful, neglects the useless and destroys the harmful. Immunol Today *10:* 57-61, 1989.
157. Wick G, Oberhuber G: Thymic nurse cells: a school for alloreactive and autoreactive cortical thymocytes? Eur J Immunol *16:* 855-858, 1986.
158. Beard J: The source of leukocytes and the true function of the thymus. Anat Anz *18:* 550-573, 1900.
159. Beard J: The origin and histogenesis of the thymus in Raja

batis. Zool Jahrbucher, Abt Ontogenie Tiere *17:* 403-480, 1902.
160. Bell ET: The development of the thymus. Amer J Anat *5:* 29-62, 1905.
161. Hammar JA: Zur Histogenese und Involution der Thymusdruse. Anat Anz *27:* 23-30 and 41-89, 1905.
162. Hammar JA: Zur Frage der Histogenese der Thymusdruse. Zentrbl Pathol *33:* 505-513, 1923.
163. Maximow AA: Untersuchungen ueber Blut und Bindegewebe. II. Ueber die Histogenese der Thymus bei Saugetieren. Arch mikrosk Anat *74:* 525-621, 1909.
164. Nicolas JF, Reano A, Kaiserlian D, Thivolet J: Epithelial cell heterogeneity in mammalian thymus: monoclonal antibody to high molecular weight keratins exclusively binds to Hassall's corpuscles. Histochemical J *21:* 357-364, 1989.
165. Norris EH: The morphogenesis and histogenesis of the thymus gland in man: in which the origin of the Hassall's corpuscles of the human thymus is discovered. Contributions to embryology No.166 pp. 193-207, 1938.
166. van Ewijk W: Cell surface topography of thymic microenvironment. Lab Invest *59:* 579-590, 1988.
167. van Ewijk W: Immunohistology of lymphoid organs. Curr Opinion Immunol *1:* 954-965, 1989.
168. van Vliet E, Melis M, van Ewijk W: Monoclonal antibodies to stromal cell types of the mouse thymus. Eur J Immunol *14:* 524-529, 1984.
169. von Gaudecker B: Functional histology of the human thymus. Anat Embryol *183:* 1-15, 1991.
170. Weller GL: Development of the thyroid, parathyroid and thymus glands in man. Contrib Embryol Carnegie Inst *24:* 95-138, 1933.
171. Jacobj W: Die Zellkerngröβe beim *Menschen.* Ein Beitrag zur quantitativen Zytologie. Zschr Mikr-Anat Forsch *38:* 161-240, 1935.
172. Downey H: Cytology of *rabbit* thymus and regeneration of its thymocytes after irradiation with notes on *human* thymus. Blood *3:* 1315-1341, 1948.
173. Törö I: The cytology of the thymus gland. Folia Biol *7:* 145-149, 1961.
174. Sainte-Marie G, Leblond CP: Cytologic features and cellular migration in the cortex and medulla of thymus in the young adult *rat.* Blood *23:* 275-299, 1964.
175. Bruijntjes JP, Kuper CF, Robinson JE, Schuurman HJ: Epithelium-free area in the thymic cortex of *rats.* Dev Immunol *3:* 113-122, 1993.
176. Smith C: Stdies on the thymus of the mammal. XV. Sites of aminopeptidase activity in the thymus of the mouse. Anat Rec *158:* 149-161, 1967.
177. 177. Wekerle H, Ketelsen UP, Ernst M: Thymis nurse cells. Lymphoepithelial cell complexes in *murine* thymuses: morphological and serological characterization. J Exp Med *151:* 925-944, 1980.
178. Wekerle H, Ketelsen UP: Thymic nurse cells. Ia-bearing epithelium involved in T-lymphocyte differentiation. Nature *283:* 402-404, 1980.
179. Jolly J, Lacassagne A: De la resistence des leucocytes du sang vis-a-vis des rayons X. Cr. Soc. Biol. Paris *89:* 379-381, 1923.
180. Prenant A, Contribution à l'étude organique et histologique du thymus de la glande thyroide et de la glande carotidienne. Cellule *10*, 1894.
181. Hammar JA: Fünfzig Jahre Thymusforschung. Ergebn Anat Entwicklunggesch *19:* 1-274, 1909.
182. Hill BH: The early development of the thymus glands in Amin calva. J Morph Physiol *57:* 61-89, 1935.
183. Bell ET: The development of the thymus. Amer J Anat *5:* 29-62, 1905.
184. Metcalf D, Ishidate M: Periodic acid-Schiff positive giant cells in the *mouse* thymus cortex. Nature *191:* 305, 1961.
185. Metcalf D, Ishidate M: PAS-positive reticulum cells in the thymus cortex of high and low leukemia strains of mice. Austr J Exp Biol Med Sci *40:* 57-72, 1962.
186. Huang RC, Bonner J: Histone, a suppressor or chromosomal RNA synthesis. Proc Nat Acad Sci (USA) *48:* 1216-1222, 1962.
187. Mandel T: Ultrastructure of epithelial cells in the medulla of the guinea pig thymus. Aust J Exp Biol Med Sci *46:* 755-767, 1968.
188. Mandel T: Ultrastructure of epithelial cells in the cortexof the guinea pig thymus. Z Zellforsch *92:*159-168, 1968.
189. Hammar JA: Die normal-morphologische Thymusforschung im letzten Vierteljahrhundert. Analyse and Syntese, Leipzig, 1936.
190. Klose W: Beiträge zur Morphologie und Histologie der Thymusdrüse und des postbrachialen Körpers von Proteus anguinen. Z Zellforsch *14:* 385-439, 1931.
191. James NS: The morphology of the thymus and its changes with age in the neotenous amphibian (Necturus Maeulosus). J Morph Physiol *64:* 455-481, 1939.
192. Herxheimer G: Fettinfiltration der Thymus. Verh Dtsch Path Ges *6*, 1903-1904.
193. Hammer JA: Über progressive und regressive Formen Hassalk schen Körprer. Z Anat *70:*466-488, 1924.
194. Holmström R: Über das Vorkommen von Fett und Fettähnliehen Substanzen im Thymus Parenchym. Arch Mikrosk Anat *77:*323-345, 1911.
195. Barbano C: Die normale Involution der Thymus. Virchows Arch *207:* 1-27, 1912.
196. Bargmann W: Der Thymus. In Handbuch der mikroskopischen Anatomie des Menschen, hrg. Möllendorf von W, Erg Bd VI/1, Springer, Berlin, pp 1-172, 1943.
197. Hart C: Thymusstudien I Über das Auftreten von Fett in der Thymus. Virchos Arsh *207*: 27-55, 1912.
198. Herrmann T: Das Auftreten des Fettegewebes im menschlichen Thymus. Anat Anz *47:* 357-359, 1914.
199. Honda I: Über Lipoide in der Thymusdrüse. Ref Jap J Med Sci *1*, 1928.
200. Kyrilov A: Über das Vorkommen und die Bedeutung von Lipoiden im Thymus. Z Konstit Lehre *10:* 460-481, 1924.
201. Bodey B, Hadjioloff AI: Thymus development, structure and function. Nature *26:* 11-19, 1977.
202. Hammar JA: Thymus. Encycl Med Technik *3:* 2161-2168, 1927.
203. Dustin A-P: A propos d' une nouvelle explicative de l' histogęnese du thymus. Arch Biol *45:* 339-352, 1934.
204. Choi MH: Experimental study about histogenesis of the *amphibian* thymus gland. Fol Anat Jpn *9:* 495-503, 1931.
205. Berblinger W: Thymus. Physiologie und Pathologie dieses Organs. N D Klinik *10:* 457-477, 1932.
206. Godwin MC: The *mammalian* thymus IV. The development in the *dog.* Am J Anat *64:* 165-201, 1939.
207. Bétance LM, de Luna J: Le réticulum thymique. C R Soc Biol *92:* 241-242, 1925.
208. Mandel T: Differentiation of epithelial cells in the *mouse* thymus. Zeitschr Zellforsch *106:* 498-515, 1970.
209. Hammar JA: Der gegenwärtige Stand der Morphologie und Physiologie der Thymusdrüse. Wien Med Wschr *2:* 2746-2750, 2795-2801, & 2910-2916, 1909.
210. Hammar JA: The new views as to the morphology of the thymus gland and their bearing on the problem of the function of the thymus. Endocrinology *5:* 543-573 & 731-760, 1921.
211. Hart C: Thymusstudien. II. Die Thymuselemente. Virchows Arch *207:* 255-277, 1912.
212. Hart C: Thymusstudien. III. Die Pathologie der Thymus.

Vichows Arch *214:* 1-83, 1913.
213. Loewenthal LA, Smith C: Studies on the thymus of *mammal*. IV. Lipid-laden foamy cells in the involuting thymus of the *mouse*. Anat Rec *112:* 1-15, 1952.
214. Bodey B, Bodey B Jr, Siegel SE, Kaiser HE: Involution of the mammalian thymus, one of the leading regulators of aging. In Vivo *11:* 421-440, 1997.
215. Gilles R: Histochemistry of Hassall corpuscles. C R Soc Biol *138:* 890-892, 1944.
216. Smith C, Parkhurts HT: Thymus of *mammal*; comparison of staining properties of Hassall's corpuscles and of thick skin of *guinea pig*. Anat Rec *103:* 649-673, 1949.
217. Rogister G, Dumoulin EL, Gerebtzoff MA: Histochemical studies on acetylcholine and choline esterases; acetylcholinesterase in thymus and biochemical measurement of its activity. Acta Anat *25:* 361-371, 1955.
218. Phillips DM, Johns EW: A study of the proteinase content and the chromatography of thymus histones. Biochem J *72:* 538-544, 1959.
219. Rasmussen PS, Murray K, Luck JM: On the complexity of *calf* thymus histone. Biochemistry *1:* 79-89, 1962.
220. Korhonen LK, Ruponen S: Leucine aminopeptidase in the thymus. Acta Anat *50:* 186-190, 1962.
221. Agostoni A, Vergani C, Villa L: Intracellular distribution of the different forms of lactic dehydrogenase. Nature *209:* 1024-1025, 1966.
222. Dioguardi N, Ideo G, Mannucci PM, Fiorelli G, Agostoni A: Multiple molecular forms of lactatedehydrogenase (LDH) of normal and leukemic cells of the myeloid line. Enzymol Biol Clin *6:* 1-9, 1966.
223. Tittobello A, Agostoni A: Leukocyte alkaline phosphatase: a quantitative and qualitative study in normal and leukemic cells. Enzymol Biol Clin (Basel) *8:* 413-420, 1967.
224. Hule V: Lactate dehydrogenase isoenzyme patterns of *human* lymphatic and myeloid leukemic leukocytes. Clin Chim Acta *17:* 349-352, 1967.
225. Hule V: Lactate dehydrogenase isoenzymes in the formed elements of human blood. Nouv Rev Fr Hematol *8:* 262-263, 1968.
226. Day AJ: Lipid metabolism by macrophages and its relationship to atherosclerosis. Adv Lipid Res *5:* 185-207, 1967.
227. Wahlqvist ML, Day AJ: Phospholipid synthesis by foam cells in *human*. Exp Mol Pathol *11:* 275-284, 1969.
228. Day AJ, Tume RK: Cholesterol-esterifying activity of cell-free preparations of *rabbit* peritoneal macrophages. Biochim Biophys Acta *176:* 367-376, 1969.
229. Fidge NH, Tume RK, Day AJ: Enzymic synthesis by glycerolipids by cell-free preparations of peritoneal macrophages. J Reticuloendothel Soc *7:* 609-616, 1970.
230. Tume RK, Day AJ: Cholesterol esterifying activity of rabbit alveolar macrophages. J Reticuloendothel Soc *7:* 338-354, 1970.
231. Smith C: Studies on the thymus of the *mammal.* XVI. Lipopigment cells in the cortex of the thymus of the mouse. Am J Anat *124:* 389-409, 1969.
232. Arvy L: Enzymology of the thymus. In Handbuch der Histochemie, VIII, Gustav Fischer, Stuttgart, 1973.
233. Smith CL, Pilarski LM, Egerton ML, Wiley JS: Nucleoside transport and proliferative rate in *human* thymocytes and lymphocytes. Blood *74:* 2038-2042, 1989.
234. Szigeti M: Age-dependent changes in the level of lysosomal enzymes in the lymphatic organs of the guinea pig. Acta Biol Acad Sci Hung *19:* 305-315, 1968.
235. Kapa E, Szigeti M, Juhasz A, Csaba G: Phylogenesis of mast cells. I. Mast cells of the *frog Rana esculenta.* Acta Biol Acad Sci Hung *21:* 141-147, 1970.
236. Timperley WR, Ahmed A: Histochemistry of the thymus and a thymoma. Arch Pathol *89:* 405-409, 1970.
237. Timperley WR, Ahmed A, Barson AJ: Histochemistry of the *human* thymus and a thymoma. J Pathol *103:* Pv, 1971.
238. Timperley WR, Barson AJ, Davies S: Enzyme histochemistry of the developing *human* thymus. Biol Neonate *19:* 42-54, 1971.
239. Timperley WR, Barson AJ, Davies S: Isoenzyme patterns of non-specific esterase and acid phosphatase in the developing brain and a variety of normal tissues. Biol Neonate *19:* 459-464, 1971.
240. Caso LV: Histochemical studies of mucosubstances and keratins of thymic corpuscles. Aust J Exp Biol Med Sci *57:* 493-496, 1979.
241. Agostoni A, Ideo G, Fiorelli G, Vergan C, Dioguardi N: Lactic dehydrogenase (LDM) molecular forms of human normal and leukemic lymphocystes. Proc X-th Congr Int Soc Saemat, Stockholm, 1964.
242. Bodey B, Bodey B Jr, Kemshead JT, Siegel SE, Kaiser HE: Identification of neural crest derived cells within the cellular microenvironment of the *human* thymus employing a library of monoclonal antibodies raised against neuronal tissues. In Vivo *10:* 39-47, 1996.
243. Bodey B, Kaiser HE: Development of Hassall's bodies of the thymus in *humans* and other *vertebrates* (especially *mammals*) under physiological and pathological conditions: Immunocytochemical, electronmicroscopic and *in vitro* observations. In Vivo *11:* 61-86, 1997.
244. Palumbi G, Millonig G: First observations on the ultrastructure of the thymus, in newborn *kittens.* Arch Ital Anat Embriol *65:* 155-167, 1960.
245. Koka T: Electron-microscopical studies on the thymus, especially on its epithelial cells. Igaku Kenkyu *30:* 309-325, 1960; and Excerpta Med *15:* 101, 1961.
246. Hoshino T: Electron microscopic studies of the epithelial reticular cells of the *mouse* thymus. Zeitschr Zellforsch *59:* 513-529, 1963.
247. Murray RG, Murray AS, Pizzo A: The fine structure of mitosis in *rat* thymic lymphocytes. J Cell Biol *26:* 601-619, 1965.
248. Lundin PM, Schelin U: Ultrastructure of the *rat* thymus. Acta Pathol Microbiol Scand *65:* 379-394, 1965.
249. Cesarini JP, Benkoel L, Bonneau H: Comparative ultrastructure of the thymus in young and adult *hamsters.* C R Seances Soc Biol Fil *162:* 1975-1979, 1969.
250. Ito T, Hoshino T: Fine structure of the epithelial reticular cells of the medulla of the thymus in the *golden hamster.* Z Zellforsch *69:* 311-318, 1966.
251. Chapman WL Jr, Allen JR: The fine structure of the thymus of the fetal and neonatal *monkey* (*Macaca mulatta*). Z Zellforsch *114:* 220-233, 1971.
252. 252. Mandel T: The development and structure of Hassall's corpuscles in the *guinea pig*. A light and electron microscopic study. Z Zellforsch Mikr Anat *89:* 180-192, 1968.
253. Gilmore RS, Bridges JB: Cystic structures in the Bursa of Fabricius of the *domestic fowl.* Ir J Med Sci *3:* 438, 1970.
254. Kameya T, Watanabe Y: Electron microscopic observations on *human* thymus and thymoma. Acta Pathol Jpn *15:* 223-246, 1965.
255. Pinkel D: Ultrastructure of *human* fetal thymus. Am J Dis Child *115:* 222-238, 1968.
256. Hirokawa K: Electron microscopic observation of the *human* thymus of the fetus and the newborn. Acta Pathol Jpn *19:* 1-13, 1969.
257. Haar JL: Light- and electron microscopy of *human* fetal thymus. Anat Rec *179:* 463-476, 1974.
258. Bartel H: Electron microscopic studies on the development of the human thymus. Acta Med Pol *18:* 277-278, 1977.

259. 259. Izard J: Ultrastructure of the thymic reticulum in *guinea pig*. Cytological aspects of the problem of the thymic secretion. Anat Rec *155:* 117-132, 1966.
260. Izard J: Ultrastructure of Hassall's corpuscles during the course of experimental thymus gland involution induced by folliculin. Zeitschr Zellforsch *66:* 276-292, 1965.
261. Ito T, Hoshino T: Histological changes of the *mouse* thymus during involution and regeneration following administration of hydrocortisone. Zeitschr Zellforsch *56:* 445-464, 1962.
262. Frazier JA: Ultrastructure of the *chick* thymus. Zeitschr Zellforsch Mikr Anat *136:* 191-205, 1973.
263. Gad P, Clark SL Jr: Involution and regeneration of the thymus in *mice*, induced by bacterial endotoxin and studied by quantitative histology and electron microscopy. Am J Anat *122:* 573-605, 1968.
264. Henry K: Mucin secretion and striated muscle in the human thymus. Lancet *1(7430)*: 183-185, 1966.
265. Pfoch M: Comparative electron microscopic investigation of entodermal reticulum cells in the thymus of newborn and aged *wistar-rats*. Zeitschr Zellforsch Mikr Anat *114:* 271-280, 1971.
266. Pfoch M, Unsicker K, Schimmler J: Quantitative electron microscopic studies on the innervation of the *human* thymus. Zeitschr Zellforsch Mikr Anat *119:*115-119, 1971.
267. Metcalf D: Functional interactions between the thymus and other organs. Wistar Inst Symp Monogr *2:* 53-73, 1964.
268. Metcalf D: The thymus. Summary of the conference. Wistar Inst Symp Monogr *2:*137-139, 1964.
269. Ropke C, Van Soest P, Platenburg PP, Van Ewijk W: A common stem cell for murine cortical and medullary thymic epithelial cells? Dev Immunol *4:* 149-156, 1995.
270. Hartmann A: Die Entwicklung der Thymus beim Kaninchen. Arch Mikr Anat *86:* 69-112, 1914.
271. Harland M: Early histogenesis of the thymus in the white rat. Anat Rec *77:* 247-262, 1940.
272. Grégoire C: Remarques au sujet du travail de M.N. de Winiwarter. Recherches sur l'évolution des dérivés branchiaux et l'histogénèse du thymus (cobaye). Arch Biol *45:* 353-354, 1935.
273. Grégoire C: Über das Verhalten des Lymphocyts bei der lymphoepithelialen Symbiose im der Thymus. Virchows Arch *303:* 457-480, 1939.
274. Grégoire C, Duchateau G: A study on lympho-epithelial symbiosis in thymus. Reactions of the lymphatic tissue to extracts and to implants of epithelial components of thymus. Arch Biol *68:* 269-296, 1956.
275. Itoh T: Establishment of an epithelial cell line from *rat* thymus. Am J Anat *156:* 99-104, 1979.
276. Itoh T, Kasahara S, Mori T: A thymic epithelial cell line, IT-45R1, induces the differentiation of prethymic progenitor cells into postthymic cells through direct contact. Thymus *4:* 69-75, 1982.
277. Itoh T, Kasahara S, Aizu S, Kato K, Takeuchi M, Mori T: Formation of Hassall's corpuscles in vitro by the thymic epithelial cell line IT-26R21 of the *rat*. Cell Tissue Res *226:* 469-476, 1982.
278. Pahwa S, Pahwa R, Incefy G, Reece E, Smithwick E, O'Reilly R, Good RA: Failure of immunologic reconstitution in a patient with the DiGeorge syndrome after fetal thymus transplantation. Clin Immunol Immunopathol *14:* 96-106, 1979.
279. Touraine JL: T-lymphocyte maturation after bone marrow, fetal liver, or thymus transplantation in immunodeficiencies. Transplant Proc *11:* 494-497, 1979.
280. Touraine JL: *Human* T-lymphocyte differentiation in immunodeficiency diseases and after reconstitution by bone marrow or fetal thymus transplantation. Clin Immunol Immunopathol *12:* 228-237, 1979.
281. Touraine JL, Touraine F: Thymic inducers of *human* T-lymphocyte differentiation *in vitro* and *in vivo*. Ann N Y Acad Sci *332:* 64-69, 1979.
282. Dardenne M, PapiernikM, Bach JF, Stutman O: Studies on thymus products. 3. Epithelial origin of the serum thymic factor. Immunology *27:* 299-304, 1974.
283. Kater L, Oosterom R, McClure J, Goldstein AL: Presence of thymosin-like factors in *human* thymic epithelium conditioned medium. Int J Immunopharmacol *1:* 273-284, 1979.
284. Köhler G, Milstein C: Continuous cultures of fused cells secreting antibody of predefined specificity. Nature *256:* 495-497, 1975.
285. Yelton DE, Scharff MD: Monoclonal antibodies. A powerful new tool in biology andmedicine. Ann Rev Biochem *50:* 657-680, 1981.
286. Sinkovics JG: Specific antibody production of lymphocyte hybridomas. Magyar Onkologia *26:* 1-6, 1982.
287. Sinkovics JG, Dreesman GR: Monoclonal antibodies of hybridomas. Rev Infect Dis *5:* 9-34, 1983.
288. Eisenbarth GS, Haynes BF, Schroer JA, Fauci AS: Production of monoclonal antibodies reacting with peripheral blood mononuclear cell surface differentiation antigens. J Immunol *124:* 1237-1244, 1980.
289. Eisenbarth GS: Review: application of monoclonal antibody techniques to biochemical research. Anal Biochem *111:* 1-16, 1981.
290. Eisenbarth GS, Jackson RA: Application of monoclonal antibody techniques to endocrinology. Endocr Rev *3:* 26-39, 1982.
291. Scearce RM, Eisenbarth GS: Production of monoclonal antibodies reacting with the cytoplasm and surface of differentiated cells. Methods Enzymol *103:* 459-469, 1983.
292. Haynes BF: The role of the thymic microenvironment in promotion of early stages of *human* T cell maturation. Clin Res *34:* 422-431, 1986.
293. Singer KH, Le PT, Denning SM, Whichard LP, Haynes BF: The role of adhesion molecules in epithelial-T-cell interactions in thymus and skin. J Invest Dermatol *94(6 Suppl):* 85S-90S, 1990.
294. Jung LK, Haynes BF, Nakamura S, Pahwa S, Fu SM: Expression of early activation antigen (CD69) during *human* thymic development. Clin Exp Immunol *81:* 466-474, 1990.
295. Patel DD, Haynes BF: Cell adhesion molecules involved in intrathymic T cell development. Semin Immunol *5:* 282-292, 1993.
296. Patel DD, Haynes BF: Stage-specific adhesion molecules in *human* intrathymic T-cell development. Res Immunol *145:* 144-146, 1994.
297. Bruggers CS, Patel DD, Scearce RM, Whichard LP, Haynes BF, Singer KH: AD2, a *human* molecule involved in the interaction of T cells with epidermal keratinocytes and thymic epithelial cells. J Immunol *154:* 2012-2022, 1995.
298. Eisenbarth GS, Walsh FS, Nirenberg M: Monoclonal antibody to a plasma membrane antigen of neurons. Proc Natl Acad Sci USA *76:* 4913-4917, 1979.
299. Nieburgs AC, Korn JH, Picciano PT, Cohen S: The production of regulatory cytokines for thymocyte proliferation by *murine* thymic epithelium in vitro. Cell Immunol 90*:* 426-438, 1985.
300. Kurtzberg J, Denning SM, Nycum LM, Singer KH, Haynes BF: Immature *human* thymocytes can be driven to differentiate into nonlymphoid lineages by cytokines from thymic epithelial cells. Proc Natl Acad Sci USA *86:* 7575-7579, 1989.
301. Haynes BF: *Human* thymic epithelium and T cell development: current issues and future directions. Thymus *16:* 143-157, 1990.

302. Shores EW, Van Ewijk W, Singer A: Disorganization and restoration of thymic medullary epithelial cells in T cell receptor-negative acid *mice*: evidence that receptor-bearing lymphocytes influence maturation of the thymic microenvironment. Eur J Immunol *21:* 1657-1661, 1991.
303. Dardenne M, Kelly PA, Bach JF, Savino W: Identification and functional activity of prolactin receptors in thymic epithelial cells. Proc Natl Acad Sci U S A *88:* 9700-9704, 1991.
304. Villa-Verde DM, de Mello-Coelho V, Farias-de-Oliveira DA, Dardenne M, Savino W: Pleiotropic influence of triiodothyronine on thymus physiology. Endocrinology *133:* 867-875, 1993.
305. Izon DJ, Jones LA, Eynon EE, Kruisbeek AM: A molecule expressed on accessory cells, activated T cells, and thymic epithelium is a marker and promoter of T cell activation. J Immunol *153:* 2939-2950, 1994.
306. Brekelmans P, van Soest P, Leenen PJ, van Ewijk W: Inhibition of proliferation and differentiation during early T cell development by anti-transferrin receptor antibody. Eur J Immunol *24:* 2896-2902, 1994.
307. Geenen V, Goxe B, Martens H, Vandersmissen E, Vanneste Y, Achour I, Kecha O, Lefebvre PJ: Cryptocrine signaling in the thymus network and T cell education to neuroendocrine self-antigens. J Mol Med *73:* 449-455, 1995.
308. Villa-Verde DM, Mello-Coelho V, Lagrota-Candido JM, Chammas R, Savino W: The thymic nurse cell complex: an in vitro model for extracellular matrix-mediated intrathymic T cell migration. Braz J Med Biol Res *28:* 907-912, 1995.
309. Haynes BF, Heinly CS: Early human T cell development: analysis of the *human* thymus at the time of initial entry of hematopoietic stem cells into the fetal thymic microenvironment. J Exp Med *181:* 1445-1458, 1995.
310. Paschke R, Geenen V: Messenger RNA expression for a TSH receptor variant in the thymus of a two-year-old child. J Mol Med *73:* 577-580, 1995.
311. Hollander GA, Wang B, Nichogiannopoulou A, Platenburg PP, van Ewijk W, Burakoff SJ, Gutierrez-Ramos JC, Terhorst C: Developmental control point in induction of thymic cortex regulated by a subpopulation of prothymocytes. Nature *373:* 350-353, 1995.
312. Savino W, Silva-Barbosa SD: Laminin/VLA-6 interactions and T cell function. Braz J Med Biol Res *29:* 1209-1220, 1996.
313. Anderson G, Anderson KL, Tchilian EZ, Owen JJ, Jenkinson EJ: Fibroblast dependency during early thymocyte development maps to the $CD25^+$ $CD44^+$ stage and involves interactions with fibroblast matrix molecules. Eur J Immunol *27:* 1200-1206, 1997.
314. Takigawa M, Imamura S, Ofuji Sh: Demonstration of epidermis-specific heteroantigens in thymic epithelial cells. Int Arch Allergy Appl Immunology *55:* 58-60, 1977.
315. Franke WW, Appelhaus B, Schmid E, Freudenstein C, Osborn M, Weber K: Identification and characterization of epithelial cells in *mammalian* tissues by immunofluorescence microscopy using antibodies to prekeratin. Differentiation *15:* 7-25, 1979.
316. Rouse RV, van Ewijk W, Jones PP, Weissman IL: Expression of MHC antigens by *mouse* thymic dendritic cells. J Immunol *122:* 2508-2515, 1979.
317. Tweel van den JG, Taylor CR, McClure J, Goldstein AL: Detection of thymosin in thymic epithelial cells by an immunoperoxidase method. Adv Exp Med Biol *114:* 511-515, 1979.
318. Monier JC, Dardenne M, Pléau JM, Schmitt D, Deschaux P, Bach J-F: Characterization of facteur thymique sérique (FTS) in the thymus. I. Fixation of anti-FTS antibodies on thymic reticuloepithelial cells. Clin Exp Immunol *42:* 470-476, 1980.
319. Janossy G, Thomas JA, Bollum FJ, Granger S, Pizzolo G, Bradstock KF, Wong L, McMichael A, Ganeshaguru K, Hoffbrand AV: The *human* thymic microenvironment: an immunohistologic study. J Immunol *125:* 202-212, 1980.
320. Ritter MA, Sauvage CA, Cotmore SF: The *human* thymus microenvironment: *in vivo* identification of thymic nurse cells and other antigenically-distinct subpopulations of epithelial cells. Immunology *44:* 439-46, 1981.
321. Dalakas MC, Engel WK, McClure JE, Goldstein A, Askanas V: Immunocytochemical characterization of thymosin in thymic epithelial cells of normal and myasthenia gravis patients and in thymic cultures. J Neurol Sci *50:* 239-247, 1981.
322. Schmitt D, Monier JC, Dardenne M, Pleau JM, Bach J-F: Location of FTS (facteur thymique serique) in the thymus of normal and auto-immune *mice*. Thymus *4:* 221-231, 1982.
323. Viac J, Schmitt D, Reano A, Thivolet J: Expression of common cytoplasmic antigens by epithelial cells of thymus and epidermis. Arch Dermatol Res *274:* 113-121, 1982.
324. Farr AG, Nakane PK: Cells bearing Ia antigens in the *murine* thymus. An ultrastructural study. Am J Pathol *111:* 88-97, 1983.
325. Haynes BF, Shimizu K, Eisenbarth GS: Identification of *human* and *rodent* thymic epithelium using tetanus toxin and monoclonal antibody A_2B_5. J Clin Invest 71*:* 9-14, 1983.
326. Haynes BF, Warren RW, Buckley RH, McClure JE, Goldstein AL, Henderson FW, Hensley LL, Eisenbarth GS: Demonstration of abnormalities in expression of thymic epithelial surface antigens in severe cellular immunodeficiency diseases. Birth Defects Orig Artic Ser *19:* 255-258, 1983.
327. Vakharia D.D., Demonstration of keratin filaments in thymic nurse cells (TNC) and alloreactivity of TNC-T cell population. Thymus *5:* 43-52, 1983.
328. Singer KH, Denning SM, Whichard LP, Haynes BF: Thymocyte LFA-1 and thymic epithelial cell ICAM-1 molecules mediate binding of activated *human* thymocytes to thymic epithelial cells. J Immunol *144:* 2931-2939, 1990.
329. Haynes BF: Phenotypic characterization and ontogeny of components of the *human* thymic microenvironment. Clin Res *32:* 500-507, 1984.
330. Bonnefoy JY, Reano A, Schmitt D, Thivolet J: Thymic Hassall's corpuscles-epidermis antigenic relations defined by a common glycoprotein in *man* (GP 37). Thymus *6:* 387-394, 1984.
331. McFarland EJ, Scearce RM, Haynes BF: The *human* thymic microenvironment: cortical thymic epithelium is an antigenically distinct region of the thymic microenvironment. J Immunol *133:* 1241-1249, 1984.
332. Haynes BF, Scearce RM, Lobach DF, Hensley LL: Phenotypic characterization and ontogeny of mesodermal-derived and endocrine epithelial components of the human thymic microenvironment. J Exp Med *159:* 1149-1168, 1984.
333. Auger C, Monier JC, Savino W, Dardenne M: Localization of thymulin (FTS-Zn) in *mouse* thymus. Comparative data using monoclonal antibodies following different plastic embedding procedures. Biol Cell *52:* 139-146, 1984.
334. Savino W, Dardenne M: Thymic hormone containing cells. VI. Immunohistologic evidence for the simultaneous presence of thymulin, thymopoietin and thymosin $\alpha 1$ in normal and pathological *human* thymuses. Eur J Immunol *14:* 987-992, 1984.
335. van Vliet E, Melis M, Ewijk W: Monoclonal antibodies to stromal cell types of the *mouse* thymus. Eur J Immunol *14:*

524-529, 1984.
336. Kruisbeek AM, Longo DL: Acquisition of MHC-restriction specificities: role of thymic stromal cells. Surv Immunol Res *4:* 110-119, 1985.
337. Lobach DF, Scearce RM, Haynes BF: The *human* thymic microenvironment. Phenotypic characterization of Hassall's bodies with the use of monoclonal antibodies. J Immunol *134:* 250-257, 1985.
338. Ritter MA, Schuurman HJ, MacKenzie WA, de Maagd RA, Price KM, Broekhuizen R, Kater L: Heterogeneity of human thymus epithelial cells revealed by monoclonal anti-epithelial cell antibodies. Adv Exp Med Biol *186:* 283-288, 1985.
339. Gromkowski SH, Krensky AM, Martz E, Burakoff SJ: Functional distinctions between the LFA-1, LFA-2 and LFA-3 membrane proteins on *human* CTL are revealed with trypsin-pretreated target cells. J Immunol *134:* 244-249, 1985.
340. de Maagd RA, MacKenzie WA, Schuurman HJ, Ritter MA, Price KM, Broekhuizen R, Kater L: The *human* thymus microenvironment: heterogeneity detected by monoclonal anti-epithelial cell antibodies. Immunology *54:* 745-754, 1985.
341. Lobach DF, Scearce RM, Haynes BF: The *human* thymic microenvironment. Phenotypic characterization of Hassall's bodies with the use of monoclonal antibodies. J Immunol *134:* 250-257, 1985.
342. Savino W, Durand D, Dardenne M: Immunohistochemical evidence for the expression of the carcinoembryonic antigen by *human* thymic epithelial cells *in vitro* and in neoplastic conditions. Am J Pathol *121:* 418-425, 1985.
343. Kendall MD: The outer and inner thymus cortex is a functional syncytium. Cell Biol Int Rep *9:* 3, 1985.
344. Kendall MD: The syncytial nature of epithelial cells in the thymic cortex. J Anat *147:* 95-106, 1986.
345. Laster AJ, Itoh T, Palker TJ, Haynes BF: The *human* thymic microenvironment. Thymic epithelium contains specific keratins associated with early and late stages of epidermal keratinocyte maturation. Differentiation *31:* 67-77, 1986.
346. Laster AJ, Haynes BF: Characterization of a monoclonal antibody, RTE-21, that binds to keratohyalin granule-associated proteins in epithelial cells of human skin and thymus. Clin Immunol Immunopathol *41:* 130-144, 1986.
347. Gaudecker von B, Steinmann GG, Hansmann ML, Harpprecht J, Milicevic NM, Müller-Hermelink HK: Immunocytochemical characterization of the thymic microenvironment. A light-microscopic and ultrastructural immunocytochemical study. Cell Tissue Res *244:* 403-412, 1986.
348. Gaudecker von B: The development of the *human* thymus microenvironment. Curr Top Pathol *75:* 1-41, 1986.
349. Lobach DF, Haynes BF: Ontogeny of the *human* thymus during fetal development. J Clin Immunol *7:* 81-97, 1987.
350. Schuurman H-J, Ritter MA, Broekhuizen R, Ladyman H, Larché M: The thymic epithelium panel of antibodies. Histologic analysis of *human* tissues. In: Leucocyte Typing III, McMichael ed, Oxford University Press, 1987, pp 259-262.
351. Schmitt D, Zambruno G, Staquet MJ, Dezutter-Dambuyant C, Ohrt C, Brochier J, Thivolet J: Antigenic thymus-epidermis relationships. Reactivity of a panel of anti-thymic cell monoclonal antibodies on *human* keratinocytes and Langerhans cells. Dermatologica *175:* 109-120, 1987.
352. Colic M, Matanovic D, Hegedis Lj, Dujic A: Immunohistochemical characterization of *rat* thymic non-lymphoid cells. I. Epithelial and mesenchymal components defined by monoclonal antibodies. Immunology *65:* 277-284, 1988.
353. Ewijk van W: Cell surface topography of thymic microenvironments. Lab Invest *59:* 579-590, 1988.
354. Nabarra B, Papiernik M: Phenotype of thymic stromal cells. An immunoelectron microscopic study with anti-IA, anti-MAC-1, and anti-MAC-2 antibodies. Lab Invest *58:* 524-531, 1988.
355. Savino W, Dardenne M: Developmental studies on expression of monoclonal antibody-defined cytokeratins by thymic epithelial cells from normal and autoimmune *mice.* J Histochem Cytochem *36:* 1123-1129, 1988.
356. Savino W, Dardenne M: Immunohistochemical studies on a *human* thymic epithelial cell subset defined by the anti-cytokeratin 18 monoclonal antibody. Cell Tissue Res 254*:* 225-231, 1988.
357. Gaudecker von B, Larché M, Schuurman H-J, Ritter MA: Analysis of the fine distribution of thymic epithelial microenvironmental molecules by immuno-electron microscopy. Thymus *13:* 187-194, 1989.
358. Kampinga J, Berges S, Boyd RL, Brekelmans P, Colic M, van Ewijk W, Kendall MD, Ladyman H, Nieuwenhuis P, Ritter MA, Schuurman H-J, Tournefier A: Thymic epithelial antibodies: immunohistological analysis and introduction of nomenclature. Thymus *13:* 165-173, 1989.
359. Colic M, Jovanovic S, Mitrovic S, Dujic A: Immunohistochemical identification of six cytokeratin-defined subsets of the *rat* thymic epithelial-cells. Thymus *13:* 175-185, 1989.
360. Picker LJ, De los Toyos J, Telen MJ, Haynes BF, Butcher EC: Monoclonal antibodies against the CD44 [In(Lu)-related p80], and Pgp-1 antigens in man recognize the Hermes class of lymphocyte homing receptors. J Immunol *142:* 2046-2051, 1989.
361. De Souza LR, Trajano V, Savino W: Is there an interspecific diversity of the thymic microenvironment? Dev Immunol *3:* 123-135, 1993.
362. Cirne-Lima EO, van Ewijk W, Savino W: Cortical and medullary phenotypes within a *mouse* thymic epithelial cell line. In Vitro Cell Dev Biol Anim *29A:* 443-445, 1993.
363. De Souza LR, Savino W: Modulation of cytokeratin expression in the *hamster* thymus: evidence for a plasticity of the thymic epithelium. Dev Immunol *3:* 137-146, 1993.
364. 364. Lannes-Vieira J, Chammas R, Villa-Verde DM, Vannier-dos-Santos MA, Mello-Coelho V, de Souza SJ, Brentani RR, Savino W: Extracellular matrix components of the *mouse* thymic microenvironment. III. Thymic epithelial cells express the VLA6 complex that is involved in laminin-mediated interactions with thymocytes. Int Immunol *5:* 1421-1430, 1993.
365. Vieira JL, de Souza LR, Savino W: Conceptual aspects of the thymic microenvironment. Immunol Today *15:* 192-193, 1994.
366. Izon DJ, Nieland JD, Godfrey DI, Boyd RL, Kruisbeek AM: Flow cytometric analysis reveals unexpected shared antigens between histologically defined populations of thymic stromal cells. Int Immunol *6:* 31-39, 1994.
367. Alves LA, Campos de Carvalho AC, Cirne Lima EO, Rocha e Souza CM, Dardenne M, Spray DC, Savino W: Functional gap junctions in thymic epithelial cells are formed by connexin 43. Eur J Immunol *25:* 431-437, 1995.
368. Freitas CS, Lyra JS, Dalmau SR, Savino W: *In vivo* and *in vitro* expression of tenascin by *human* thymic microenvironmental cells. Dev Immunol *4:* 139-147, 1995.
369. Patel DD, Whichard LP, Radcliff G, Denning SM, Haynes BF: Characterization of *human* thymic epithelial cell surface antigens: phenotypic similarity of thymic epithelial cells to epidermal keratinocytes. J Clin Immunol *15:* 80-92, 1995.
370. Patel DD, Hale LP, Whichard LP, Radcliff G, Mackay CR, Haynes BF: Expression of CD44 molecules and CD44

ligands during *human* thymic fetal development: expression of CD44 isoforms is developmentally regulated. Int Immunol *7:* 277-286, 1995.

371. Patel DD, Wee SF, Whichard LP, Bowen MA, Pesando JM, Aruffo A, Haynes BF: Identification and characterization of a 100-kD ligand for CD6 on *human* thymic epithelial cells. J Exp Med *181:* 1563-1568, 1995.
372. Bodey B, Bodey B Jr, Hall FL: Expression of Transforming Growth Factor-β type II receptors in the cells of the *human* thymic microenvironment during ontogenesis. *In:* Molecular Biology of Hematopoiesis. Vol. 5. (NG Abraham, S Asano, G Brittinger, R. Shadduck, eds) New York, Plenum Press, Chapter 78, 1996, pp. 645-658.
373. Saunders D, Lucas K, Ismaili J, Wu L, Maraskovsky E, Dunn A, Shortman K: Dendritic cell development in culture from thymic precursor cells in the absence of granulocyte/macrophage colony-stimulating factor. J Exp Med *184:* 2185-2196, 1996.
374. Vicente A, Varas A, Sacedón R, Zapata AG: Histogenesis of the epithelial component of *rat* thymus: an ultrastructural and immunohistological analysis. Anat Rec *244:* 506-519, 1996.
375. Rinke de Wit TF, Izon DJ, Revilla C, Oosterwegel M, Bakker AQ, van Ewijk W, Kruisbeek AM: Expression of tyrosine kinase gene in *mouse* thymic stromal cells. Int Immunol *8:* 1787-1795, 1996.
376. Kranz A, Kendall MD, von Gaudecker B: Studies on *rat* and *human* thymus to demonstrate immunoreactivity of calcitonin gene-related peptide, tyrosine hydroxylase and neuropeptide Y. J Anat *191:* 441-450, 1997.
377. Gaudecker von B, Kendall MD, Ritter MA: Immuno-electron microscopy of the thymic epithelial microenvironment. Microsc Res Tech *38:* 237-249, 1997.
378. 378. Markert ML, Watson TJ, Kaplan I, Hale LP, Haynes BF: The *human* thymic microenvironment during organ culture. Clin Immunol Immunopathol *82:* 26-36, 1997.
379. Bodey B, Bodey B Jr, Kaiser HE: Dendritic type, accessory cells within the mammalian thymic microenvironment. Antigen presentation in the dendritic neuro-endocrine-immune cellular network. In Vivo *11:* 351-370, 1997.
380. 380. Lucas K, Vremec D, Wu L, Shortman K: A linkage between dendritic cell and T- cell development in the *mouse* thymus: the capacity of sequential T-cell precursors to form dendritic cells in culture. Dev Comp Immunol. *22:* 339-349, 1998.
381. Imami N, Ladyman HM, Vincents B, Al-Tubuly A, Freysdottir J, Sedibane ML, Taylor-Fishwick DA, Foxwell BM, Ritter MA: K21-antigen: a molecule shared by the microenvironments of the *human* thymus and germinal centers. Dev Immunol *6:* 41-52, 1998.
382. Anderson KL, Moore NC, McLoughlin DE, Jenkinson EJ, Owen JJ: Studies on thymic epithelial cells in vitro. Dev Comp Immunol. *22:* 367-377, 1998.
383. Porritt HE, Anderson KL, Suniara RK, Jenkinson EJ, Owen JJ: Cellular interactions in the thymus regulate the protein kinase C signaling pathway. Eur J Immunol *28:* 1197-1203, 1998.
384. Miralles GD, Smith CA, Whichard LP, Morse MA, Haynes BF, Patel DD: $CD34^+CD38^-lin^-$ cord blood cells develop into dendritic cells in *human* thymic stromal monolayers and thymic nodules. J Immunol *160:* 3290-3298, 1998.
385. Alves LA, de Carvalho AC, Savino W: Gap junctions: a novel route for direct cell-cell communication in the immune system? Immunol Today *19:* 269-275, 1998.
386. Calvo W, Fliedner TM, Herbst EW, Hugl E, Bodey B: Degenerative changes and recovery of the thymus of lethally irradiated *dogs*, rescued by transfusion of cryopreserved autologous blood leukocytes. Exp Hemat *15:* 1171-1178, 1987.
387. Vakharia DD: Demonstration of keratin filaments in thymic nurse cells (TNC) and alloreactivity of TNC-T cell population. Thymus *5:* 43-52, 1983.
388. Wijngaert van de FP, Rademakers LHPM, Schuurman HJ, de Weger RA, Kater L: Identification and in situ localization of the "thymus nurse cell" in man. J Immunol *130:* 2348-2351, 1983.
389. van Vliet E, Melis M, van Ewijk W: Immunohistology of thymic nurse cells. Cell Immunol *87:* 101-109, 1984.
390. Schuurman HJ, Kater L: Relevance of "nurse cells' in histophysiology of lymphoid tissues. Thymus *7:* 317-322, 1985.
391. Villa-Verde DM, Savino W: Thymic "nurse" cells express extracellular matrix proteins. Mem Inst Oswaldo Cruz 86 *Suppl 3:* 109-110, 1991.
392. Penninger J, Rieker T, Romani N, Klima J, Salvenmoser W, Dietrich H, Stossel H, Wick G: Ultrastructural analysis of thymic nurse cell epithelium. Eur J Immunol *24:* 222-228, 1994.
393. Villa-Verde DM, Lagrota-Candido JM, Vannier-Santos MA, Chammas R, Brentani RR, Savino W: Extracellular matrix components of the *mouse* thymus microenvironment. IV. Modulation of thymic nurse cells by extracellular matrix ligands and receptors. Eur J Immunol *24:* 659-664, 1994.
394. Rieker T, Penninger J, Romani N, Wick G: *Chicken* thymic nurse cells: an overview. Dev Comp Immunol *19:* 281-289, 1995.
395. Drenckhahn D, von Gaudecker B, Müller-Hermelink HK, Unsicker K, Groschel-Stewart U: Myosin and actin containing cells in the *human* postnatal thymus. Ultrastructural and immunhistochemical findings in normal thymus and in myasthenia gravis. Virchows Arch B Cell Pathol *32:* 33-45, 1979.
396. 396. Boyd RL, Tucek CL, Godfrey DI, Izon DJ, Wilson TJ, Davidson NJ, Bean AG, Ladyman HM, Ritter MA, Hugo P: The thymic microenvironment. Immunol Today *14:* 445-459, 1993.
397. Bekkum DW van, Betel I, Blankwater MJ, Kruisbeek AM, Swart AC: Biologic activities of various thymus preparations. Transplat Proc *9:* 1197-1199, 1977.
398. Comsa J: Thymus hormones. Facts and problems. Med Welt *31:* 533-536, 1980.
399. Low TL, Goldstein AL: Thymic hormones: an overview. Methods Enzymol *116:* 213-219, 1985.
400. Oosterom R, Kater L: Effects of conditioned media of *human* thymus epithelial cultures on T-lymphocyte functions. Ann N Y Acad Sci *332:* 113-122, 1979.
401. Sunder-Plassmann P: Problem of thymus and Basedow's disease: further studies on biology, pathology and clinical aspects of neurohormonal cell system. Deutsche Ztsschr Chir *255:* 523-615, 1942.
402. Constant GA, Porter EL, Andronis A, Rider JA: Effect of thymic extracts on neuromuscular response. Texas Rep Biol Med *7:* 350-354, 1949.
403. Selye H, Marion D: Direct thymolytic effect of hexestrol (estrogen). Acta Anat *23:* 180-184, 1955.
404. Savino W, Cirne-Lima EO, Soares JF, Leite-de-Moraes M do C, Ono IP, Dardenne M: Hydrocortisone increases the numbers of $KL1^+$ cells, a discrete thymic epithelial cell subset characterized by high molecular weight cytokeratin expression. Endocrinology *123:* 2557-2564, 1988.
405. Goya RG, Sosa YE, Quigley KL, Reichhart R, Meites J: Homeostatic thymus hormone stimulates corticosterone secretion in a dose- and age-dependent manner in *rats*. Neuroendocrinology *51:* 59-63, 1990.
406. Timsit J, Savino W, Safieh B, Chanson P, Gagnerault MC,

Bach J-F, Dardenne M: Growth hormone and insuline-like growth factor-I stimulate hormonal function and proliferation of thymic epithelial cells. J Clin Endocrinol Metab *75:* 183-188, 1992.
407. Villa-Verde DM, Defresne MP, Vannier-dos-Santos MA, Dussault JH, Boniver J, Savino W: Identification of nuclear triiodothyronine receptors in the thymic epithelium. Endocrinology *131:* 1313-1320, 1992.
408. Wick G, Hu Y, Schwarz S, Kroemer G: Immunoendocrine communication via the hypothalamo-pituitary-adrenal axis in autoimmune diseases. Endocr Rev *14:* 539-563, 1993.
409. Anselmino KJ: Thymotropic hormone of anterior pituitary and thymus hormone. Klin Wchnschr *21:* 611-612, 1942.
410. Sacks J: Thytuitary (thymus-posterior pituitary preparation). JAMA *122:* 1083-1084, 1943.
411. Geenen V, Legros JJ, Franchimont P, Baudrihaye M, Defresne MP, Boniver J: The neuroendocrine thymus: coexistence of oxytocin and neurophysin in the *human* thymus. Science *232:* 508-511, 1986.
412. Weusten JJ, Blankenstein MA, Gmelig-Meyling FH, Schuurman HJ, Kater L, Thijssen JH: Presence of oestrogen receptors in *human* blood mononuclear cells and thymocytes. Acta Endocrinol *112:* 409-414, 1986.
413. Geenen V, Legros JJ, Franchimont P, Defresne MP, Boniver J, Ivell R, Richter D: The thymus as a neuroendocrine organ. Synthesis of vasopressin and oxytocin in *human* thymic epithelium. Ann N Y Acad Sci *496:* 56-66, 1987.
414. Le PT, Kurtzberg J, Brandt SJ, Niedel JE, Haynes BF, Singer KH: *Human* thymic epithelial cells produce granulocyte and macrophage colony-stimulating factors. J Immunol *141:* 1211-1217, 1988.
415. Geenen V, Robert F, Defresne MP, Boniver J, Legros JJ, Franchimont P: Neuroendocrinology of the thymus. Horm Res *31:* 81-84, 1989.
416. Le PT, Lazorick S, Whichard LP, Yang YC, Clark SC, Haynes BF, Singer KH: *Human* thymic epithelial cells produce IL-6, granulocyte-monocyte-CSF, and leukemia inhibitory factor. J Immunol *145:* 3310-3315, 1990.
417. Le PT, Vollger LW, Haynes BF, Singer KH: Ligand binding to the LFA-3 cell adhesion molecule induces IL-1 production by *human* thymic epithelial cells. J Immunol *144:* 4541-4547, 1990.\
418. Le PT, Lazorick S, Whichard LP, Haynes BF, Singer KH: Regulation of cytokine production in the *human* thymus: epidermal growth factor and transforming growth factor α regulate mRNA levels of interleukin 1 α (IL-1 α), IL-1 β, and IL-6 in human thymic epithelial cells at a post-transcriptional level. J Exp Med *174:* 1147-1157, 1991.
419. Argiolas A, Gessa GL, Melis MR, Stancampiano R, Vaccari A: Effects of neonatal and adult thyroid dysfunction on thymic oxytocin. Neuroendocrinology *52:* 556-559, 1990.
420. Argiolas A, Melis MR, Stancampiano R, Mauri A, Gessa GL: Hypothalamic modulation of immunoreactive oxytocin in the rat thymus. Peptides *11:* 539-543, 1990.
421. Robert F, Geenen V, Schoenen J, Burgeon E, De Groote D, Defresne MP, Legros JJ, Franchimont P: Colocalization of immunoreactive oxytocin, vasopressin and interleukin-1 in *human* thymic epithelial neuroendocrine cells. Brain Behav Immun *5:* 102-115, 1991.
422. al-Shawaf AA, Kendall MD, Cowen T: Identification of neural profiles containing vasoactive intestinal polypeptide, acetylcholinesterase and catecholamines in the *rat* thymus. J Anat *174:* 131-143, 1991.
423. Melis MR, Stancampiano R, Argiolas A: Oxytocin- and vasopressin-like immunoreactivity in the *rat* thymus: characterization and possible involvement in the immune response. Regul Pept *45:* 269-272, 1993.
424. Robert FR, Martens H, Cormann N, Benhida A, Schoenen J, Geenen V: The recognition of hypothalamo-neurohypophysial functions by developing T cells. Dev Immunol *2:* 131-140, 1992.
425. de Leeuw FE, Jansen GH, Batanero E, van Wichen DF, Huber J, Schuurman HJ: The neural and neuro-endocrine component of the *human* thymus. I. Nerve-like structures. Brain Behav Immun *6:* 234-248, 1992.
426. Batanero E, de Leeuw FE, Jansen GH, van Wichen DF, Huber J, Schuurman HJ: The neural and neuro-endocrine component of the *human* thymus. II. Hormone immunoreactivity. Brain Behav Immun 1992 *6:* 249-264, 1992.
427. Caux C, Dezutter-Dambuyant C, Schmitt D, Banchereau J: GM-CSF and TNF-α cooperate in the generation of dendritic Langerhans cells. Nature *360:* 258-261, 1992.
428. Goya RG, Gagnerault MC, De Moraes MC, Savino W, Dardenne M: *In vivo* effects of growth hormone on thymus function in aging *mice*. Brain Behav Immun *6:* 341-354, 1992.
429. Martin-Fontecha A, Broekhuizen R, de Heer C, Zapata A, Schuurman HJ: The neuro-endocrine component of the rat thymus: studies on cultured thymic fragments before and after transplantation in congenitally athymic and euthymic *rats*. Brain Behav Immun *7:* 1-15, 1993.
430. Dardenne M, Savino W: Control of thymus physiology by peptidic hormones and neuropeptides. Immunol Today *15:* 518-523, 1994.
431. Savino W, Dardenne M: Immune-neuroendocrine interactions. Immunol Today *16:* 318-322, 1995.
432. Duijvestijn AM, Kohler YG, Hoefsmit ECM: Interdigitating cells and macrophages in the acute involving *rat* thymus. An electro-microscopic study on phagocytic activity and population development. Cell Tissue Res *224:* 291-301, 1982.
433. Deman J, Van Meurs M, Claassen E, Humblet C, Boniver J, Defresne MP: In vivo expression of interleukin-1 β (IL-1 β), IL-2, IL-4, IL-6, tumour necrosis factor-α and interferon-γ in the fetal *murine* thymus. Immunology *89:* 152-157, 1996.
434. Ewijk van W, de Kruif J, Germeraad WT, Berendes P, Ropke C, Platenburg PP, Logtenberg T: Subtractive isolation of phage-displayed single-chain antibodies to thymic stromal cells by using intact thymic fragments. Proc Natl Acad Sci USA *94:* 3903-3908, 1997.
435. 435. Vanneste Y, Thome AN, Vandersmissen E, Charlet C, Franchimont D, Martens H, Lhiaubet AM, Schimpff RM, Rostene W, Geenen V: Identification of neurotensin-related peptides in *human* thymic epithelial cell membranes and relationship with major histocompatibility complex class I molecules. J Neuroimmunol *76:* 161-166, 1997.
436. Savino W, Villa-Verde DM, Alves LA, Dardenne M: Neuroendocrine control of the thymus. Ann N Y Acad Sci 840*:* 470-479, 1998.
437. Geenen V, Martens H, Vandersmissen E, Achour I, Kecha O, Franchimont D: Cellular and molecular aspects of thymic T-cell education in neuroendocrine self principles. Implications for autoimmunity. Ann N Y Acad Sci *840:* 328-337, 1998.
438. Martens H, Kecha O, Charlet-Renard C, Defresne MP, Geenen V: Neurohypophysial peptides stimulate the phosphorylation of pre-T cell focal adhesion kinases. Neuroendocrinology *67:* 282-289, 1998.
449. de Mello-Coelho V, Gagnerault MC, Souberbielle JC, Strasburger CJ, Savino W, Dardenne M, Postel-Vinay MC: Growth hormone and its receptor are expressed in *human* thymic cells. Endocrinology *139:* 3837-3842, 1998.
450. Bodey B, Bodey B Jr, Siegel SE, Kaiser HE: The role of zinc in pre- and postnatal *mammalian* thymic immunohistogenesis. In Vivo, *12:* 695-722, 1998.

Chapter 3

Neuroendocrine Influence on Thymic Hematopoiesis Via the Reticulo-Epithelial (Re) Cellular Network.

Abstract: The thyrnus provides an optimal cellular and humoral microenvironment for a cell line committed differentiatiation of pluripotent hematopoietic stem cells. The thymic RE cells are functionally specialized based on their intrathymic location and this differentiation is modulated by various interaction signals of differentiating thymocytes and other nonlymphatic, hematopoietic stem cells. Thus, although subcapsular, cortical, and medullary RE cells are derived from a common, endodermal in origin epithelial precursor cell, their location within the gland determines their specialization in terms of their IP and *in situ* physiological properties. The subcapsular, endocrine, RE cell layer (also named, giant or nurse cells-TNCs) is comprised of cells filled with PAS positive granules, which also express A_2B_5/TE4 cell surface antigens and MHC Class I (HLA A, B, C) molecules. In contrast to the medullary RE cells, these subcapsular TNCs also produce thymosins β3 and β4. The TNCs display a neuroendocrine cell specific IP: Thy-1^+, $A_2B_5^+$, TT^+, $TE4^+$, UJ13/A^+, UJ127.11^+, UJ167.11^+, UJ181.4^+, and presence of common leukocyte antigen (CLA^+). Cortical RE cells express a surface antigen, gp200-MR6 that plays a significant role of thymocyte differentiation. Medullar RE cells display MHC Class II (HLA-DP, HLA-DQ, HLA- DR) molecule restriction. These cells also contain transforming growth factor (TGF)-β type II receptors and are involved in the positive selection of T cells. Transmission electron microscopic (TEM) observations have defined four, functional subtypes of medullary RE cells: undifferentiated, squamous, villous, and cystic. All subtypes were connected with desmosomes. The secreted thymic hormones, thymulin, thymosin-α 1 and thymopoietin (its short form, thymopentin or TP5) were detected IC to be produced by RE cells. Thymic RE cells also produce numerous cytokines including IL-1, IL-6, G-CSF, M-CSF, and GM-CSF molecules that likely are important in various stages of hematopoietic cell activation and differentiation. The co-existence of pituitary hormone and neuropeptide secretion [oxytocin (OT), growth hormone (GH), growth hormone secretagogue (GHS), prolactin (PRL), adrenocorticotropic hormone (ACTH), thyroid stimulating hormone (TSH), triiodothyronine (T3), somatostatin, follicle stimulating hormone (FSH), luteinizing hormone (LH), arginine vasopressin (AVP), growth hormone releasing hormone (GHRH), corticotropin releasing hormone (CRH), nerve growth factor (NGF), vasoactive intestinal peptide (VIP), pro-enkephalin (pro-enk), and β-endorphin (β-end)], as well as the production of a number of interleukins and growth factors and expression of receptors for all, by RE cells is an unique molecular biological phenomenon. This repertoire of polypeptides belongs to a number of neuroendocrine protein families (such as the neurohypophysial, tachykinin, neurotensin and insulin families). Thymic neuroendocrine polypeptides are also the source of self-antigens presented by the MHC molecules to differentiating hematopoietic stem cells. Furthermore, it is quite possible that even on the level of individual RE cells, the numerous projections associated with a single cell, which engulf developing lymphocytes, nurturing and guiding them in their maturation, may differ in their hormone production and/or hormone receptor expression profile, thus allowing a single cell to be involved in distinct, separate steps of the T cell and other hematopoietic cell maturation process. I suggest that the thymic RE cells represent an important cellular and humoral network within the thymic microenvironment and are involved in the homeopathic regulation mechanisms of the multicellular organism, in addition to the presentation of various antigens to developing hematopoietic cell lines, and providing growth regulatory signals which may range from stimulatory to apoptotic signaling within the thymus. In my understanding, the intrathymic T lymphocyte selection is also a complex, multi-step process, influenced by several functionally specialized RE cells and under immuno-neuroendocrine regulation control reflecting the dynamic changes of the mammalian organism.

Key words: Stem cell factor (SCF); Thymic hormones; Pluripotent Hematopoietic Stem Cells (PHSCs); Surface antigenic markers; Androgen receptor; Erythropoietic activity; Erythropoietin; Activin A; Granulocytopoiesis; Hematopoietic growth factor KL (HGF-KL); Multipotential Colony Stimulating Factor (Multi-CSF); Granulocyte CSF (G-CSF); Granulocyte-Macrophage CSF (GM-CSF); Macrophage CSF (M-CSF); Leukemia-inhibitory factor (LIF); Neurotensin (neuromedin); Eosinophilic cells; Thymic reticulo-epithelial (RE) cells; Major histocompatibility complex (MHC); Thymic cellular microenvironment; Thymic eosinophilia; Hassall's bodies (HBs); Mast cells; Intrathymic pituitary hormones; Intrathymic neuropeptides; Immuno-neuroendocrine regulation; Neuroendocrine protein families (such as the neurohypophysial, tachykinin, neurotensin and insulin families); Hypothalamo-hypophyseal-thymic axis; Immunocytochemistry (IC); Immunophenotype (IP): Monoclonal antibody (MoAB).

1. INTRODUCTION

"The presence in the thymus of hemopoietic cells other than thymocytes has been known for many years, but the extent of the hemopoietic activity of the thymus and the possible functional implications have only recently begun to receive much attention."

Marion D. Kendall (1)

It has become clear that thymus is the primary and major site of T lymphocyte development and functional maturation, as well as of positive and negative selection of immunocompetent T lymphocytes. According to Manley (2), in the complex process of thymic organogenesis, the progenitor, endodermal epithelial cells proliferate into a three-dimensional thymic stromal environment under the continuous adhesion signals of differentiating, already committed T lymphocyte progenitors. The functionally developed thymus provides an optimal cellular-humoral microenvironment for the development of immunocompetent T cells and other cell lines of non-lymphatic hematopoiesis (3, 4, 5). An antigen ChT1, a member of the Ig superfamily with one V-like and one C2-like domain, was identified on avian early T cell progenitors to play a significant role in the early differentiation of cortical thymocytes (6). Anti-ChT1 MoAB inhibits T cell development in embryonic thymic organ cultures and in lymphatic progenitors cocultures on thymic RE cells.

During the last twenty years, advances in the fields of biochemistry and molecular biology have led to the scientific concept of early endocrinization of the embryo (7-11). During the course of embryogenesis, the hormonal microenvironment and the development and differentiation of cell membrane receptors mediate the various functional expressions of different tissues. An increasing number of growth factors have been shown to be responsible for the proliferation, survival, and enhanced function of many cell types within the hematopoietic system. The action of these hematopoietic growth factors in stimulating cell growth and survival applies both to cells within the progenitor (PHSC) compartment, as well as mature cells. The process of mature blood cell formation, by which a small number of self-renewing PHSCs give rise to lineage committed hematopoietic progenitor cells that subsequently proliferate and differentiate is regulated by a group of glycoprotein hormone growth factors, such as HGF-KL, encoded by the Sl locus and a ligand to the c-kit receptor (12-14). These hormones have been termed colony-stimulating factors (CSFs) (15), a designation derived from the in vitro observation that CSFs stimulate hemopoietic progenitor cells of different cell lineages to produce colonies of recognizable maturing cells (16). The different factors have been operationally characterized by defining the major cell type produced in the colony (e.g. Multi-CSF; G-CSF; GM-CSF; M-CSF and erythropoietin, as well as commitment factor activin A) (17, 18).

It is well known that exposure of the mammalian organism to ionizing radiation causes a drastic decrease in immunity and nonspecific resistance, as well as an impairment in hematopoiesis (19, 20). In animal experiments, peptide preparations from thymic (thymalin) and bone marrow (haemalin) cells were capable of normalizing the immunological reactivity and hematopoiesis of irradiated animals. Moreover, their mixture compound (thymohaemin) favored the post-irradiation recovery of hematopoiesis and immunogenesis. Lymphopoiesis stimulation in the bone marrow was provoked by thymalin and granulopoiesis stimulation by haemalin. These results strongly suggest an important regulatory role of the thymic cellular microenvironment on mammalian hematopoiesis.

2. HEMATOPOIESIS DURING MAMMALIAN ONTOGENESIS

Hematopoiesis in invertebrates does not include the development of lymphocytes, nor of other cell lineages of blood cells. Primitive blood cells (hemocytes), produced in the embryonal coelomic cavity retain their generalized functional capabilities during the entire life span of the organism (21).

The thymocytes in the early larvae of Xenopus laevis have been shown to be derived from precursor cells immigrating interstitially through the mesenchyme into the organ rudiments at 3-4 days of age (Nieuwkoop and Faber stages 42-45) (22). Orthotopic grafting of diploid tissues onto triploid stage 22 embryos followed by ploidy analyses of their hematopoietic cells revealed that both thymocytes and erythrocytes in early larvae are derived from the ventral blood islands (VBI), whereas those in late larvae and adults come mainly from the dorsolateral plate (DLP). The authors produced a number of interspecific chimeras in X. laevis and X. borealis embryos in order to study how the VBI cells of embryos at stage 22 participate in hemopoiesis. Sections of the chimeras at various developmental stages were examined by employing the unique stainability of X. borealis nuclei to quinacrine as a marker and the results showed that the VBI-derived cells enter into the circulation around stage 35/36, and that some of them leave the blood vessels to migrate interstitially through the mesenchyme toward the thymic rudiment during stages 43-45. A minor population of the VBI-derived cells was also found extravascularly in the mesonephric primordia. In contrast to the VBI, the DLP-derived cells contributed to the hematopoietic cell population not in early larvae, but in late ones as a major constituent in the mesonephros, thymus, liver, and peripheral blood. There is no need for de novo formation of blood cells during their postnatal histogenesis (23, 24). Again, postnatal hematopoiesis, as a continuous production of erythrocytes and leukocytes, does not exist in the least developed invertebrates.

It is generally accepted that modern birds evolved from reptiles and that their histogenetical evolutionary line, as the second class of warm-blooded vertebrates, is far from the mammalian line (25-30). The emergence of the bursa of Fabricius in birds and the development of lymph nodes first in crocodiles, a few birds (e.g. the ostrich and the duck), and most abundantly in mammals, as specialized lymphocytopoietic tissue organizations, predicted the future dichotomy of lymphopoiesis and hematopoiesis in mammals (30). The sites of formation of erythrocytes, granulocytes, lymphocytes, and thrombocytes have been described by Machotka (31). The microenvironment of the yolk sac is permissive only for erythropoiesis, which proceeds synchronously and may be not under the regulation of the erythropoietin (32-34). The differentiated, mature cells are nucleated red cells. This phenomenon is characteristic of primitive erythropoiesis. Intravascular erythropoiesis is present in birds where mature cells need not travel through the vascular wall to become enucleated. This early hematopoiesis is no longer detectable by the 10th fetal week of intrauterine life.

During phylogenetical development, it is only in mammals that all hematopoietic and mature blood cells are formed extravascularly (35, 36). The migratory nature of mammalian hematopoiesis during prenatal ontogenesis has been well established (14, 21). The phenomenon is first extra-embryonic, developing in blood islands of the yolk sac. The precursors of these islands migrate from the primitive streak region of blastoderm into the developing area opaca vasculosa (37). Cell proliferation observations indicate that the blood island forming cells, known as hemangioblasts, invaginate through the primitive streak and migrate laterally to produce the mesodermal layer (38, 39). Pluripotent hematopoietic stem cells (PHSCs) are also derived from these hemangioblasts, and thus the entire stem cell pool is composed of cells of mesodermal origin. PHSCs not only have the capacity to generate all of the hematopoietic cell types, but also have extensive self-renewal abilities (40, 41).

The development of hematopoiesis in rodents, particularly in the Mongolian gerbil (Meriones unguiculatus), was conducted by Smith and Glomski in order to determine the temporal sequence, the organs involved, and the cytology of blood cell formation in this species (42). Hematopoiesis in the intrauterine life of the gerbil was divided into four consecutive stages based on the site of blood cell formation:

1. vitelline stage;
2. hepatic stage, including thymic histogenesis;
3. splenic stage; and
4. the medullary stage, with the development of secondary lymphoid tissues.

At the onset of each of these phases, a blast-like cell was identifiable in each hematopoietic organ which, because of its morphology and its presumed multipotentiality was classified as a "lymphoid cell". During the yolk sac stage (gestational day 12), two generations of erythrocytes, a primitive and a definitive, are formed. The liver is by day 15 erythropoietic and megakaryopoietic, but later, a few granulocytes are also found in its extravascular compartment. The thymus is exclusively lymphopoietic from the appearance of its earliest cells on day 15. Splenic hematopoiesis is initiated by the presence of lymphoid cells (day 20 of gestation), followed later by the appearance of morphologically identifiable blood cell lines. Early normoblastic and granulocytic activity begins in the marrow cavities on day 23, though the marrow is not considered to be a source of circulating blood cells during fetal life. Lymph node histogenesis occurs during the last four days of gestation, first in the cervical region and then in other parts of the body. The finding of undifferentiated lymphoid cells in all organs at the initiation of hematopoiesis and in the peripheral blood throughout gestation was discussed in light of the migratory nature of hematopoiesis. This migratory pattern is basically similar in all mammals. Unfortunately, it still remains unclear how one hematopoietic cellular microenvironment within the embryo (fetus) loses its hematopoietic potential, while this potential is gained by a tissue very different in structure.

It is well known that the shift in the site of erythropoiesis (from intravascular to extravascular) is coincident with the development of non-nucleated red cells (43). The loss of the nucleus results from cell migration across the vascular endothelium. Hematopoiesis during prenatal development moves to intraembryonic sites, first to the liver and spleen, and finally to the bone marrow. Recent morphological observations have also defined well-developed island-like hematopoiesis in other developing human organs (lung, thymus, etc.). In our earlier works, we have also described pulmonal and intrathymic stages of prenatal hematopoiesis (44).

Hematopoiesis in the liver is detectable in humans by the 6th week of the pregnancy (45). Doubling time at first is just 8 hours, ensuring that the hematopoietic tissue expands exponentially. After a few days, however, the doubling time stabilizes at two days (46). The liver remains the major hematopoietic organ during fetal life until the first postnatal week. Thus, the cellular microenvironment in the liver is primarily inductive for erythroid differentiation, although it is not impermissive for other cell lineages. Erytropoiesis here is extravascular.

Splenic hematopoiesis in humans is absent at birth, but the spleen can easily regain its hematopoietic function under abnormal situations.

The final hematopoietic organ during fetal development is the bone marrow. The development of bone marrow requires the penetration of avascular cartilage by perichondrial mesenchymal cells and their associated blood vessels. This leads to calcification of cartilage. The vascular mesenchyme by week 20 of intrauterine life forms the reticular cell network upon which PHSCs can seed and proliferate. Hematopoiesis is heralded by the appearance of undifferentiated basophilic cells in dilated sinuses of the bone marrow. The seeding of the PHSCs occurs by immigration from the liver via the bloodstream. In humans, prenatal bone marrow is mainly erythropoietic from the very beginning of life.

To establish a hypothesis of molecular events of hematopoiesis regulation, it would be helpful to identify physiologically relevant function-associated molecules on human PHSCs. Therefore, a library of MoABs was utilized, with immunoreactivities against CD2, CD3, CD7, CD11, CD15, CD16, CD19, CD33, CD34, CD38, CD56, CD69, CD71 etc., to study the IP characteristics of PHSCs (47-51). The most often detected cell membrane antigens, representing the IP of human PHSCs, were $CD34^+$, $CD43^+$, HLA-DR^+ (CD7, CD19, CD43, and CD69 antigens were also detected). Moore and co-workers (41) reported their detailed examination of CD43 expression on murine and human PHSCs. Mouse stem cells were found within the $Ly6^+Lin^-$ $CD43^{high}$ subpopulation of bone marrow. These cells, upon transfer into SCID mice, caused rapid repopulating of the thymus, spleen, and bone marrow. Retransfer of bone marrow cells from primary SCID recipients of $Ly6^+Lin^-CD43^{high}$ cells into secondary recipients resulted in repopulation of the lymphohematopoietic cells. All $Ly6^+Lin^-$ $CD43^{high}$ cells were found to express high levels of c-kit. In contrast, $Ly6^+Lin^-CD43^{high}$ cells caused limited and variable thymic and splenic repopulation. These cells failed to repopulate the marrow cavity and did not contain re-transplantable stem cells. These data indicate that murine PHSCs express high levels of CD43. Although the exact function of CD43 is still not clear, it may play a crucial role in the regulation of hematopoiesis, because it appears that intercellular adhesion molecule-1 (ICAM-1), an important cell adhesion molecule expressed on stromal cells, is a ligand to CD43 (52-54). It appears that ICAM-1 binds to LFA-1 and CD43, both present on PHSCs. Adhesion molecules expressed on stromal cells probably regulate the cell to cell interaction between stromal elements and PHSCs. CD43 may act as a signal transduction molecule.

Observations of human fetal bone marrow cells revealed a subpopulation of $CD34^+CD38^-CD43^+$ cells. This IP represents the morphologically described large basophilic cells, regularly detected in the epithelial thymic rudiment during the first inflow of PHSCs (55-58). When single sorted cells with this IP were cultured in vitro, they were able to produce colonies with a dispersed growth pattern. Cells with this growth pattern have previously been shown to have myeloid and lymphoid growth potentials and extensive self-renewal capacity. Furthermore, $CD34^+CD38^-$HLA-DR^+ cells have recently been shown to be highly enriched in stem cell activity, expressing relatively high levels of CD43. Testi and co-workers (59) recently submitted that CD69, which is transiently present during lymphocyte activation, is expressed on a variety of hematopoietic stem cells at various stages during their maturation.

2.1 Molecular basis of cell lineage formation and cell homing in hematopoiesis

Recent advances in our knowledge about the molecular biological level of cell homing and growth factor regulation have provided new directions. The most noteworthy is the identification of the Steel gene product (60). This hematopoietic factor is produced by stromal cells, defines a multipotent growth factor, well known as stem cell factor (SCF) or HGF-KL, encoded by the Sl locus and serves as the ligand for the c-kit receptor, which is a tyrosine kinase and is the gene product of the W locus expressed in stem cells (13, 61, 62). The highly glycosylated molecule of SCF is present in both a soluble, as well as a cell membrane-bound form, and during the embryogenesis it is associated with both the migratory pathway and the homing of PHSCs (63). Another receptor, a membrane located lectin heterodimer with a molecular weight of 110kD is involved in the homing of PHSCs to bone marrow (64). The recognition is keyed in two carbohydrate residues (galactosyl and mannosyl) of a yet undetermined membrane glycoconjugate (65-67).

In addition, characteristic molecules of the extracellular matrix, particularly fibronectin, are known to be involved in surrounding tissue recognition (68, 69). Most of these molecular mechanisms are regulated genetically during the advancement of histogenesis and are likely to be involved in the receptiveness of one or the other organ/tissue microenvironment for the task of hematopoiesis during ontogenesis (21, 70).

3. THYMIC RE CELL MESHWORK AND ITS ROLE IN THYMIC HEMATOPOIESIS

After the classical embryological, and morphological descriptions of histogenesis, thymic research, has increasingly moved towards a better understanding of intrathymic hematopoietic stem cell maturation and differentiation, one of the main functions of the thymic stromal microenvironment (71-78). Our literature searches concur with Trainin (79), Maximov (57, 80, 81) was the first morphologist to mention a close cellular interaction between thymic RE cells and maturating thymocytes. Cortical RE cells express a 200000 mol. weight protein called gp200-MR6 (82). This antigen is also present on some epithelial neoplasms and is the same protein that inhibits the proliferation of an established colon carcinoma cell line, HT29. This antigen plays a significant role in the differentiation pathway of undifferentiated cortical thymocytes.

For a further discussion on thymic RE cells, please refer to Chapter 2. In general, thymic RE cells provide the main cellular and humoral microenvironment for intrathymic T lymphocyte differentiation, which manifests itself into a number of regulatory pathways:

1. secretion of a great number of polypeptides, including thymic hormones and cytokines;
2. cell-cell interactions through adhesion molecules and those between major histocompatibility complex (MHC) class I and class II proteins and the TRC within the context of CD4 and CD8 molecules, respectively (83, 84); and
3. RE cells can bind and interact with differentiating T lymphocytes using extracellular matrix (ECM) ligands and their respective receptors. Cytokine activities has been analyzed in the supernatant of human thymic RE cell cultures by employing the respective cytokine-dependent cell lines and by neutralization with specific MoABs (85). Three cytokine activities were identified: IL-6, G-CSF, and M-CSF, a thymocyte related one and two acting on myeloid/monocyte cell lines. The supporting effect on Concavalin-A induced proliferation of murine thymocytes has been completely eliminated employing anti-IL-6 MoABs.

In the last two decades, the potent immunomodulatory effects of extrinsic factors, such as hormones, neuropeptides, cytokines and growth factors have been investigated thoroughly. Consequently, numerous research groups have observed the actions of steroid hormones and their effects on the thymus have been reviewed (86-90). The cells of the RE network have been shown to express receptors for and produce a wide array of pituitary hormones and neuropeptides *in vivo* and *in vitro* including, OT, GH, GHS, PRL, ACTH, TSH, T_3, somatostatin, FSH, LH, AVP, GHRH, CRH, NGF, VIP, pro-enk, and β-end (91-104). This repertoire of polypeptides belongs to a number of neuroendocrine protein families, including the neurohypophysial, tachykinin, neurotensin and insulin families (105). These self-antigen molecules modulate the expression of major histocompatibility complex (MHC) gene products through the use of

RE cells and the extracellular matrix (ECM) mediated cell interactions, producing an enhanced thymocyte adhesion to the RE cells (105,106).

The regulation of cytokine production by RE cells in the thymus is under coordinated and temporal control and is important for the development of T lymphocytes and non-lymphatic hematopoietic cells. Cells of the RE network in the thymus express TGF-βR (107) and epidermal growth factor (EGF) receptor, and produce TGF-β3 *in vitro* and *in vivo*. Furthermore, EGF has been shown to increase IL-1α, IL-1β, IL-6 mRNA and protein levels in human RE cells. It has been shown that individual RE cells may concomitantly express both EGF receptor and TGF-βR. TGF-β plus EGF synergistically increase leukemia-inhibitory factor (LIF), additively increase IL-6, but have little effect on IL-1α and IL-1β mRNA levels. In contrast, TGF-β alone has been shown to increase LIF and IL-6, exert little effect on IL-1α, and cause a slight decrease of IL-1β mRNA levels. The increases in LIF and IL-6 mRNA levels by TGF-β plus EGF correlate with the increases in LIF and IL-6 concentrations in RE cell culture supernatants as detected by ELISA. TGF-β plus EGF does not affect transcription of the LIF and IL-6 genes, suggesting that the increases in the steady state levels of cytokine mRNA were mediated post-transcriptionally, most likely at the level of mRNA stability. TGF-β produced *in situ* possibly plays a role in thymocyte development by directly affecting thymocyte differentiation and by indirectly modulating cytokine production by the cells of the RE network (108).

In serum-free human fetal thymic RE cell cultures, IL-3 and IL-7 production has not been detected. Lack of IL-7 and presence of thymosin-α1 and of the surface molecule TE-4 identifies these cells as subcapsular/medullary RE cells. RE cells constitutively secrete IL-6, IL-8, fibronectin and thymosin-α1, but not GM-CSF of LIF. Recombinant IL-1, anti-ICAM-1, and anti LFA-3 alone and collectively induce modest, but significant increases in the secretions of thymic peptides, as well as IL-6 and IL-8, and this increase in secretion is not inhibited by hydrocortisone (HC) treatment. Recombinant IL-1, but neither anti-ICAM-1 nor anti-LFA-3 induces GM-CSF and LIF. HC has been shown to inhibit the secretion of GM-CSF and LIF induced by IL-1. None of the above stimulants augment the constitutive secretion of fibronectin or thymosin-α1, while HC inhibits thymosin-α1 secretion (109). Thus, it is quite clear that the secretion of cytokines by the cells of the thymic RE network is under very complex regulation, and is modulated by hormones (derived from neighboring cells of the RE network, other stromal cells, and/or endocrine glands of the nervous system), growth factors, and other possible mediators (neuropeptides, thymic peptides, etc.).

It appears that both the other stromal elements of the thymic microenvironment and the differentiating hematopoietic cells significantly influence the functional state of thymic RE cells, thus establishing the bi-directional intrinsic regulatory circuit. The expression of a panel of cytokines in thymic RE cells and $CD4^-8^-$ double negative thymocytes has been explored following cell to cell lymphostromal interaction, in an experimental model which enhances *in vitro* thymocyte maturation. Since retinoic acid (RA) has been previously shown to be an inhibitor of thymocyte maturation process in this model, cytokine expression in double negative thymocytes and thymic RE cells was assessed following the RA-induced impairment of *in vitro* thymocyte maturation. Cell to cell lymphostromal interaction results in increased IL-2 and decreased IL-7 expression in thymocytes while the expression of IL-1β and IL-7 increased and decreased, respectively, in RE cells. Addition of RA to lympho-stromal cell co-culture results in the enhancement of IL-4 and IL-7 expression in thymocytes while in thymic RE cells IL-1α decreased and IL-6 and IL-7 increased. These data indicate that discrete patterns of cytokine expression are present in thymocyte precursors and in the cells of the thymic RE network during *in vitro* T-cell development. Furthermore, specific cytokine modulation might contribute to the RA-induced impairment of thymocyte differentiation (110). A ganglioside of the neural crest and neuroendocrine cells (A_2B_5) was identified *in vivo* on TNC in the human thymus and recently, also *in vitro* throughout the adult rat thymus (111). The described diversity of cellular immunoreactivity in cultures of rat thymuses suggests that a variety of neural and endocrine factors may influence thymic development and function.

4. INTRATHYMIC THYMOCYTE-RE CELL TO CELL INTERACTIONS

Significant cell to cell contacts between T lymphocytes and RE cells are essential for normal T cell development, and interactions between T cells and skin epidermal keratinocytes occur in the context of inflammatory skin diseases and cutaneous T cell malignancies (112). On the basis of observations that the T cell ALL cell line, HSB, bound to IFN-γ activated epidermal keratinocytes, whereas the CTCL cell line H9 bound poorly. MoAB 13H12 could identify a 36 kDa molecule, termed AD2, that was highly expressed on HSB but not on H9 T cells. MoAB 13H12 inhibited the binding of HSB T cells to IFN-γ-treated epidermal keratinocytes, inhibited the binding of peripheral blood T cells to IFN-γ-treated EK, and inhibited the binding of IFN-γ-treated human thymic RE cells to thymocytes. Although AD2 was expressed at a low level on all T cells, AD2 was highly expressed on the $CD3^-4^-8^-$, $CD3^-4^{low+}8^-$, and $CD3^-4^+8^+$ subsets of immature thymocytes in the thymic subcapsular and inner cortex. These data suggest a role for the AD2 molecule in interactions of T cells with ecto- and endodermal epithelial cells of skin and thymus (112).

A novel thymic stromal cell antigen, named HS9, is a potent immunoregulatory molecule that participates in intrathymic T cell development. HS9 antigen is expressed on cortical thymic stromal cells but not on thymocytes. The cDNA (N14) encoding the HS9 Ag has been identified and characterized. Sequencing analysis of N14 cDNA revealed it to be a novel one without any significant homology to previously reported functional molecules. COS7 cells transfected with expression vectors harboring N14 cDNA became reactive with HS9-specific MoAB. Tissues that are positive for HS9 MoAB expressed N14 mRNA. To examine the role of this molecule in T cell development, the authors generated transgenic mice. The transgene was significantly overexpressed on both cortical and medullar thymic stromal cells but not on thymocytes. Flow cytometric analyses showed that the percentages of mature $CD4^-8^+$ or $CD4^+8^-$ thymocytes in transgenic mice were approximately twice and triple, respectively, those in the control littermates. Moreover, substantial $CD4^+8^+$ thymocytes appeared to have high levels of TCR compared with peripheral T cells. Histologic examination revealed that transgenic mice had thin cortex and relatively developed medulla. These data indicate the critical role of the N14 gene in T lymphocytes development (113).

Intrathymic development of T lymphocytes progresses as a consequence of both TCR-mediated and non-TCR-mediated interactions between thymocytes and stromal cells. As relβ -deficient mice appear to lack thymic medullary RE cells and mature dendritic cells, the effect of such a "cortex-only" thymus on T cell development was observed by DeKoning and co-workers (114). Two major consequences were detected. First, in both relβ mutant and TCR transgenic/relβ mutant mice, positive selection of both TCR αβ and δγ T cells appeared to proceed normally, with export of fully functional T cells to the periphery, suggesting that the thymic medullary stromal cells are not required for full maturation of T cells nor is an organized medullary compartment required for accumulation of mature single positive CD4 and CD8 T cells. Second, thymic negative selection was impaired, as evidenced by significant autoreactive proliferative responses to normal spleen stimulators. Peripheral T cells in relβ mutant mice showed an unusually high proportion of $CD69^+$ and $CD44^{high}$ cells. While some of these cells may be autoreactive T cells, most of the cells appeared to be activated by cytokines produced by relβ mutant non-lymphoid cells, as the effect is minimized in relβ mutant bone marrow chimeras. In conclusion, while the TCR-mediated steps in T lymphocyte maturation require both thymic cortex and medulla (RE and dendritic cells) for normal positive and negative selection of the repertoire, non-TCR-mediated interactions in the thymic cortex alone are sufficient to generate mature, immunocompetent T cells. Events following the initiation of positive selection were investigated by Hare and co-authors (115) using re-aggregate organ cultures to follow the maturation of purified $CD4^+8^+69^+$ thymocytes. These thymocytes represent a subpopulation that has already received positive selection signals. Using a dilution analysis of an FITC-based membrane-binding dye, 5-(and -6)-carboxyfluorescein diacetate succinimidyl ester, to allow a quantitative measure of proliferation, the authors show that while newly selected $CD4^+$ and $CD8^+$ cells are nondividing, both subsets subsequently undergo a wave of postpositive selection proliferation involving multiple cell divisions. Moreover, in the presence of fetal stromal cells, postselection expansion is more extensive in newborn thymocytes compared with adult thymocytes, suggesting that this phase of expansion is developmentally regulated. Proliferation of $CD4^+$ and $CD8^+$ cells is seen in re-aggregates of purified MHC class II^+ thymic RE cells, while $CD4^+$ and $CD8^+$ cells generated from bcl-2 transgenic $CD4^+8^+69^+$ thymocytes in the absence of stromal cell support survive but do not proliferate; this observation indicates that MHC class II^+ thymic RE cells are both necessary and sufficient to mediate this wave of cell division. The maturation of $CD4^+8^+69^+$ thymocytes and the subsequent proliferation of $CD4^+$ and $CD8^+$ cells occur in the presence of MHC-mismatched thymic stromal cells, suggesting that the later stages of positive selection and the associated postselection events do not depend on interactions with the same peptide/MHC complexes responsible for initiation (115).

The role of medullary thymic RE cells in tolerance induction to MHC class I restricted self peptides was analyzed by studying the β-galactosidase (β-gal)-specific cytotoxic T cell response of a transgenic mouse expressing β-gal in the thymus, skin, and central nervous system (Tg β-gal mouse). β-gal expression in the thymus was limited in to a subpopulation of medullary RE cells and Tg β-gal mice did not mount an anti-β-gal CTL response even in the presence of exogenous IL-2, while Tg β-gal-->B6 chimeras responded to β-gal as strongly as NTg β-gal mice. Furthermore, Tg β-gal mice did not generate CTL against the immunodominant Kb-restricted β-gal 497-504 peptide, and tolerance was due to the thymic RE cells that expressed β-gal because nude mice grafted with thymus from Tg β-gal mice were also unable to respond to β-gal. The Tg β-gal mouse-derived β-gal^+ medullary RE cells, the X10 cell line was found to present the Kb-restricted β-gal 497-504 epitope. These findings demonstrate that medullary thymic RE cells induce a complete tolerance towards class I-restricted self-peptides presented on their own surface (116). It has also been shown that cells of the medullary RE cellular network, but not thymic bone marrow-derived cells can use a MHC class II endogenous presentation pathway to induce tolerance to nuclear proteins (117).

Apoptosis of normal thymocytes has been shown to be triggered by several mechanisms (e.g. glucocorticoids, γ-irradiation). Thymocyte apoptosis induced by thymic RE cells have also been reported. The thymocytes undergo a massive apoptotic death within 24 hours of cocultivation with thymic RE cell monolayers derived from primary cultures or from a thymic RE cell line. Non-thymic epithelial monolayers were inactive. Induction of T cell apoptosis required direct contact between the thymocytes and the thymic RE monolayer and can be blocked by anti-CD2 and anti-LFA-1 MoABs. The immature $CD3^-/^{+dull}$ $CD4^+8^+$ thymocytes were the cells which undergo apoptosis. This massive apoptotic process of immature cells occurs in the absence of exogenous antigen, suggesting that it involves nonselected thymocytes, although apoptosis is certainly involved in the intrathymic selection processes. The apoptotic pathway selected by thymocytes following their culturing on RE cell lines involves p53 expression. Indeed, it was found that RE cell-induced apoptosis *in vitro*, led to the accumulation of p53 protein that preceded the step of DNA fragmentation in freshly isolated thymocytes, as well as in a glucocorticoid resistant thymoma cell line. Since glucocorticoid-induced thymocyte apoptosis is p53-independent, glucocorticoids are conceivably not involved in RE cell-induced thymocyte death (118).

The *in vitro* effect of thymic RE cells on dexamethasone-induced DNA fragmentation of thymocytes as a parameter for apoptosis has also been investigated. By co-culture of thymocytes with an RE cell line, at the condition without cell-cell contact, the DNA fragmentation of thymocytes caused by physiological level of glucocorticoid was significantly suppressed. The same effect was detected when treated the thymocytes with the supernatant from the cultures of the RE cell line. Thus, humoral factors released by RE cells can rescue thymocytes from the glucocorticoid-induced apoptosis (119).

5. RE CELLS: KEY CELLS OF IMMUNE-NEUROENDOCRINE REGULATION

It is now largely established that the immune and neuroendocrine systems cross talk by using similar ligands and receptors. In this context, the hypothalamo-hypophyseal-thymic axis can be regarded as a paradigm of connectivity in both normal and pathological conditions (120). It is well known that cytokines and thymic hormones modulate hypothalamic-pituitary functions:

1. IL-1 seems to upregulate the production of corticotropin-releasing factor and by adrenocorticotropin by hypothalamic neurons and pituitary cells, respectively;
2. thymulin enhances LH secretion. Conversely, a great deal of data strongly indicates that the hypothalamo-pituitary axis plays a role in the control of thymic physiology. Growth hormone (GH) for example, enhances thymulin secretion by RE cells, both *in vivo* and *in vitro*, also increasing extracellular matrix-mediated RE cell-thymocyte interactions. Additionally, gap junction-mediated cell coupling among RE cells is upregulated by ACTH. In a second vein, it

was shown that GH injections in aging mice increased total thymocyte numbers and the percentage of CD3-bearing cells, as well concanavalin-A mitogenic response and IL-6 production. In addition to mutual effects, thymus-pituitary similarities for cytokine and hormone production have been demonstrated. Cytokines such as IL-1, IL-2, IL-6, IFN-γ, TGF-β and others can be produced by hypothalamic and/or pituitary cells. Conversely, hormones including GH, PRL, LH, oxytocin, vasopressin and somatostatin can be produced intrathymically (120). Moreover, receptors for a number of cytokines and hormones are expressed in both the thymic and the hypothalamus/pituitary cells. It is also noteworthy that a thymus-pituitary connectivity can also be seen under pathological situations. In this regard, an altered HPA axis has been reported in AIDS, human falciparum malaria and murine rabies, that also show a severe thymic atrophy (120).

Cortical, undifferentiated thymocytes undergo a complex process of maturation, largely dependent on interactions with the thymic microenvironment, a tridimensional cellular network formed by RE cells, macrophages, dendritic cells, and fibroblasts. Lympho-stromal interactions in the thymus crucially determine the fate of developing T cells. Endodermal RE cells, interdigitating reticular cells, macrophages and fibroblasts all play a role in the shaping of the T lymphocyte repertoire. Recently published evidence shows that lympho-stromal interaction is bi-directional (121). Developing T cells themselves, at different stages of maturation, also control the microarchitecture of thymic cellular microenvironments, a phenomenon designated as 'crosstalk' (121). One essential cellular interaction involves the TCR-CD3 complex expressed by thymocytes with MHC-peptide complexes present on microenvironmental cells. Additionally, thymic RE cells interact with thymocytes via soluble polypeptides such as thymic hormones and interleukins, as well as through extracellular matrix (ECM) ligands and receptors (101). Such types of heterotypic interactions are under neuroendocrine control. For example, thymic endocrine function, represented by thymulin production, is up regulated, both *in vivo* and *in vitro*, by thyroid and pituitary hormones, including prolactin and growth hormone. These peptides also enhance the expression of ECM ligands and receptors, as well as the degree of RE cell-thymocyte adhesion. In addition, intrathymic T-cell migration is also hormonally regulated as ascertained by the thymocyte entrance into and exit from the TNC lymphoepithelial complexes. Taken together these data clearly illustrate the concept that neuroendocrine circuits exert a pleiotropic control on thymus physiology. The intrathymic production of hormones such as prolactin and GH suggests that, in addition to endocrine circuits, paracrine and autocrine interactions mediated by these peptides and their respective receptors may exist in the thymus, thus influencing both lymphoid and microenvironmental compartments of the organ.

The expression of extracellular matrix ligands and receptors by cultured thymic RE cells is upregulated by prolactin and GH, the latter apparently occurring *via* insulin-like growth factor 1 (IGF-1). GH enhances immune responses (122). Recently, a small mol. weight compound, GHS was identified as an inducer of the GH production by the pituitary gland. In young mice, GHS enhanced the number of peripheral blood lymphocytes, but there was no increase in T and B cell proliferative responses. Suprisingly, in old mice (16 to 24 months of age), GHS treatment for three weeks induced significant regeneration of the thymic architectonic through the increase in cell proliferation and differentiation. These treated old mice also demonstrated significant resistance to the development of tumors, induced by a transplantable lymphoma cell line, EL4. Generation of cytotoxic lymphocyte lines against the tumor cells also was promoted by GHS treatment (122). These experimental results ascribed to GHS, suggest a possible, new therapeutical approach for aging, AIDS and transplant recipients, whose immune functions are weak. Thymocyte release from the TNC lymphoepithelial complexes, as well as the reconstitution of these complexes is enhanced by PRL, GH, GHS, and IGF-1. Treatment of a mouse RE cell line with these hormones has been shown to induce an increase in thymocyte adhesion, an effect significantly prevented in the presence of antibodies to fibronectin, laminin or their respective receptors VLA-5 and VLA-6 (123).

The three well defined and biochemically characterized thymic hormones, Facteur Serique Thymique [FTS, later named thymulin (124, 25)], thymosin-α and thymopoietin (126, 27) have also been detected in the cells of RE network and in HBs. These results allowed us to suggest that HBs possess an active secretory role and thereby may participate in the intrathymic T lymphocyte differentiation (128). Furthermore, it is significant that immunocytological evidence for the simultaneous presence of all three thymic hormones in the same

RE cells of the human thymic microenvironment has also been determined (129). In a recent study, Gaudecker and co-investigators (130) used a library of MoABs together with gold-coupled reagents in immuno-electron microscopy to study the fine cellular distribution of the molecules to which the antibodies bind. Anti-cytokeratin antibodies were employed to identify the epithelial nature of all thymic RE cells, while the distribution of MHC class II molecules was revealed with MoABs to shared nonpolymorphic determinants. MoAB MR6 showed strong surface labeling of cortical RE cells and subcapsular TNCs and very weak surface staining of thymocytes, medullary macrophages, and interdigitating cells. MoAB 8.18 also labeled an another cell surface molecule; this was present on HBs and associated, hypertrophied medullary RE cells. The two molecules detected by MoABs MR6 and 8.18 are therefore located in a position where they are available to interact with external cellular and soluble signals within the thymus. In contrast, MoABs MR10 and MR19 recognized intracellular molecules within subcapsular, perivascular, and medullary RE cells. A similar distribution pattern was seen with MoAB 4β, developed against the thymic hormone thymulin, although, in addition to the expected intracellular RE cell staining, large lymphoblasts in the subcapsular zone showed surface positivity, indicating the presence of thymulin bound to surface receptors on these early thymoblasts. As expected, MHC class II molecules were expressed on some medullary and essentially all cortical RE cells. However, although most subcapsular epithelium was MHC class II-negative, some cells did express these MHC molecules on their apical surface and on the surface of their cytoplasmic extensions into the cortex. Interestingly, some cortical RE cells surrounding capillaries expressed MR6 and MR10, MR19 antigens, and presence of thymulin was revealed. Double-labeling experiments, using both MR6 and MR19 MoABs simultaneously, revealed the existence of a double-positive perivascularly located RE cell subpopulation in the thymic cortex. It remains a possibility that these "double positive" RE cells represent a common thymic RE cell progenitor.

Similar results were published after immunocytochemistry on RE cells *in vitro* from thymic tissue cultures (131). By the use of rapidly expanding cultures of thymic RE cells from 14 to 16 day-old murine fetuses and by specific antibodies against cortical and medullary located RE cells, respectively, the authors were able to demonstrate a small subpopulation of double-labeled RE cells in the cultures. These cells were not present in RE cell cultures initiated from thymuses of neonatal mice. Double-labeled RE cells were also detected in prenatal thymic tissue sections. These findings may indicate that the "double positive" RE cell subpopulation represent a common epithelial in nature stem cell for the thymic RE networks in the cortex and the medulla.

Amongst the most important signal transduction molecules involved in regulating growth and differentiation are the protein tyrosine kinases (PTK). Since intrathymic T lymphocyte maturation and differentiation is a consequence of cellular and humoral interactions between thymic stromal cells (TSC) and thymocytes, identification of the PTK in both compartments is required to dissect this process. PTK was observed in mouse TSC, using RT-PCR to survey the repertoire of PTK mRNAs expressed in a freshly isolated TSC preparation (132, 133). Ten different PTK cDNAs were identified among the 216 cDNAs sequenced. Transcripts of three of those (*ufo*, *fyn* and *fer*) were widely expressed among a large panel of immortalized thymic RE cell lines and in primary cultures of TSC. Of the other seven, none were expressed in established RE cell lines but, instead, displayed distinct expression patterns in cell types likely to have contaminated the fresh TSC preparation, i.e., macrophages, B cells, T cells and fibroblasts. Among the three PTK expressed in RE cell lines, only one, *ufo*, exhibited expression exclusively in cells of non-hemopoietic origin. Although expression of *ufo* (also named *tyro 7*, *axl* or *ark*) is not thymus specific, it is also expressed in cell types of mesodermal origin, placed in other tissues, its presence in thymic RE cells suggests a role for *ufo* in differentiation of the TSC microenvironment. Consistent with this notion, high-level expression of this receptor PTK at the protein level could be documented in every thymic RE cell line observed, as well as in fresh, frozen thymus tissue sections. These results provided the initial example of expression a receptor PTK in TSC and opened a new direction in study of the molecular regulation of thymic stromal microenvironment differentiation (133).

Previous studies have shown that a single dose of hydrocortisone in mice was able to transiently upregulate the expression of cytokeratins (CKs) 3 and 10 by medullary RE cells of the mouse thymus. It has also been found that single versus repeated exposure to high doses of glucocorticoid hormone may trigger different circuits regulating intrathymic CK expression (134). The possible role of glucocorticoids (GCs) in the maturation of thymic dendritic cells (DCs) during early ontogeny was analyzed by Sacedon *et al.* in the progeny of adrenalectomized pregnant rats (Adx fetuses) (135). In the absence of maternal GCs, there was a high percentage of DCs, many of them exhibiting a mature phenotype, demonstrating that as for other stromal cellular components of the rat thymus, DC maturation is accelerated in an early fetal microenvironment devoid of glucocorticoids.

Murine thymic RE cells express the cytochrome P450 hydroxylases Cyp11A1, Cyp21, and Cyp11B1. These enzymes, in combination with 3β-hydroxysteroid dehydrogenase (3βHSD), convert cholesterol into corticosterone, the major GC in rodents. In addition, when RE cells are cocultured with 'reporter cells' containing the glucocorticoid receptor (GR) and a GR-dependent reporter gene, a specific induction of reporter gene activity was observed. Induction of reporter gene activity was blocked when the RE cells and reporter cells were incubated in the presence of the Cyp11B1 inhibitor metyrapone or the 3βHSD inhibitor trilostane, as well as by the GR antagonist RU486. Coculturing of RE cells with thymocytes induced apoptosis in the latter, which was partially blocked by the enzyme inhibitors and RU486 (136).

Parathyroid hormone-related protein (PTHrP), a cytokine-like peptide, has been reported to affect the proliferation of lymphocytes *in vitro*. Antibodies directed against 2 different PTHrP epitopes, PTHrP(1-34) and PTHrP (34-53), demonstrated prominent specific PTHrP immunoreactivity in both subcapsular and medullary RE cells. In addition, faint but specific staining for PTHrP was described in the cortex, interdigitating between cortical lymphocytes while sparing epithelial-free subcapsular areas, thus suggesting that cortical RE cells could also be a source of PTHrP immunoreactivity. In contrast, PTH/PTHrP receptor immunoreactivity was only seen in medullary and occasionally in septal RE cells, while no evidence of cortical or lymphocytic PTH/PTHrP receptor immunoreactivity was detected. Cultured cytokeratin-positive rat thymic RE cells were also immunoreactive for both PTHrP and PTH/PTHrP receptor. Thus, it seems that PTHrP may exert significant intrathymic effects mediated by direct interactions with RE cells, or through indirect effects on PTH/PTHrP receptor-negative thymocytes (137).

The molecular and functional expression of serpentine membrane receptors for vasoactive intestinal peptide (VIP), calcitonin gene-related peptide (CGRP), a neuropeptide and calcitonin (CT) were characterized in human thymus and thymomas from myasthenia gravis (MG) patients and thymic RE cells either in primary culture (PREC) or transformed by the simian virus 40 large T (SV40LT) oncogene (LT-REC) by Marie and co-workers (138). Receptor (-R) transcripts for VIP, CGRP, and CT were identified in the thymus from control subjects and MG patients with either hyperplasia or thymoma. Similar expressions of the VIP- and CGRP-R transcripts were observed in PREC, whereas the CT-R message was not detected. In LT-REC, the signals for VIP-R, CGRP-R, and CT-R transcripts were seen with a lower intensity than those in control and MG thymus. VIP seems to be the most potent peptide among VIP-related peptides (VIP > PACAP > PHM > PHV) to stimulate cAMP production through specific type 1 VIP receptors in both PREC and LT-REC. cAMP generation was induced by CGRP in PREC and by CT in LT-REC. Type 1 VIP-R, CGRP-R, and CT-R are expressed on thymic RE cells. The transcription of the acetylcholine receptor α subunit (the main autoantigen in MG) was induced by CGRP and CT in PREC and LT-REC, respectively. These data suggest that the neuroendocrine peptides VIP, CGRP, and CT may be associated with MG and malignant transformation of the human thymus. Proliferation of thymic RE cells increases in response to addition of CGRP and histamine, but is decreased by isoproterenol, 5-HT, and acetylcholine. The percentage of multinucleate cells decreased following isoproterenol, and increase after 5-HT, GABA, and histamine treatment. Compared to controls, thymulin in the supernatant was decreased after incubation with acetylcholine, isoproterenol, 5-HT, and histamine (139). Although the results presented by Head and co-authors (139) demonstrate direct effects of neuropeptides and neurotransmitters on rat thymic RE cells, it is very important to realize that these data were obtained based on observations in cell culture, and serve more to demonstrate the physiological flexibility of thymic RE cells, than to delineate any *in vivo* mechanisms involved in the regulation of thymic lymphopoiesis.

Catecholaminergic nerve fibers have been detected in close connection to thymic RE cells, which therefore may represent preferred target cells. β1- and β2-adrenoceptor mRNA was detected in cultured rat thymic RE cells. Adrenaline, noradrenaline or the β-adrenoceptor agonist, isoprenaline, were found to increase intracellular cAMP levels 25 fold in thymic RE cells *in vitro* in a dose-dependent manner, and the response was mediated by receptors of the β1- and the β2-

subtypes. The increase in cAMP was downregulated by preincubation with glucocorticoids, such as dexamethasone or cortisol which also changed the relative importance of β1-/β2-adrenoceptors to the response. Incubation with isoprenaline or the adenylate cyclase activator forskolin decreased basal and serum-stimulated proliferation of thymic RE cells. However, adrenergic stimulation of RE cells did not induce IL-1 production. The ability of cells of the RE cellular network of the thymic microenvironment to respond to catecholamines suggests that adrenergic stimulation may be modulate the regulation of immune functions (140).

Somatostatin (SS) and its analogs exert inhibitory effects on secretive and proliferative processes of various cells via high affinity SS receptors (SS-R). SS analogs bind with different affinity to the five cloned SS-R subtypes. Octreotide, an octapeptide SS analog, binds with high affinity to the SS-R subtype 2 (sst2). SS-R has been demonstrated *in vivo* and *in vitro* on cells of the endocrine and immune systems. Among the lymphatic tissues, the thymus has been shown to contain the highest amount of SS, suggesting a local functional role of the peptide (141). The autocrine and/or paracrine role of SS and its receptors in the regulation of cell growth and differentiation within the thymic microenvironment has yet to be clarified.

Sex hormone receptors and thymulin are co-localized in thymic RE cells, but not in T lymphocytes. Furthermore, production/expression of thymic factors (thymulin, thymosin-α1) are remarkably inhibited by sex hormone treatment, and sex hormones cause changes in T cell subpopulations in the thymus. Sex hormones also strongly influence the development of thymus tumors in spontaneous thymoma BUF/Mna rats through their receptor within the tumor cells. The proliferation/maturation of thymic RE cells is mediated through protein kinase C activity induced by sex hormones and sex hormones can directly influence DNA synthesis and cdc2 kinase activity (142). The long-term effects of sex steroids have been well documented for thymocytes and cells of the thymic microenvironment. The rapid actions of progesterone upon aspects of thymic RE cell physiology were examined in a recent study (143). Progesterone (0.1-10 μm) was applied to cultured thymulin-secreting RE cells and changes in transmembrane potential, transmembrane current, intracellular calcium levels and thymulin secretion were assessed. Rapid changes in electrophysiology and intracellular calcium provide evidence for a membrane-bound progesterone receptor in these cells, in addition to classical cytoplasmic receptors. Application of progesterone to these RE cells caused electrophysiological changes in a majority of the cells, activating an inward current and dose-dependent depolarization. Intracellular calcium levels increased within seconds of progesterone application. Progesterone increased thymulin levels in supernatant above the levels in the pre-application period. This effect was reduced in the presence of cobalt chloride, which blocks voltage-dependent calcium channels. In addition, RE cells in culture were immunoreactive to antibody AG7. This antibody was raised to a membrane-bound antigen involved in calcium influx subsequent to progesterone binding in sperm (143). Higher incidents of autoimmune diseases in females were the main suggestion that gonadal steroid hormones may also modulate the human immune response (144, 145). The effects of androgens have been observed on the differentiation of lymphocytes in the thymus and bone marrow. The expression of androgen receptor, a ligand activated transcription factor that mediates hormone actions, has been detected on lymphatic and nonlymphatic cells of the thymic microenvironment and in bone marrow. Hormonal treatments suggest that thymic RE cells and bone marrow stromal cells are the mediators of androgenic effects on immature T and B lymphocytes.

Oxytocin (OT) has been shown to be the dominant peptide of the neurohypophysial family expressed by thymic RE and nurse cells in various species (146). Ontogenetic observations defined that thymic presence of OT gene precedes the hypothalamic one. Thymic OT is not secreted but, after translocation of a hybrid neurophysin/MHC class I protein, is integrated within the plasma membrane of RE cells, thus allowing its presentation to pre-T cells. Immunoreactive-OT (ir-OT), ir-IL-1β, ir-IL-6 and ir-LIF have been detected in the thymic RE cell cultures (147). ir-OT was restricted to thymic RE cells, while some ir-IL-6 and ir-LIF were occasionally detected in fibroblasts. In basal conditions, ir-IL-6 and ir-LIF (but neither ir-OT nor ir-IL-1β) were detected in the supernatants of human RE cell cultures. MoABs to OT induced a marked increase in ir-IL-6 and ir-LIF secretion in RE cell cultures.

6. CONCLUSIONS

During mammalian ontogenesis, the thymus is a hematopoietic organ. In the past, it was believed that after birth, the thymus loses this function of hematopoiesis. However, as shown by numerous studies and researchers over the last half century, lymphatic and nonlymphatic hematopoiesis remains one of the main functions of the cellular microenvironment. It appears that both the other stromal elements of the thymic microenvironment and the differentiating hematopoietic cells significantly influence the functional state of thymic RE cells, thus establishing the bi-directional intrinsic regulatory circuit. I hope that future functional observations will be carried out employing single-chain antibodies specific for evolutionary conserved RE, microenvironmental interspecific molecules. These studies will detect shared, rodent/human antigens and will allow us new, functional, and comparative analyses.

Certainly, the thymus remains the first, central and most important organ of T lymphopoiesis. I think that the different stages of this intrathymic T lymphocyte maturation and differentiation is fully described throughout scientific literature.

We paid much attention to the thymic reticulo-epithelial (RE) cellular network as a cellular level of neuro-endocrine regulation in this chapter. These libraries of polypeptides belong to a number of neuroendocrine protein families (such as the neurohypophysial, tachykinin, neurotensin and insulin families). These neuroendocrine polypeptides are self-antigens and via the thymic RE cells are able to modulate the hematopoietic stem cell differentiation in different cell lines. The thymic RE cells play a key role in the induction of the self-tolerance to neuroendocrine polypeptide families. I agree with Geenen and co-workers (105) that the tolerogenic properties of the self-antigens of the thymic stromal microenvironment could be used for vaccination to prevent the development of autoimmune diseases.

Recently a small mol. weight compound, a GHS was identified as an inducer of the GH production by the pituitary gland. In young mice, GHS enhanced the number of peripheral blood lymphocytes, but there was no increase in T and B cell proliferative responses. Suprisingly, in old mice (16 to 24 months of age) GHS treatment for only three weeks induced significant increase in thymic cellularity and cell differentiation. These treated old mice also demonstrated significant resistance to the initiation of tumors by a transplantable lymphoma cell line, EL4. Generation of cytotoxic lymphocytes against the tumor cells also was promoted by GHS. I agree with Koo and co-workers (122) that the actions ascribed to GHS suggest a possible, new therapeutical approach for the aging population, AIDS patients and transplant recipients, whose immune functions are weak.

Androgen receptor expression on thymic RE cells and bone marrow stromal cells also modulate the T and B lymphocyte maturation. Recent castration or androgen resistant testicular feminization experiments in normal male rodents produced significant enlargement of the thymus, and androgen treatment reverses these changes. Probably therefore is so high the incidence of autoimmune disorders in females.

All of this demonstrates and puts forth crucial evidence that the thymus is a very important organ of hematopoiesis throughout the entire life of a mammalian organism.

REFERENCES

1. Kendall MD: Hemopoiesis in the thymus. Dev Immunol *4:* 157-168, 1995.
2. Manley NR: Thymus organogenesis and molecular mechanisms of thymic epithelial cell differentiation. Semin Immunol *12:* 421-428, 2000.
3. Furusawa T, Yanai N, Hara T, Miyajima A, Obinata M: Integrin-associated protein (IAP, also termed CD47) is involved in stroma-supported erythropoiesis. J Biochem *123:* 101-106, 1998.
4. Fisher RC, Scott EW: Role of PU.1 in hematopoiesis. Stem Cells *16:* 25-37, 1998.
5. Anderson G, Harman BC, Hare KJ, Jenkinson EJ: Microenvironmental regulation of T cell development in the thymus. Semin Immunol *12:* 457-464, 2000.
6. Katevuo K, Imhof BA, Boyd R, Chidgey A, Bean A, Dunon D, Gobel TW, Vainio O: ChT1, an Ig superfamily molecule required for T cell differentiation. J Immunol *162:* 5685-5694, 1999.
7. Hakanson H, Larsson LI, Sundler F: Peptide and amine producing endocrine-like cells in the chicken thymus. A chemical, histochemical and electron microscopic study. Histochemistry *39:* 25-34, 1974.
8. Caso LV: Some endocrine aspects of the thymus gland. Japan J Med Sci Biol *29:* 289-321, 1976.
9. Peschle C: Human ontogenic development: studies on the hemopoietic system and the expression of homeo box genes. Ann NY Acad Sci *511:* 101-116, 1987.
10. de Pablo F, Roth J: Endocrinization of the early embryo: an emerging role for hormones and hormone-like factors. Trends Biochem Sci *15:* 339-342, 1990.
11. Murray MJ: Hypothyroidism and respiratory insufficiency in a neonatal foal. J Am Vet Med Assoc *197:* 1635-1638, 1990.
12. Sieff CA, Niemeyer CM, Nathan DG, Ekern SC, Bieber FR, Yang YC, Wong G, Clark SC: Stimulation of human hematopoietic colony formation by recombinant gibbon multi-colony-stimulating factor or interleukin 3. J Clin Invest *80:* 818-823, 1987.
13. Huang E, Nocka K, Beier DR, Chu TY, Buck J, Lahm HW, Wellner D, Leder P, Besmer P: The hemopoietic growth factor KL is encoded by the Sl locus and is the ligand of the c-kit receptor, the gene product of the W locus. Cell *61:* 953-963, 1990.
14. Tavassoli M: Embryonic and fetal hemopoiesis: an overview. Blood Cells *17:* 269-281, 1991.
15. Metcalf D, Nicola NA: The regulatory factors controlling murine erythropoiesis in vitro. Prog Clin Biol Res *148:* 93-105, 1984.
16. Ni K, O'Neill HC: Hemopoiesis in long-term stroma-dependent cultures from lymphoid tissue: production of cells with myeloid/dendritic characteristics. In Vitro Cell Dev Biol Anim *34:* 298-307, 1998.
17. Danova M, Aglietta M: Cytokine receptors, growth factors and cell cycle in human bone marrow and peripheral blood hematopoietic progenitors. Haematologica *82:* 622-629, 1997.
18. Shiozaki M, Kosaka M, Eto Y: Activin A: a commitment factor in erythroid differentiation. Biochem Biophys Res Commun *242:* 631-635, 1998.
19. Khmel'nitskii OK, Khavinson VKh, Zhukhina GE, Kotov VA, Seryi SV: The cytomorphological characteristics of hemopoiesis and of the organs of immunogenesis in exposure to thymus and bone marrow peptides during hematopoietic depression. Arkh Patol *51:* 39-44, 1989.
20. Khavinson VKh, Seryi SV, Malinin VV: Correction with thymic and bone marrow peptides of radiation injuries of immuno- and hemopoiesis. Radiobiologiia *31:* 501-505, 1991.
21. Tavassoli M: Embryonic origin of hematopoietic stem cells. Exp Hematol *22:* 7, 1994.
22. Maeno M, Tochinai S, Katagiri C: Differential participation of ventral and dorsolateral mesoderms in the hemopoiesis of Xenopus, as revealed in diploid-triploid or interspecific chimeras. Dev Biol *110:* 503-508, 1985.
23. Cowden RR: The embryonic origin of blood cells in the tunicate Clavelina picta. Trans Am Microsc Soc *87:* 521-524, 1968.
24. Dorn A: Ultrastructure of differentiating hemocytes in the embryo of Oncopeltus fasciatus dallas (insecta, heteroptera). Cell Tissue Res *187:* 79-488, 1978.
25. Ward P, D'Cruz D: Seasonal changes in the thymus gland of a tropical bird. Ibis *110:* 203-205, 1968.
26. Anderson WL: Seasonal changes of thymus weights in ring-necked pheasants. Condor *72:* 205-208, 1970.
27. Kendall MD, Ward P: Erythropoiesis in an avian thymus. Nature *249:* 366-367, 1974.
28. Kendall MD, Frazier JA: Ultrastructural studies on erythropoiesis in the avian thymus. Cell Tissue Res *199:* 37-61, 1979.
29. Fonfria J, Barrutia MG, Garrido E, Ardavin CF, Zapata A: Erythropoiesis in the thymus of the spotless starling Sturnus unicolor. Cell Tissue Res *32:* 445-455, 1983.
30. Kaiser HE: Comparative importance of the lymphatic system during neoplastic progression: lymphohematogenous spreading. In: Fundamental Aspects of Cancer. (Goldfarb RH, ed) Cancer Growth and Progression. (Kaiser HE, series ed), v.1, ch.12, Kluwer Academic Publishers, Dordrecht, 1989, p.126.
31. Machotka SV: Lymphocytic neoplasms in reptiles and fish. In: Comparative Aspects of Tumor Development. (Kaiser HE, ed) Cancer Growth and Progression. (Kaiser HE, series ed), v.5, ch.11, Kluwer Academic Publishers, Dordrecht, 1989, p.89.
32. Cole RJ, Paul J: The effects of erythropoietin on haem synthesis in mouse yolk sac and cultured foetal liver cells. J Embryol Exp Morphol *15:* 245-260, 1966.
33. Fantoni A, Bank A, Marks PA: Globin composition and synthesis of hemoglobins in developing fetal mice erythroid cells. Science *157:* 1327-1329, 1967.
34. Rich IN, Kubanek B: The ontogeny of erythropoiesis in the mouse detected by the erythroid colony-forming technique. II. Transition in erythropoietin sensitivity during development. J Embryol Exp Morphol *58:* 143-155, 1980.
35. Metcalf D, Moore MAS: Haemopoietic cells. London: North-Holland Publishing Company, 1971.
36. Christensen RD: Hematopoiesis in the fetus and neonate. Pediatr Res *26:* 531-535, 1989.
37. Tavassoli M, Friedenstein A: Hemopoietic stromal microenvironment. Am J Hematol *15:* 195-203, 1983.
38. Kau CL, Turpen JB: Dual contribution of embryonic ventral blood island and dorsal lateral plate mesoderm during ontogeny of hemopoietic cells in Xenopus laevis. J Immunol *131:* 2262-2266, 1983.
39. Smith PB, Turpen JB: Differential contribution of dorsal and ventral lateral plate mesoderm to hemopoiesis during Rana pipiens embryogenesis. Dev Biol *104:* 497-499, 1984.
40. Ogawa M: Differentiation and proliferation of hematopoietic stem cells. Blood *81:* 2844-2853, 1993.
41. Moore T, Huang S, Terstappen LW, Bennett M, Kumar V: Expression of CD43 on murine and human pluripotent hematopoietic stem cells. J Immunol *153:* 4978-4987, 1994.
42. Smith RA, Glomski CA: Embryonic and fetal hemopoiesis in the Mongolian gerbil (Meriones unguiculatus). Anat Rec

189: 499-517, 1977.

43. Kendall MD, Singh J: The presence of erythroid cells in the thymus gland of man. J Anat *130:* 183-189, 1980.
44. Bodey B: Histomorphology and histochemistry of the human thymus during the prenatal development. D. Sc. Thesis, Sofia, pp. 1-360, 1977.
45. Petti S, Testa U, Migliaccio AR, Mavilio F, Marinucci M, Lazzaro D, Russo G, Mastroberardino G, Peschle C: Embryonic hemopoiesis in human liver: morphologic aspects at sequential stages of ontogenic development. Prog Clin Biol Res *193:* 57-71, 1985.
46. Rencricca NJ, Howard D, Kubanek B, Stohlman F: Erythroid differentiation of fetal, newborn and adult haemopoietic stem cells. Scand J Haematol *16:* 189-195, 1976.
47. Andrews RG, Singer JW, Bernstein ID: Precursors of colony-forming cells in humans can be distinguished from colony-forming cells by expression of the CD33 and CD34 antigens and light scatter properties. J Exp Med (1989) 169:1721-1731.
48. Andrews RG, Singer JW, Bernstein ID: Human hematopoietic precursors in long-term culture: single CD34$^+$ cells that lack detectable T cell, B cell, and myeloid cell antigens produce multiple colony-forming cells when cultured with marrow stromal cells. J Exp Med *172:* 355-358, 1990.
49. Verfallie C, Blakolmer K, McGlave P: Purified primitive human hematopoietic progenitor cells with long-term in vitro repopulating capacity adhere selectively to irradiated bone marrow stroma. J Exp Med *172:* 509-602, 1990.
50. Terstappen LW, Huang S, Safford M, Lansdorp PM, Loken MR: Sequential generations of hematopoietic colonies derived from single nonlineage-committed CD34$^+$CD38$^-$ progenitor cells. Blood *77:* 1218-1227, 1991.
51. Lansdorp PM, Dragowska W: Long-term erythropoiesis from constant numbers of CD34$^+$ cells in serum-free cultures initiated with highly purified progenitor cells from human bone marrow. J Exp Med *175:* 1501-1509, 1992.
52. Rosenstein Y, Park JK, Hahn WC, Rosen FS, Bierer BE, Burakoff SJ: CD43, a molecule defective in Wiskott-Aldrich syndrome, binds ICAM-1. Nature *354:* 233-235, 1991.
53. Pedraza-Alva G, Merida LB, Burakoff SJ, Rosenstein Y: CD43-specific activation of T cells induces association of CD43 to Fyn kinase. J Biol Chem *271:* 27564-27568, 1996.
54. Pedraza-Alva G, Merida LB, Burakoff SJ, Rosenstein Y: T cell activation through the CD43 molecule leads to Vav tyrosine phosphorylation and mitogen-activated protein kinase pathway activation. J Biol Chem *273:* 14218-14224, 1998.
55. Hammar JA: Zur Histogenese und Involution der Thymusdrüse. Anat Anz *27:* 23-30 and 41-89, 1905.
56. Hammar JA: Zur Frage der Histogenese der Thymusdrüse. Zentrbl Pathol *33:* 505-513, 1923.
57. Maximov AA: Untersuchungen über Blut und Bindegewebe. II. Über die Histogenese der Thymus bei Säugetieren. Arch mikrosk Anat *74:* 525-621, 1909.
58. Le Douarin NM, Jotereau FV: Tracing of cells of the avian thymus through embryonic life in interspecific chimeras. J Exp Med *142:* 17-40, 1975.
59. Testi R, D'Ambrosio D, DeMaria R: The CD69 receptor: A multipurpose cell-surface trigger for hematopoietic cells. Immunol Today *15:* 479-483, 1994.
60. Witte ON: Steel locus defines new multipotent growth factor. Cell *63:* 5-6, 1990.
61. Moore NC, Anderson G, Smith CA, Owen JJ, Jenkinson EJ: Analysis of cytokine gene expression in subpopulations of freshly isolated thymocytes and thymic stromal cells using semiquantitative polymerase chain reaction. Eur J Immunol *23:* 922-927, 1993.
62. Moore TA, Zlotnik A: Differential effects of Flk-2/Flt-3 ligand and stem cell factor on murine thymic progenitor cells. J Immunol *158:* 4187-4192, 1997.
63. Matsui Y, Zsebo KM, Hogan BL: Embryonic expression of a haematopoietic growth factor encoded by the Sl locus and the ligand for c-kit. Nature *347:* 67-669, 1990.
64. Tavassoli M, Hardy CL: Molecular basis of homing of intravenously transplanted stem cells to the marrow. Blood *76:* 1059-1070, 1990.
65. Aizawa S, Tavassoli M: In vitro homing of hemopoietic stem cells is mediated by a recognition system with galactosyl and mannosyl specificities. Proc Natl Acad Sci USA *84:* 4485-4489, 1987.
66. Matsuoka T, Hardy C, Tavassoli M: Characterization of membrane homing receptors in two cloned murine hemopoietic progenitor cell lines. J Clin Invest *83:* 904-911, 1989.
67. Matsuoka T, Tavassoli M: Purification and partial characterization of membrane-homing receptors in two cloned murine hemopoietic progenitor cell lines. J Biol Chem *264:* 20193-20198, 1989.
68. Yokota T, Oritani K, Mitsui H, Aoyama K, Ishikawa J, Sugahara H, Matsumura I, Tsai S, Tomiyama Y, Kanakura Y, Matsuzawa Y: Growth-supporting activities of fibronectin on hematopoietic stem/progenitor cells in vitro and in vivo: structural requirement for fibronectin activities of CS1 and cell-binding domains. Blood *91:* 3263-3272, 1998.
69. Schofield KP, Humphries MJ, de Wynter E, Testa N, Gallagher JT: The effect of α4β1-integrin binding sequences of fibronectin on growth of cells from human hematopoietic progenitors. Blood *91:* 3230-3238, 1998.
70. Tavassoli M, Minguell JJ: Homing of hemopoietic progenitor cells to the marrow. Proc Soc Exp Biol Med *196:* 367-373, 1991.
71. Sato VL, Waksal SD, Herzenberg LA: Identification and separation of pre T cells from nu/nu mice by preculture with thymic reticuloepithelial cells. Cell Immunol *24:* 173-185, 1976.
72. Stutman O: Intrathymic and extrathymic T cell maturation. Immunol Rev *42:* 138-184, 1978.
73. Kruisbeek AM: Thymic factors and T cell maturation in vitro: a comparison of the effects of thymic epithelial cultures with thymic extracts and thymus dependent serum factors. Thymus *1:* 163-185, 1979.
74. Jason JM, Janeway CA: In vitro cultivation of nonlymphoid thymic cells: morphological and immunological characterization. Clin Immunol Immunopathol *30:* 214-226, 1984.
75. Schuurman HJ, van de Wijngaert FP, Huber J, Schuurman RK, Zegers BJ, Roord JJ, Kater L: The thymus in "bare lymphocyte" syndrome: significance of expression of major histocompatibility complex antigens on thymic epithelial cells in intrathymic T-cell maturation. Hum Immunol *13:* 69-82, 1985.
76. Singer KH, Wolf LS, Lobach DF, Denning SM, Tuck DT, Robertson AL, Haynes BF: Human thymocytes bind to autologous and allogeneic thymic epithelial cells in vitro. Proc Natl Acad Sci USA *83:* 6588-6592, 1986.
77. Kurtzberg J, Denning SM, Nycum LM, Singer KH, Haynes BF: Immature human thymocytes can be driven to differentiate into nonlymphoid lineages by cytokines from thymic epithelial cells. Proc Natl Acad Sci USA *86:* 7575-7579, 1989.
78. Zuniga-Pflucker JC, Jones LA, Longo DL, Kruisbeek AM:

Both TCR/MHC and accessory molecule/MHC interactions are required for positive and negative selection of mature T cells in the thymus. Cold Spring Harb Symp Quant Biol 54 Pt *1:* 153-158, 1989.
79. Trainin N: Thymic hormones and the immune response. Physiol Rev *54:* 272-315, 1974.
80. Maximov AA: Untersuchungen über Blut und Bindegewebe. IV. Über die Histogenese der Thymus bei Amphibien. Arch mikrosk Anat *79:* 560-611, 1912.
81. Maximov AA: Untersuchungen über Blut und Bindegewebe. V. Über die embryonale Entwicklung der Thymus bei Selachiern. Arch mikrosk Anat *80:* 39-88, 1912.
82. Palmer DB, Crompton T, Marandi MB, George AJ, Ritter MA: Intrathymic function of the human cortical epithelial cell surface antigen gp200-MR6: single-chain antibodies to evolutionary conserved determinants disrupt mouse thymus development. Immunology *96:* 236-245, 1999.
83. Viret C, Janeway CA Jr: MHC and T cell development. Rev Immunogenet *1:* 91-104, 1999.
84. Viret C, Sant'angelo DB, He X, Ramaswamy H, Janeway CA Jr: A role for accessibility to self-peptide-self-MHC complexes in intrathymic negative selection. J Immunol *166:* 4429-4437, 2001.
85. Meilin A, Shoham J, Sharabi Y: Analysis of thymic stromal cell subpopulations grown in vitro on extracellular matrix in defined medium. IV. Cytokines secreted by human thymic epithelial cells in culture and their activities on murine thymocytes and bone marrow cells. Immunology *77:* 208-213, 1992.
86. Sunder-Plassmann P: Problem of thymus and Basedow's disease: further studies on biology, pathology and clinical aspects of neurohormonal cell system. Deutsche Ztsschr Chir *255:* 523-615, 1942.
87. Selye H, Marion D: Direct thymolytic effect of hexestrol (estrogen). Acta Anat *23:* 180-184, 1955.
88. Savino W, Cirne-Lima EO, Soares JF, Leite-de-Moraes Mdo C, Ono IP, Dardenne M: Hydrocortisone increases the numbers of KL1+ cells, a discrete thymic epithelial cell subset characterized by high molecular weight cytokeratin expression. Endocrinology *123:* 2557-2564, 1988.
89. Timsit J, Savino W, Safieh B, Chanson P, Gagnerault MC, Bach JF, Dardenne M: Growth hormone and insuline-like growth factor-I stimulate hormonal function and proliferation of thymic epithelial cells. J Clin Endocrinol Metab *75:* 183-188, 1992.
90. Wick G, Hu Y, Schwarz S, Kroemer G: Immunoendocrine communication via the hypothalamo-pituitary-adrenal axis in autoimmune diseases. Endocr Rev *14:* 539-563, 1993.
91. Geenen V, Legros JJ, Franchimont P, Baudrihaye M, Defresne MP, Boniver J: The neuroendocrine thymus: coexistence of oxytocin and neurophysin in the human thymus. Science *232:* 508-511, 1986.
92. Le PT, Lazorick S, Whichard LP, Yang YC, Clark SC, Haynes BF, Singer KH: Human thymic epithelial cells produce IL-6, granulocyte-monocyte-CSF, and leukemia inhibitory factor. J Immunol *145:* 3310-3315, 1990.
93. Argiolas A, Gessa GL, Melis MR, Stancampiano R, Vaccari A: Effects of neonatal and adult thyroid dysfunction on thymic oxytocin. Neuroendocrinology *52:* 556-559, 1990.
94. Robert F, Geenen V, Schoenen J, Burgeon E, De Groote D, Defresne MP, Legros JJ, Franchimont P: Colocalization of immunoreactive oxytocin, vasopressin and interleukin-1 in human thymic epithelial neuroendocrine cells. Brain Behav Immun *5:* 102-115, 1991.
95. Martin-Fontecha A, Broekhuizen R, de Heer C, Zapata A, Schuurman HJ: The neuro-endocrine component of the rat thymus: studies on cultured thymic fragments before and after transplantation in congenitally athymic and euthymic rats. Brain Behav Immun *7:* 1-15, 1993.
96. Dardenne M, Savino W: Control of thymus physiology by peptidic hormones and neuropeptides. Immunol Today *15:* 518-523, 1994.
97. Savino W, Dardenne M: Immune-neuroendocrine interactions. Immunol Today *16:* 318-322, 1995.
98. Deman J, Van Meurs M, Claassen E, Humblet C, Boniver J, Defresne MP: In vivo expression of interleukin-1 β (IL-1 β), IL-2, IL-4, IL-6, tumour necrosis factor-α and interferon-γ in the fetal murine thymus. Immunology *89:* 152-157, 1996.
99. van Ewijk W, de Kruif J, Germeraad WT, Berendes P, Ropke C, Platenburg PP, Logtenberg T: Subtractive isolation of phage-displayed single-chain antibodies to thymic stromal cells by using intact thymic fragments. Proc Natl Acad Sci USA *94:* 3903-3908, 1997.
100. Vanneste Y, Thome AN, Vandersmissen E, Charlet C, Franchimont D, Martens H, Lhiaubet AM, Schimpff RM, Rostene W, Geenen V: Identification of neurotensin-related peptides in human thymic epithelial cell membranes and relationship with major histocompatibility complex class I molecules. J Neuroimmunol *76:* 161-166, 1997.
101. Savino W, Villa-Verde DM, Alves LA, Dardenne M: Neuroendocrine control of the thymus. Ann NY Acad Sci *840:* 470-479, 1998.
102. Geenen V, Martens H, Vandersmissen E, Achour I, Kecha O, Franchimont D: Cellular and molecular aspects of thymic T-cell education in neuroendocrine self principles. Implications for autoimmunity. Ann NY Acad Sci *840:* 328-337, 1998.
103. Martens H, Kecha O, Charlet-Renard C, Defresne MP, Geenen V: Neurohypophysial peptides stimulate the phosphorylation of pre-T cell focal adhesion kinases. Neuroendocrinology *67:* 282-289, 1998.
104. de Mello-Coelho V, Gagnerault MC, Souberbielle JC, Strasburger CJ, Savino W, Dardenne M, Postel-Vinay MC: Growth hormone and its receptor are expressed in human thymic cells. Endocrinology *139:* 3837-3842, 1998.
105. Geenen V, Kecha O, Brilot F, Charlet-Renard C, Martens H: The thymic repertoire of neuroendocrine-related self antigens: biological role in T-cell selection and pharmacological implications. Neuroimmunomodulation *6:* 115-125, 1999.
106. Savino W, Dardenne M: Neuroendocrine control of thymus physiology. Endocr Rev *21:* 412-443, 2000.
107. Bodey B, Bodey B Jr, Hall FL: Expression of Transforming Growth Factor-β type II receptors in the cells of the human thymic microenvironment during ontogenesis. In: Molecular Biology of Hematopoiesis, Volume 5 (N.G. Abraham, S. Asano, G. Brittinger, G.J.M. Maestroni and R. Shadduck, eds.) New York: Plenum Press, Chapter 78, 1996, pp. 645-658.
108. Schluns KS, Cook JE, Le PT: TGF-β differentially modulates epidermal growth factor-mediated increases in leukemia-inhibitory factor, IL-6, IL-1 α, and IL-1 β in human thymic epithelial cells. J Immunol *158:* 2704-2712, 1997.
109. Shu S, Naylor P, Touraine JL, Hadden JW: IL-1, ICAM-1, LFA-3, and hydrocortisone differentially regulate cytokine secretion by human fetal thymic epithelial cells. Thymus *24:* 89-99, 1996.
110. Napolitano M, Bellavia D, Maroder M, Farina M, Vacca A, Frati L, Gulino A, Screpanti I: Modulation of cytokine gene expression by thymic lympho-stromal cell to cell interaction: effect of retinoic acid. Thymus *24:* 247-258, 1997.
111. Botham CA, Jones GV, Kendall MD: Immuno-characterisation of neuroendocrine cells of the rat thymus

gland in vitro and in vivo. Cell Tissue Res *303:* 381-389, 2001.

112. Bruggers CS, Patel DD, Scearce RM, Whichard LP, Haynes BF, Singer KH: AD2, a human molecule involved in the interaction of T cells with epidermal keratinocytes and thymic epithelial cells. J Immunol *154:* 2012-2022, 1995.
113. Takeuchi T, Kuro-o M, Miyazawa H, Ohtsuki Y, Yamamoto H: Transgenic expression of a novel thymic epithelial cell antigen stimulates abberant development of thymocytes. J Immunol *159:* 726-733, 1997.
114. DeKoning J, DiMolfetto L, Reilly C, Wei Q, Havran WL, Lo D: Thymic cortical epithelium is sufficient for the development of mature T cells in relB-deficient mice. J Immunol *158:* 2558-2566, 1997.
115. Hare KJ, Wilkinson RW, Jenkinson EJ, Anderson G: Identification of a developmentally regulated phase of postselection expansion driven by thymic epithelium. J Immunol *160:* 3666-3672, 1998.
116. Oukka M, Cohen-Tannoudji M, Tanaka Y, Babinet C, Kosmatopoulos K: Medullary thymic epithelial cells induce tolerance to intracellular proteins. J Immunol *156:* 968-975, 1996.
117. Oukka M, Colucci-Guyon E, Tran PL, Cohen-Tannoudji M, Babinet C, Lotteau V, Kosmatopoulos K: CD4 T cell tolerance to nuclear proteins induced by medullary thymic epithelium. Immunity *4:* 545-553, 1996.
118. Schreiber L, Sharabi Y, Schwartz D, Goldfinger N, Brodie C, Rotter V, Shoham J: Induction of apoptosis and p53 expression in immature thymocytes by direct interaction with thymic epithelial cells. Scand J Immunol *44:* 314-322, 1996.
119. Gao Y, Kinoshita Y, Hato F, Tominaga K, Tsuji Y: Suppression of glucocorticoid-induced thymocyte apoptosis by co-culture with thymic epithelial cells. Cell Mol Biol *42:* 227-234, 1996.
120. Savino W, Arzt E, Dardenne M: Immunoneuroendocrine connectivity: the paradigm of the thymus-hypothalamus/pituitary axis. Neuroimmunomodulation *6:* 126-136, 1999.
121. van Ewijk W, Wang B, Hollander G, Kawamoto H, Spanopoulou E, Itoi M, Amagai T, Jiang YF, Germeraad WT, Chen WF, Katsura Y: Thymic microenvironments, 3-D versus 2-D? Semin Immunol *11:* 57-64, 1999.
122. Koo GC, Huang C, Camacho R, Trainor C, Blake JT, Sirotina-Meisher A, Schleim KD, Wu TJ, Cheng K, Nargund R, McKissick G: Immune enhancing effect of a growth hormone secretagogue. J Immunol *166:* 4195-4201, 2001.
123. de Mello-Coelho V, Villa-Verde DM, Dardenne M, Savino W: Pituitary hormones modulate cell-cell interactions between thymocytes and thymic epithelial cells. J Neuroimmunol *76:* 39-49, 1997.
124. Dalakas MC, Engel WK, McClure JE, Goldstein AL, Askanas V: Immunocytochemical characterization of thymosin in thymic epithelial cells of normal and myasthenia gravis patients and in thymic cultures. J Neurol Sci *50:* 239-247, 1981.
125. Schmitt D, Monier JC, Dardenne M, Pleau JM, Bach JF: Location of FTS (facteur thymique serique) in the thymus of normal and auto-immune mice. Thymus *4:* 221-231, 1982.
126. van den Tweel JG, Taylor CR, McClure J, Goldstein AL: Detection of thymosin in thymic epithelial cells by an immunoperoxidase method. Adv Exp Med Biol 114*:* 511-515, 1979.
127. Monier JC, Dardenne M, Pleau JM, Schmitt D, Deschaux P, Bach JF: Characterization of facteur thymique sérique (FTS) in the thymus. I. Fixation of anti-FTS antibodies on thymic reticuloepithelial cells. Clin Exp Immunol *42:* 470-476, 1980.
128. Bodey B, Kaiser HE: Development of Hassall's bodies of the thymus in humans and other vertebrates (especially mammals) under physiological and pathological conditions: immunocytochemical, electronmicroscopic and in vitro observations. In Vivo *11:* 61-86, 1997.
129. Savino W, Dardenne M: Thymic hormone containing cells. VI. Immunohistologic evidence for the simultaneous presence of thymulin, thymopoietin and thymosin $\alpha 1$ in normal and pathological human thymuses. Eur J Immunol *14:* 987-992, 1984.
130. von Gaudecker B, Kendall MD, Ritter MA: Immuno-electron microscopy of the thymic epithelial microenvironment. Microsc Res Tech *38:* 237-249, 1997.
131. Ropke C, van Soest P, Platenburg PP, van Ewijk W: A common stem cell for murine cortical and medullary thymic epithelial cells? Dev Immunol *4:* 149-156, 1995.
132. Boyd RL, Tucek CL, Godfrey DI, Izon DJ, Wilson TJ, Davidson NJ, Bean AG, Ladyman HM, Ritter MA, Hugo P: The thymic microenvironment. Immunol Today *14:* 445-459, 1993.
133. Rinke de WIT TF, Izon DJ, Revilla C, Oosterwegel M, Bakker AQ, van Ewijk W, Kruisbeek AM: Expression of tyrosine kinase gene in mouse thymic stromal cells. Int Immunol *8:* 1787-1795, 1996.
134. Cirne-Lima EO, Savino W: Repeated in vivo hydrocortisone treatment promotes a dual modulation of cytokeratin expression by mouse thymic epithelial cells. Neuroimmunomodulation *5:* 328-331, 1998.
135. Sacedon R, Vicente A, Varas A, Jimenez E, Zapata AG: Early differentiation of thymic dendritic cells in the absence of glucocorticoids. J Neuroimmunol *94:* 103-108, 1999.
136. Pazirandeh A, Xue Y, Rafter I, Sjovall J, Jondal M, Okret S: Paracrine glucocorticoid activity produced by mouse thymic epithelial cells. FASEB J *13:* 893-901, 1999.
137. Funk JL, Jones GV, Botham CA, Morgan G, Wooding P, Kendall MD: Expression of parathyroid hormone-related protein and the parathyroid hormone/parathyroid hormone-related protein receptor in rat thymic epithelial cells. J Anat *194:* 255-264, 1999.
138. Marie J, Wakkach A, Coudray A, Chastre E, Berrih-Aknin S, Gespach C: Functional expression of receptors for calcitonin gene-related peptide, calcitonin, and vasoactive intestinal peptide in the human thymus and thymomas from myasthenia gravis patients. J Immunol *162:* 2103-2112, 1999.
139. Head GM, Mentlein R, von Patay B, Downing JE, Kendall MD: Neuropeptides exert direct effects on rat thymic epithelial cells in culture. Dev Immunol *6:* 95-104, 1998.
140. Kurz B, Feindt J, von Gaudecker B, Kranz A, Loppnow H, Mentlein R: β-adrenoceptor-mediated effects in rat cultured thymic epithelial cells. Br J Pharmacol *120:* 1401-1408, 1997.
141. Ferone D, van Hagen PM, van Koetsveld PM, Zuijderwijk J, Mooy DM, Lichtenauer-Kaligis EG, Colao A, Bogers AJ, Lombardi G, Lamberts SW, Hofland LJ.: In vitro characterization of somatostatin receptors in the human thymus and effects of somatostatin and octreotide on cultured thymic epithelial cells. Endocrinology *140:* 373-380, 1999.
142. Seiki K, Sakabe K: Sex hormones and the thymus in relation to thymocyte proliferation and maturation. Arch Histol Cytol *60:* 29-38, 1997.
143. Head GM, Downing JE, Brucker C, Mentlein R, Kendall MD: Rapid progesterone actions on thymulin-secreting epithelial cells cultured from rat thymus. Neuroimmunomodulation *6:* 31-38, 1999.
144. Olsen NJ, Olson G, Viselli SM, Gu X, Kovacs WJ:

Androgen receptors in thymic epithelium modulate thymus size and thymocyte development. Endocrinology *142:* 1278-1283, 2001.
145. Olsen NJ, Kovacs WJ: Effects of androgens on T and B lymphocyte development. Immunol Res *23:* 281-288, 2001.
146. Geenen V, Martens H, Brilot F, Renard C, Franchimont D, Kecha O: Thymic neuroendocrine self-antigens. Role in T-cell development and central T-cell self-tolerance. Ann NY Acad Sci *917:* 710-723, 2000.
147. Martens H, Malgrange B, Robert F, Charlet C, De Groote D, Heymann D, Godard A, Soulillou JP, Moonen G, Geenen V: Cytokine production by human thymic epithelial cells: control by the immune recognition of the neurohypophysial self-antigen. Regul Pept *67:* 39-45, 1996.

Chapter 4

Development of Lymphopoiesis as a Function of the Thymic Microenvironment

Abstract: Previous thymic studies detected a subcapsular $A_2B_5^+$ and Thy-1$^+$, TE4$^+$, Vimentin$^+$, Cytokeratin$^+$, so-called "endocrine" RE cell or nurse cell (TNC) subpopulation within the cortical RE cell network. Secretion of multiple *in situ* active, autocrine growth factors and a humoral chemotactic factor by the cells of ectomesenchymal origin allows the commencement of immigration of hematopoietic stem cells. The thymic lymphopoiesis is initiated by the immigration of pluripotent (with cellular immunophenotype TdT$^+$, Ki67$^+$, CD3-, CD7$^+$, CD34$^+$, CD38$^+$, CD44$^+$, CD45$^+$ or T200$^+$), but already to T lymphocyte cell lineage committed hematopoietic stem cells during the 6-7th week of ontogenesis. CD2, a 50-55 kD, glycoprotein is the first intrathymic, early differentiation antigen expressed during the 8-9th ontogenetic weeks. This antigen also serves as a cell surface component of the alternative or antigen independent pathway of thymocyte activation. The 10th week is defined as the first expression of CD4 and CD8 antigens, which determine the basic, characteristic dichotomy of the T lymphocytes. The induction of the initial proliferative wave of immature cortical thymocytes is carried out by the LFA-3 (CD58) adherence molecules, the receptors of CD2 antigens located on reticulo-epithelial cells. Because of the extremely high proliferation rate the thymic mass markedly expands in all dimensions, numerous microlobules are formed. Between the 13th to 16th week, the typical thymic cell environment is formed and the first Hassall's bodies are developed. The outer layer of the Hassall's bodies contains hypertrophized TE8$^+$, TE16$^+$ and TE19$^+$ reticulo-epithelial cells, with an active secreting cytoplasmic structure.

Key words: Antigen presenting cell (APC); Growth factor; Interdigitating cell (IDC); Maturation; Langerhans cell (LC); Neural crest; Proto-oncogene; Organogenesis; Reticulo-epithelial cellular network; Thymus; T-lymphocyte; T-lymphocyte receptor; Thymic cellular microenvironment (TCM)

1. BIOLOGICAL REGULATION OF CELL DIFFERENTIATION AND THE PROLIFERATION DURING EARLY ONTOGENESIS

Cellular differentiation is a biological phenomenon and by definition is "a process by which a cell acquires or displays a new stable phenotype without changing its genotype" (1-4). This process involves *induction* (activation) of genes that characterize the differentiated state and the *repression* of their independent suppressor genes (5-9). The control of cell differentiation is integrally coordinated with the process of cell proliferation (10-20). A defective regulation of these processes may result in the development of carcinogenesis (2, 21-23). Pluripotent hematopoietic cells, termed "stem cells" capable of fast cell growth and rapid cell divisions, have the potential for self-renewal or differentiation in two or more morphologically various cells within one tissue type (24-33). Infection of long-term marrow cultures with *src*, a molecular recombinant of Rous' sarcoma virus and murine amphotropic leukemia virus, led to continuous *in vitro* production of multipotential stem cells (7). Adipocytes, chondrocytes, muscle cells, and macrophages all represent derivatives of a special, common type mesenchymal stem cell.

The term "stem cell" is also applied to multipotential somatic cells; "progenitor cell" represents a stem cell that is already differentiated and committed in one direction and its proliferation and differentiation are more limited, compared to the original mother stem cell, but not its growth properties (34-49). A granulocyte progenitor cell is programmed to differentiate only toward granulocytes and not towards mesenchymal cells. Progenitor cells usually involve only a single cell lineage. A very important question is whether yolk sac or fetal liver-derived stem cells become committed to T cell lineage within the yolk sac or liver, or the commitment develops once the stem cells have entered the thymic microenvironment (50-52). These progenitor cells proliferate further to produce the functionally mature, post-mitotic cells required to replace those lost through natural cell death.

At the same time, "genetic programs" control the destiny of cells during ontogenetic development and differentiation (10, 24, 40, 53–66), with the mechanism of restriction to the number of cell lineages that stem cells have the potential to form. Usually at this level the terms *commitment* and *determination* are used (31, 67-72). The process of intrathymic T lymphocyte maturation is marked with activation or inactivation of a set of genes such as those encoding components of T cell receptors for antigen (TCR α,β, γ, δ) (1, 17, 21, 22, 58, 73-78) and those coding for the T3 chains necessary for both the expression of T lymphocyte receptor on the cell surface and the antigen dependent activation of T lymphocytes (3, 5, 17, 34, 41, 47, 48, 55, 79-114).

1.1 Role of c-myc and c-fos oncogene expression in growth and differentiation embruogenesis

Expression of viral oncogenes by transforming retroviruses infected cells alters the proliferation and differentiation rate in these target cells, suggesting that the cellular homologues of viral oncogenes, the proto-oncogenes, have a significant role in normal cell proliferation and differentiation (115, 116). The *c-onc* genes, particularly those encoding nuclear antigens, might be differentially regulated at the transcriptional level during cell-cycle progression (117-119). Several of these genes encode protein products that are homologous to either a growth factor or a growth factor receptor. Therefore *c-myc* activation in thymocytes is mediated by growth factors, supporting the notion that *c-myc* play an important role in the control of cell proliferation. Since freshly isolated immature thymocytes express low levels of *c-myc* mRNA it is possible that IL-2 or IL-7 alone and Con A together with the tumor promoter 12–O – tetradecanoylphorbol –13-acetate (TPA) in dose 1ng/ml induces *c-myc* production in at least a subpopulation of thymocytes during the embryonal ontogenesis (120, 121). Other data suggest an IL-2 induced increase in *c-myc* expression in lectin-activated T lymphocytes *in vitro*. Thymocyte activation not only increases *c-myc* expression but also induces immature thymocytes to produce this oncogene *de novo*. The majority of cortical thymocytes are $CD3^{+}CD6^{+}$ cells, and have been experimentally shown that even $CD3^{-}$ cells could be induced to synthesize lymphokines, autocrine regulatory factors, and express their receptors such as IL-2R (120). The *sis* oncogene of the simian sarcoma virus demonstrates a homology with the platelet-derived growth factor (PDGF), which stimulates the growth of connective tissue cells (7).

1.2 Growth factors secreted during early organogenesis

Some growth factors are unique, appearing only during early embryonic development; others, the so-called "defined" growth factors, derive from adult, partly differentiated tissues (2). Four groups of growth factors are involved in early ontogenesis:

1. Transforming growth factor α and β (TGF-α and TGF-β) (2, 122, 123);
2. Platelet-derived growth factor (PDGF-α and PDGF-β) (3, 124);
3. Epidermal growth factor (EGF); and
4. Insulin-like growth factor I and II (IGF).

The target cells for growth factors are characterized by the expression of specific transmembrane receptors, chemically binding the factor and stimulating the cellular response (27). Only within 15 minutes of initial stimulation (0-24 hours after serum stimulation), nuclei isolated from BALB/c-3T3 cells demonstrated a striking increase in the transcription of two genes, *c-fos* and actin. The *c-fos* transcription increased more than 15-fold and very rapidly, but returned to its initial level in 30 minutes. The nuclear protein, encoded by *c-fos* has multifunctional properties (7, 117). *C-fos* expression interferes with thymic development in transgenic mice (125). Japanese authors have already compiled a list of the various growth factors involved in lymphocyte differentiation (126). *Src*-infected lymphocyte cultures generated a continuous transformation into multipotential stem cells (127). At the same time human immature lymphocytes were driven to differentiate into non-lymphatic cells by the use of thymic RE cell-derived cytokines (128). The Thy-1 receptor on human T lymphocytes and transfected B cells functions also as a signal transduction molecule (129).

A transgenic H2-*c-fos* mouse, expressing *c-fos* from the H2-K promoters in several organs has been developed (125). These transgenic mice have enlarged spleens and hypertrophied thymuses, which contain increased numbers of RE cells. These results suggested that the prenatal presence of *c-fos* specifically increased the proliferation of thymic RE cells. However, the functional analyses of T and B lymphocytes demonstrated the presence of a deficient, immunophysiologic status of these cells in transgenic mice. T lymphocytes differentiation is altered in H-Y-specific T cell receptor α/β transgenic mice (89, 130-132). The positive selection was diminished, but not the negative in MHC class II transgenic mice which demonstrates that the place of MHC class II restriction is at the level of interdigitating (ID) cells (133).

One of the main physiologic role of the thymus, as a special cellular and humoral microenvironment is the full development of the thymus-dependent T lymphocyte production, depends upon the successful completion of a number of well coordinated cellular interactions and growth factors related inductions between cells of widely divergent, primordial tissues (epithelium - primitive pharynx; ectomesenchyme - neural crest; and pluripotent or already cell lineage committed hemopoietic stem cells - yolk sac, fetal liver) during the embryonal ontogenesis (13, 14, 26, 27, 31, 53, 128, 134-140). The major steps in the differentiation and maturation of lymphatic precursor cells to immunocompetent T lymphocytes occur through the secretion of factors with biological response modification activity (141-143). In the adult thymus, the majority of thymocytes are at a non-proliferating end stage, during which 3% of all $CD4^+CD8^+$ cortical thymocytes are selected on the basis of appropriate T-cell antigen receptor (TcR) specificity to become $CD4^-CD8^+$ or $CD4^+CD8^-$ mature T lymphocytes (144-146). These selected mature cells gradually emigrate to the secondary lymphatic tissue. The result of this differentiation, several distinct thymocyte subpopulations can be distinguished in the mammalian thymus and peripheral blood.

2. HYPOTHESES ABOUT THE ORIGIN OF THYMOCYTES

Historical overviews of functional hypotheses, considering the cell origin of thymocytes are always interesting and reflect the logical way of research thinking of the first classical era of thymology: The theory of "pure transformation" of the RE cells into lymphatic elements (24, 53, 54). Some ideas from the work of Koelliker (1852) are cited in German: "Koelliker (1852), der die Beziehung der Thymusanlagen zu den Schlundspalten zuerst erkannte war der Vater, der s.g. "Transformationslehre", also für direkte Verwandlung der Epithelzellen in die lymphatischen Elemente. Die weitere Entwicklung der ursprünglichen Änsicht Koelliker's war die Theorie der Transformation, der eine Einwanderung fremder Elemente in die epitheliale Anlage nicht sicher konstatieren konnte. Unten ihren Vertretern waren sehr berühmte Morphologisten wie Tourneux und Herrmann (147), Maurer (148), Prenant (149), und Bell (137). Alle these Autoren leugnen auf, daβ die Einwanderung von lymphatische Elemente hat die Richtung von Bindegewebe in die reine epitheliale Thymusanlagen. In Bindegewebe und den Gefaβen soll es zur zeit des Beginns der lymphoiden Verwandlung der Thymus überhaupt keine Wanderzellen geben also können auch keine solchen ins Epithel einwandern. Das Auftreten von lymphoiden Zellen in der Thymusanlage hangt einzis und allein von einer besonderen Weiterentwickhung der Epithelzellen Selbst ab. Sie sollen sich durch schnelle Proliferation vermehren und in einen grossen Teil der Tochterzellen sollen sich dabei die ursprünglich groβen, flesigen, helle Kerne verkleinern, ein kompaktes, dunkles Aussehen bekommen durch Protoplasma Reduzirung, so daβ echte, freie Lymphoblasten oder Lymphozyten entstehen. Prenant (149) beschreibt nicht nur mitotische Figuren, sondern noch die s.g. Stenose, eine besondere Art von Amitose und ihre Rolle. Jadenfalls ist es sehr wichtig hervorzuheben, daβ alle Autoren (Anhanger der Transformationstheorie) die "kleinen Rundzellen" der Thymus trotz ihrer epithelialen Abstammung doch als echte, richtige Lymphozyten betrachten, die also auch in das Bindegewebe und in das Blut als cleren vollwertige Bestandteile übertraten können. Prenant spricht über Lymphoblasten und Lymphozyten (Bell auch) und vergleicht soger der Thymus mit Lymphknoten, während er das Mark für eine typ Keimzentrum erklert.

Stöhr (150) was the pioneer of the hypothesis of "paratransformation" of the epithelial cells into small lymphocytes. The transformed RE cells were called free-epithelial cells, and the thymocytes were characterized as not real lymphocytes. In original German notes: Stöhr (1906, 1910) Vater des "modifizierte Transformationstheorie": "sie sind Abkömmlinge von Epithelzellen und bleiben Epithelzellen, solange sie bestehen". Diese Autoren laβt die Rundzellen der Thymus ebenfalls aus dem Epithel der ursprünglichen Anlagen entstehen, aber sie erkloren these Zellen als frei gewordere Epithelzellen, also bloss für lymphozytenähnliche und nicht für echte Lymphozyten. Diese Zellen mit den Lymphozyten des Blutes auβer einer rein auβerlichen, morphologischen Ähnlichkeit nichts gemeinsames haben und auch nicht aus der Thymus ins Blut ausgeschwemmt werden. Also der Thymus hat keiner Beziehung zur blutbildendes Organen. Stöhr nach seinen Untersuchungen die an anussen Amphibien und an erschiedenen Säugetieren ausgeführt wurden, ist der Thymus, ein rein epitheliales bebilde. Lymphatische Zellen warden in der Thymus nicht erzeugt. Die "kleinen Zellen" entstehen *in situ* durch vielfoche Teilung (mitose) der Epithelzellen. Aderseits können sie sich durch eine Wachtum in groβe wieder in typische Epithelzellen zurückverwandeln. Es wandern ebensowenig Leukozyten in der Thymus ein, wie etwas minimale Menge "kleinkernigen Zellen" der Thymus verlosst, auch die von den Thymus wegführenden Blutgefäβe sollen keine solchen Zellen enthalten. Aber wir missen sagen in der Untersuchungen von Stöhr von Menschen nur ziemlich spate Stadien vorlagen, sein frühester Fötus war drei Monate alt. In dem Mesenchym sah Stöhr nur sehr sparliche Saxersche Wanderzellen während in der Thymus Leukozyten schon massenhaft vorkenden waren, und Zellgrapper mit Mitosen und auch Amitosen fehlten. Aber auch Stöhr in gewissen Grade laβt doch Immigration von echter Leukozyten in die rein epitheliale Anlage zu. Grade in Mark sollen die echten Leukozyten sehr zahlreich sein. Dies soll aber ernst dann eintreten, wenn das ursprunglich von der finde ganz umschloβene Mark sich spater an vielen Stellen entbloβt, was meistens in der Tiefe der die Lappchen trennenden Einschnitte geschieht. Es muss notient werden, daβ aus Stöhr's Theorie nicht klar hervorgeht, was eigentlich der morphologische Unterschied der ins Mark eindringenden "echten" Leukozyten und Lymphozyten und der in der Rinde entstandenen lymphozytenähnlichen Zellen sein soll und wie es neiglich ist, folls solche Unterschiede überhaupt nicht existeren, einen verschiedenen Ursprung für die beiden hart nebeneinander liegenden und ganz gleich aussehenden Zellarten anzunehmen. In genau denselben Sinne wie Stöhr schreiben auch Marcus (1907, 1908), Cheval (1908) und Gamburzeff (1908). Marcus untersuchte allerdings nicht der Säugerthymus, sondern das Organ von Gymnophionen und bildet die Meinung, “die Thymusrundzellen gingen infolge einer Störung der Kern-Plasmarelation aus den Epithelzellen hervor”.

The investigation and analysis of population changes in mature thymus glands was not able to supply valuable information about the origin of the thymic "small cells", the thymocytes (151-154).

Sainte-Marie & Leblond (151) found four main cell types (reticular cells, large, medium, and small lymphocytes) in the cortex of rat thymus. Mitoses of reticular cells (epithelial or mesenchymal in origin) would yield an equal number of large lymphocytes and of new reticular cells. The number of large lymphocytes in medulla was only 0.025 per field, but at the same time in the cortex, there were 2.61 per field, i.e. about 100 times more. The very important morphogenetic opinion: "Thus, only 1% of reticular cell mitoses produce large lymphocytes, the other 99% is yielding new reticular cells instead." The large lymphocytes would then begin a series of successive divisions, so that a total of eight generations of small lymphocytes would be produced and thus each initial large lymphocyte would yield 128 "mature" small lymphocytes. The presence of numerous small lymphocytes in medulla, suggested that the small lymphocytes are moving into from the cortex. Indeed, diapedesis of small lymphocytes was commonly seen across the walls of the perivascular lymphatic channels and blood vessels in medulla but not in cortex. An initial immigration of lymphocytes into the thymic cortex through the medulla must be postulated to allow then to reach the peripheral circulation.

Auerbach (24, 53, 54) employing *in vitro* experimental techniques of tissue culture and organ transplantation was able to describe the origin of the thymic lymphocytes, as one of the main events of thymic ontogenesis: the development was seen to depend on a cellular interaction between mesenchyme and epithelial tissue. The same interaction could be reproduced in vitro between tissues separated with Millipore filter. While *in vitro* morphogenesis included lobulation and growth, the explants did not become lymphoidal in these studies, but transplantation of rudiments or explants to the anterior chamber of adult mouse eyes gave rise morphologically characteristic lymphoidal thymus grafts. Experimental combinations involving mouse and chicken tissues have been useful in the study of cell origins owing to the ready distinction between mouse and chicken cells in size and staining reactions. They demonstrate that under a variety of different experimental observations the epithelial component of the thymic rudiment is capable of forming the lymphoidal tissue of the thymus, and that neither immigrating cells of the host, nor thymic mesoderm, nor generalized mesenchyme contributes significantly to the initial lymphoid cell population of the mouse thymus. The 12 days thymic rudiments were separated into epithelial and mesenchymal components by the use of trypsin. The vesicular nature of the epithelium in this developmental stage is disappearing, but the rudiment has not yet reached its definite position, has not yet lobulated and does not yet appear lymphoidal. Studies have demonstrated that mesenchyme, from a variety of sources served us inducers of lobulation and growth, but no attempt was made to carry analysis to morphogenesis of the lymphatic elements.

It has been shown that the *in vitro* inductive effect leading to lobulation could be exerted across Millipore filter barriers, but here too, extension of the experiments to histogenesis in transplant situations was not attempted. For example, mouse lung mesenchyme was separated from the isolated thymic epithelium by a 20 TH filter barrier. After four days the epithelia opposite mesenchyme had grown and lobulated, where as control epithelia had become reduced in size and had undergone no morphogenesis.

Auerbach (24, 53, 54) mentioned in his *in vitro* experiments a very important morphogenetic note: "the lymphoid development can occur in the absence of any direct cellular contribution by mesenchyme." In fact, in other experiments mouse thymus rudiments were grown on chorioallantoic membranes of chicken embryos, for seven days. All transplanted mouse thymuses underwent typical histogenetic changes toward lymphatic histo-organization. The cells of the grafts were of mouse origin, chicken lymphocytes being visible only at the periphery of the grafts.

Other experimental conditions allowed, for mouse thymic epithelium and chicken nine-day bursal mesenchyme to fuse on filter membranes. All cultures, after five days demonstrated presence of lymphatic cells and in all cases the lymphatic cells were predominantly of mouse origin. Finally, Auerbach concluded: "The studies demonstrate that lymphoid tissue of the developing thymus originates in the epithelial component of the early thymic rudiment." Auerbach's observations did not, however, exclude the possibility that lymphocyte precursors or lymphoid stem cells might have already penetrated the epithelial thymic rudiment before it was explanted *in vitro* and *in vivo*. Additionally these experiments represented *in vivo* and *in vitro* results to the authors, supporting the "pure transformation" of RE cells into lymphatic elements in the middle of the 20th century. After all, we are not surprised that Tachibana and co-workers (155, 156) described the "transitional forms" between epithelial cells and lymphoblasts in detailed cytological studies of serial sections on mouse and chicken embryonal thymuses. During the transformation, there is an increase in cellular and nuclear size, a reduction in the amount of rough endoplasmic reticulum, an increase in the number of polysomes and frequently an enlargement of the Golgi apparatus. The most important histogenetic note: "Desmosomes are often seen between lymphoblasts and adjacent epithelial cells but are also present between the lymphoblasts. Desmosome-gearing transitional forms between epithelial cells and lymphoblasts can be recognized. In the developing chicken thymus as well it is possible to follow the process of the development of epithelial cells comprising the thymic primordium into lymphoblasts or lymphocytes on the basis of the presence of desmosomes and the increase and decline of cytoplasmic organelles."

By experiments carried out in chicken embryo, Moore & Owen (157-159) demonstrated the vascular immigration pathway of chromosomally labeled lymphoid stem cells invading the thymus in seven to eight day old embryos. According to them, the thymus receives an inflow of lymphoid stem cells, presumably originating from the blood islets of the yolk sac, which proliferate and differentiate into thymocytes after reaching the thymus (see Fig. 1 to Fig. 5) (160).

Le Douarin (161-165) devised a biological cell marking technique, which can be employed in the study of the embryological histogenesis of the thymus, during cell immigrations. Her embryological implantation technique, was based on structural differences in the interphase nucleus of two species of birds the Japanese quail (Coturnix coturnix japonica) and the chicken (Gallus gallus). In the first, a large amount of heterochromatic DNA is associated to the nucleolus, resulting in a hypertrophy of this organelle. This feature is not usual in birds or mammals, especially not in the chicken (the chromatin is dispersed in the nucleoplasm with some small chromocenters and participates only little to nuclear structure). Using these heterospecific combinations between quail and chicken thymic rudiments, the respective contribution of endodermal epithelium, thymic mesenchyme and eccentric hemopoietic stem cells to the histogenesis of the thymus were studied. It has been possible to investigate in these observations the chronology of the pluripotent hematopoietic stem cell inflows in the thymus during development. The lateroventral wall of the embryonic pharynx from quail embryos at 15-30 somite stages (after treatment with trypsin) were transplanted in the sometopleura of a 3-3.5 day old chicken between anterior and posterior limb buds. The transplants were in contact with the sometopleural mesenchyme, which participate to thymic histogenesis, the stimulating influence on growth and differentiation. The host embryo was sacrificed 12 days after the transplantation procedure. It was demonstrated that the whole lymphoid population of the thymus is derived from immigrant hematopoietic stem cells, which are chemically attracted by the endoderm of the third and fourth pharyngeal pouch (Le Douarin & Jotereau, 1975). The lymphoid population of the thymuses should be of host origin when they have been transplanted before the normal time of hematopoietic stem cell invasion from the yolk sac. On the contrary, if they are taken after the stem inflow, donor type lymphocytes can be found in the developing thymus. Moreover, with the evaluation of the proportion of host and donor lymphoid cells is a permanent or transitory process during morphogenesis may be determined. In the first experiments (transplantation of quail thymus rudiments into chicken), the lymphatic cells remained of host origin, coming from four to five days old chicken embryos. If the rudiments are taken at five to five and a half days, they contain both quail and chicken lymphoid cells. When the thymuses were transplanted from six to eight day old quail embryos the lymphocytes which they thereafter contain were either all, or a large percent of quail origin. The transplantation of the chicken thymus rudiments showed the same experimental results. These results suggested that the lymphatic stem cells were present in the thymus from the stages of five days in quail and six and a half in chicken embryos. This stage coincides with the time at which basophilic cells become detectable in the thymuses of both species. The most important question arose: Whether these cells are the real precursors of thymic lymphocytes? Six days old chicken thymic rudiments were grafted for three to four days into three days old quail embryos. In these conditions, the epithelial chicken thymic rudiments contained a number of cells with a strongly basophilic cytoplasm also showing the quail marker.

The hypothesis of Moore and Owen (157, 158) that the first detectable lymphoid elements of the developing thymus are cells with a large nucleus and a strongly basophilic cytoplasm is thus confirmed. The precise stage of the inflowing of the stem cells is relatively short, about 24 hours in the quail (during the sixth day of incubation) and 36 hours in the chicken (during the second half of the seventh day and the whole eighth day of incubation). However, stem cells with a capacity to differentiate into thymic lymphocytes were searched for in the embryo at stages before and after that of the normal thymic invasion. In order to know whether thymic lymphocyte stem cells are present in the embryo the following experiment was carried out: thymic rudiments were dissected from a four day old quail and grafted for two days in a three day old chicken somatopleura and then retransplanted into a three day old quail for eight days. In these cases, the lymphatic population of the thymic tissue showed the characteristic nuclei of the chicken, while RE cells and connective tissue elements belonged the quail species. These results demonstrated that stem cells able to respond to thymic endoderm attraction (humoral chemotaxis) and induction are available in the embryo as soon as the fourth day of incubation.

Clearly, thymic lymphocytes of the leopard frog (Rana pipiens) are derived from immigrating, committed, lymphopoietic stem cells (extrinsic hypothesis) rather than from lymphoid cells indigenous to the thymus. This thesis in amphibians has gained strong support from experimental observations in an urodele (166, 167) and two anuras, Xenopus (168-170) and Rana (171, 172). The results of the Volpe's group demonstrated:

1. localization of stem cells in the region of the developing thymus does not occur until after 120 hours of embryonic differentiation;
2. immigration of stem cells takes place after the establishment of full body circulation; and
3. most likely occurs concomitant with the differentiation of the thymus as an independent organ;
4. stem cells most likely immigrate during a developmental period between 200 and 300 hours after fertilization at 18 °C. Volpe (171, 172) reported: "The thymic lymphocytes of the leopard frog (Rana pipiens) are derived exclusively from migratory stem cells that enter the thymic rudiment during a developmental period when the rudiment undergoes definitive, structural differentiation."

The lymphocyte population changes were studied also with an autoradiographic experimental technique. Borum (173) reported in CBA mice an uptake of tritiated-thymidine labeled lymphocytes to the periphery of the thymic cortex and to a lesser extent to the medulla, using only a single intraperitoneal injection. Immediate incorporation was seen in the RE cells and in the large and medium sized lymphocytes, but not in the small ones. As time elapsed more cells became labeled and the immigration of labeled cells from the cortex to the medulla took place. Autoradiographs of smears of whole thymus revealed that 13% of the cells were labeled one hour after the treatment with tritiated-thymidine. The percentage of labeled cells increased with time and reached a maximum of 37% three days after a single dose. Calculations on the basis of mitotic counts with and without colchicine pretreatment, gave an average mitotic rate in thymic cortex of 3.7% per hour, which renders an average renewal time of 27 hours by cell division for thymic cortex. The straightest interpretation of cell movement is that the large and medium sized lymphocytes which are generated in the cortex, especially in the subcapsular area, immigrate to the medulla and leave the organ from there. Hemmingsson and Alm (174) had the hemopoietic cells of the chicken's yolk sac (16 day old fetus) locally labeled with tritiated-thymidine. With autoradiographic technique, heavily labeled hematopoietic cells derived from the yolk sac were demonstrated in the thymus, both in the cortex and the medulla, and in the bursa of Fabricius twenty-four hours later. The injection of the tritiated-thymidine directly into the yolk sac wall resulted also in the labeling of the cells outside the yolk sac. These immigrant cells might be the precursor cells of the thymic and bursal lymphoid cells. The results therefore support the hypothesis that the yolk sac is the origin of at least some of the cells entering the embryonic thymic and bursal anlagen (157-159). The experimental injection of in vitro tritiated-thymidine labeled cells from the yolk sac and other embryonic hemopoietic organs into chicken embryos resulted in an accumulation of labeled cells in the thymus and the bursa of Fabricius (175). These observations suggest that these cells are at home in these primary lymphopoietic organs.

The results of *in situ* labeling experiments demonstrated that this technique may be used to detect an inflow of stem cells from the yolk sac to the bursa of Fabricius.

Most of the thymic lymphocytes reside in the cortex, which contains a subcapsular band of large lymphoblasts and a major population of smaller, immunoincompetent midcortical and juxtomedullary thymocytes, whereas only 5-10% of the thymic cells are immunocompetent medullary thymocytes. The humoral regulatory factor for homing, the differentiation and maturation of prothymocytes into immunocompetent T lymphocytes is presumed to be produced by the cells of RE cellular network (see Fig.5 to Fig 7). Previous studies have shown that the humoral products (cytokines) of these cells have capacity to induce the differentiation of stem cells into lymphocytes that bear recognizable T cell typical surface antigens in mouse (176-179), and in man (180-182). The in vitro studies of the rate of the DNA synthesis, as measured by 4 hours tritiated-thymidine incorporation assay, indicated that human prothymocytes display an active DNA synthesis. The studies using mouse bone marrow derived prothymocytes (stem cells) suggested that the production and proliferation of stem cells that precede intrathymic prothymocytes is controlled by homeostatic and genetic regulation of lymphopoiesis (183-185).

Sainte-Marie & Leblond (151) found, in young adult rats that the cortex is composed of variously sized thymocytes and scattered reticular cells. The thymocytes showed a well outlined pale blue cytoplasm, which varied in amount with the size of the nucleus. A frequency curve on randomly measured nuclear diameters revealed that the diameter of the largest thymocyte was 8.6 μm and in the smallest the diameter was 3.5 μm. The curve showed dips at 4.6 μm and 5.9 μm. Nevertheless, these two values were used as guides to classify thymocytes. Those with a nuclear diameter greater than 5.9 μm, were called large; those with a nuclear diameter less than 4.6 μm, were called small; and all of the ones with a diameter between 4.6 μm and 5.9 μm, were called medium thymocytes. This method of defining the thymocytes was practical. The large thymocytes (nuclear diameter greater than 5.9 μm) usually had a round to oval nucleus with colorless sap and prominent chromatin. The range in size was wide (5.9 μm to 8.6 μm). These large thymoblasts may be seen in a group under the capsule, and the members of the group had approximately the same size. The medium thymocytes (nuclear diameter 4.6 μm to 5.9 μm) had lesser cytoplasm and nuclei with a more compact appearance than in large. The small thymocytes (nuclear diameter smaller than 4.6 μm) had a very small amount of basophilic cytoplasm, which was hardly visible in sections. The nucleus usually round, contained very dark masses of chromatin. Degenerating small thymocytes were occasionally observed with karyolytic or more commonly pyknotic nuclei. The pyknotic figures were usually round, regularly outlined and appeared uniformly dark, although a pale, eccentric, crescent-shaped zone could be seen. Mitotic figures were found usually in the peripheral layers of the cortex, to the corticomedullary junction, and their frequency decreased with the proximity of the medulla. Some mitotic figures exhibited the basophilic, well outlined cytoplasm. Since the prophases of the cells showed the same variation in the amount of cytoplasm and in size of nucleus (3.8 μm to 8.8 μm), our opinion is that large, medium, and even small thymocytes may enter mitosis. When the thymocyte mitosis was classified according to the scale of nuclear diameters (the diameters of prophase and metaphase nuclei were already measured), difficulties arose in the case of anaphase and telophase figures. At prophase the cytoplasm was unchanged, the nucleus showed prominent chromatin dots of uniform size and a lighter sap than in the "resting" cells. At the end of this phase, the dots transformed into chromosomes. At metaphase, the cytoplasm was less blue than at prophase and appeared vesicular. Individual chromosomes were distinct throughout the metaphase plate in large thymoblasts and mainly at the margin in medium thymocytes, but made up a compact dark mass in small lymphocytes. Equatorial views of anaphases showed the two separating groups of chromosomes as trapezoids followed by chromosome legs. Polar views of anaphases showed the chromosome groups as dark rings with a light center and a rugged margin. At early telophase, the two chromosomal groups were trapezoids as at anaphase, but with no chromosomes protruding. In the medulla, almost all cells than RE or accessory cells were thymocytes. Their nuclei were slightly larger and paler than those of small cortical thymocytes and often had an irregular shape. Long nuclear processes, often containing chromatin were commonly recognized in thymocytes migrating through the walls of blood vessels. The nucleolus can be demonstrated in narrow portions of nuclei. Mitoses of thymocytes were found regularly in the cortex in the medullary region were few in number.

Casali and co-workers (186) investigated the proliferation and differentiation of thymocytes in bursa Fabricii and thymus using cytophotometric techniques in White Leghorn (WL) and Red Rhode Island (RRI) chickens between 20th day of incubation to the 100th day after hatching. It was suggested that the cortical thymocytes are different from the medullary located ones by means of their nuclear volume and nuclear content in DNA, in total weight of nucleic acids, and in expressions of histone proteins. These differences furthermore appear analogous in the organ, independently from the fact that she produced precursors of T lymphocytes. The cytophotometric techniques provide a basic approach to the cell function since they quantitate the nuclear components that are directly involved in proliferation and differentiation. Lymphocyte proliferation and maturation, within the bursal and thymic environment has also been evaluated in cell kinetic studies (187). Adult thymocytes showed a higher nuclear content of DNA and total nucleic acids in the cortex than in medulla (6.99 resp. 3.60) and in addition, the cortical ones had a higher nuclear volume (30.65 resp. 21.23 μm). The submitted differences observed between cortical and medullary thymocytes represent the various stages and pathways in the process of maturation of the cell types.

3. ELECTRON MICROSCOPY

The earliest transmission electron microscopic (TEM) investigations of the developed thymus were submitted in the 1950s. We can find reports on the fine structure of thymocytes in many original works (188-190). Murray and co-workers (188) described small thymocytes in the cortex of rats, with nuclei from 3.5 to 4.5 μm in diameter, and found there are in the vast majority. The thymocytes showed a primitive, undifferentiated structure. There are no protoplasmic bridges between thymocytes and reticulo-epithelial (RE) cells. The nuclei in TEM contain typical pores and in many cases a clearly defined, finely granular nucleolus. The thymocytes are closely applied to one another, with a relatively uniform intercellular space of approximately 150 A where the separation is regular. Occasionally the space between the thymocytes might contain profiles of cytoplasmic processes of the RE cells. The actual shape of a thymocyte profile is round or polygonal in response to the pressure of adjacent cells. The nuclear diameters (a largely random selection of profiles of 1000 cells) show a peak at 3.9 μm and the vast majority are below 5.0 μm. In the plasma membrane, although minor irregularities were observed, no projections which could be classed as microvilli were noted. Throughout the cytoplasm of the thymocytes, an extensively accumulated finely fibrillar component can be found at the plasma membrane. The widest component of the cytoplasm is an angular granule, which has a very large density and measures from 160 to 200 Å. From their appearance and characteristic distribution, we have identified them as ribosomes.

It appears as if most of the cytoplasmic granules in the thymocytes are ribosomes, and that there is no clear evidence of the presence of glycogen. The endoplasmic reticulum in the thymocytes is represented by an occasional cysterna studded with free ribosomes, and a few scattered vesicles without attached ribosomes. In the majority of the cases, the nucleus shows a flattening or slight indentation on one side, associated with an accumulation of cytoplasm. Almost invariably, this area of the cytoplasm demonstrates a so-called grouping of membrane-bound vesicles and cysternae (which clearly represents the well-known Golgi apparatus). The paired membranes have a relatively uniform minimal spacing, this lies in the range of 600-800 A. Small vesicular components devoid of internal content are common, in the range of 800-1000 A, with occasional larger ones. In the Golgi region, usually on the side toward the plasma membrane, one or two centrioles might be found, which show the characteristic nine radially arranged triple-tubular filaments. Although the mitochondria never penetrates the zone of the cell center, they are numerous in its immediate vicinity. They are of all shapes, but near-circular profiles are in the large majority. They may be as long as 2 μm, very rarely appear branched, and measure about 0.25 μm in their shortest dimension. As many as twelve have been counted in one cell. Between the usually transversely oriented cristae, there is a moderate density consisting of vague, amorphous or finely fibrillar material, much like that recognized in the cytoplasm.

Raitsina and Kalinina (191) investigated the thymocytes of human fetuses by TEM. The thymocytes of 7.5-8 weeks old fetuses showed that the cells contained irregularly shaped or round nuclei with 1-3 nucleoli and that heterochromatin or compact chromatin was not present in the nuclei of 97-97.4% of the cells. The electron-dense cytoplasm of these cells contained a few number of mitochondria and many polysomes. Most of the cells had cytoplasmic outgrowths. Thymocytes of this stage had virtually no receptors for SRCB and the number of rosette-forming cells (RFC) did not exceed 4% and averaged 1.3%. In all dilution tested the antiserum had no cytotoxic action on these thymocytes. Thymocytes from 9-10 week human fetuses were almost indistinguishable in structure from thymocytes from 7.5-8 week fetuses. A few clumps of chromatin could be seen in the nuclei near the nuclear membrane, but the number of RFC in the thymus averaged 23%, and 24% of the cells were lyzed by antiserum in a dilution of 1:10. Thymocytes of 11-12 week fetuses were reduced in size, they were rounder in shape, and regions of compact chromatin (granules near the nuclear membrane) appeared in the nuclei of 90-97% of the cells. The nucleoli were reduced in size, the outgrowths disappeared and the number of polysomes was reduced. Rosettes with SRCB were formed by 79% of the thymocytes and 60% of these cells were lyzed by antiserum in a dilution of 1:10. The thymocytes of 17-24 week old fetuses were small, with a narrow rim of electron-dense cytoplasm, containing a few mitochondria and polysomes. Compact chromatin occupied a large part of the cell nuclei, in which it formed a curious and complex pattern. Receptors for SRBC were found on 80% of the thymocytes, and 70% of the cells were lyzed by antiserum in a dilution of 1:10. This investigation suggested a clear and regular order of development and maturation of the thymocytes during thymic histogenesis. The stem cells (enter the thymus of 7.5-8 week fetuses) are characterized by absence of compact chromatin in the nuclei and of T-antigen and receptors for SRBC on the cell surface.

In his systematic study of the thymic histogenesis in human fetuses, Haar (192) described large thymocytes with irregular shapes and often presented blunt pseudopodia-like cytoplasmic processes or more slender cytoplasmic extensions. They also often possessed numerous elongated mitochondria, a large Golgi complex, and a strongly basophilic cytoplasm. However, these large thymocytes were not attached to the RE cells by desmosomes, although some of the cytoplasmic processes from them were in association with extensions from the RE cells. The medium sized thymocytes were round at all stages and had fewer mitochondria than the large ones. The large thymocytes had only occasional short profiles of rough endoplasmic reticulum and a nucleus with a fairly dense band of chromatin. These cells are mostly found near the outer surface of the thymus and are separated from the basal lamina by only a thin projection of cytoplasm from a RE cell.

A topographic correlation between the blood vessels, RE framework and lymphoid cells of the thymus is essential for a better understanding of the maturation processes and the migration of thymocytes. The use of the scanning electron microscope (SEM) permits the visualization of the HBs (Fig. 16) and the three dimensional complexity of the supporting framework (193). The thymus, of rat, consists of three parts:

1. an outer cortex containing mainly lymphoblasts (large thymocytes);
2. an inner cortex showing mainly small thymocytes and
3. the medulla, characterized by numerous RE cells.

The cortex of the thymus is supplied by arcades of capillaries, which at the corticomedullary junction are connected to postcapillary venules and arterioles. The vascular supply of thymus was studied in detail by Raviola and Karnovsky (194). Capillaries arising from the arterioles at the boundary of cortex and medulla ascend into the cortex, undergo branching, and finally, curve back toward the medulla to join the postcapillary venules. The thymoblasts form a layer three to four cells thick. These cells have an average diameter of 7 μm (7.164 +/- 0.278). The nucleus is round, with a diameter of 4.7 μm (4.73 +/- 0.5794). There is a clumping of heterochromatin near the membrane. The rest is filled mainly with euchromatin. One or two nucleoli were present and the cytoplasm contains numerous free ribosomes, which often form aggregates (groups). The majority of them appear as polyhedral smooth surfaced cells and do not show any surface variations in different regions (195). The cell surfaces in the medulla are relatively variable with regards to their shape, slight undulations, short protuberances or pseudopods. This pheomorphism which might be indicative of motility is often observed, especially in the areas adjacent to the postcapillary venules. Thymocytes are demonstrated in all three parts of the thymus, naturally, mainly in the inner cortex. They have an average diameter of 5 μm (4.957 +/- 1.684). The slightly indented nucleus, with a diameter of 3.3 μm (3.83 +/- 0.006), contains a large amount of heterochromatin, which accounts for the electron-dense appearance of the nucleus. Very few profiles of rough endoplasmic reticulum and few mitochondria are present. These cells by SEM have partly smooth surface. In the postcapillary venule, the thymocytes (mature, immunocompetent T lymphocytes) are frequently seen migrating through the endothelial wall, reentering the peripheral blood circulation. An observation of this process of migration indicates that the thymocytes follow an intercellular pathway by squeezing in between and pushing apart the adjoining endothelial cells. Microvilli or other surface specializations become evident as soon as the progressing regions of the thymocytes appear in the capillary lumen. In the human material of Bhalla and Karnovsky (195) only 19% of the migrating T lymphocytes had microvilli. It has also been suggested that T and B cells can be distinguished morphologically only by their surface structure (196). The development of villous surface can be also a result of biochemical activation processes. Stinson and Glick (196), using SEM, studied lymphocytes from the bursa, spleen and thymus in 2-7 weeks old chickens. The results were scored as presence of rough (villous), intermediate or smooth lymphocyte surface. In contrast to other systems, both T and B lymphocytes in the chicken belonged to the smooth type, regardless of the tissue origin and the age of the bird (thymus -98%; bursa - 96%; spleen - 95%). Virtually no cells with rough surface were observed in the thymus (only approximately 1%). It is known today that only SEM investigations cannot distinguish between T and B lymphocytes.

4. INTRATHYMIC IMMUNOLOGICAL MATURATION OF LYMPHOCYTES

Studies of the last two decades, particularly in mice have clarified certain main aspects of thymic ontogenesis and the role of the thymus in the generation of functional T lymphocyte subsets (141, 142, 197, 198). It has been an attractive postulate, that the so-called "thymic hormones", produced by the RE cells, have influence on the thymocyte maturation by direct cell to cell, receptor to receptor contacts (199-202). The symmetrical cell interaction system as a model for the T cell development, also cooperated with the MHC regions and their products, may be postulated:

1. Different MHC regions in the "inducer, accessory cells" as well as specific and non-specific humoral factors can regulate the intra- and extrathymic T lymphocyte histogenesis, and that such signals are necessary for functional differentiation of T lymphocyte subsets; and
2. the "inducer" cells can impart three types of signals: one for irreversible commitment to the T cell lineage, second for the elaboration of the appropriate dictionary of cell interaction molecules and the third one for specialization into a determined subclass of T lymphocytes.

The first two signals are intrathymic events (142). Recent results suggest that the immature thymocytes "learn" to recognize the MHC antigens within the thymus and during the process of differentiation into mature, immunocompetent T lymphocytes, having arrived at the periphery or secondary lymphatic organs. The thymocytes in the cortex and medulla show different features (203). The medullary cells are highly immunocompetent and similar to peripheral T lymphocytes (204, 205). In contrast, the cortical thymocytes are immunologically incompetent, express cell surface antigens not found on other cell types (TL antigen in mouse, HTA-1 in man and CD-differentiation antigens) (204, 205) and contain a nuclear DNA polymerase enzyme, terminal deoxynucleotydil transferase-TdT (206, 207). TdT adds nucleotides to single stranded DNA and many contribute to thymocyte differentiation or even to diversification. During fetal thymus organogenesis there is a

6. FIGURES

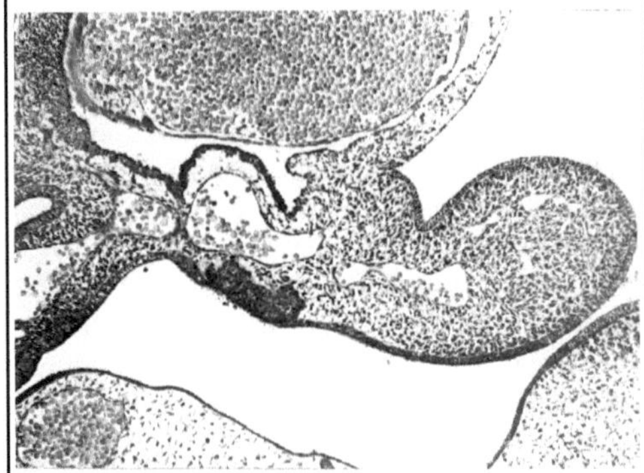

Figure 1. Dog embryo (23 days of gestation). Pure epithelial thymic anlage. Methacrylate embedding; 3 μm thin section. Schaffer fixation. Giemsa staining. Magnification 64x.

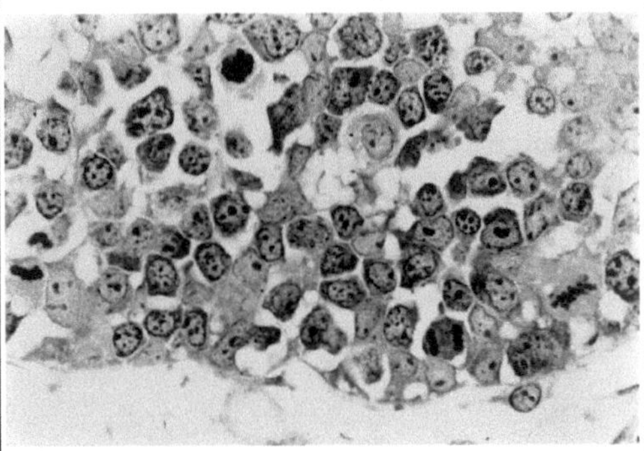

Figure 2. Dog fetus in 36th day of gestation. Thymic cortex. Early stage of lymphopoietic thymus, with numerous lymphatical and RE cell mitoses. Well developed RE cell processes. Methacrylate embedding; 3 μm thin section. Schaffer fixation. Giemsa staining. Magnification 500x.

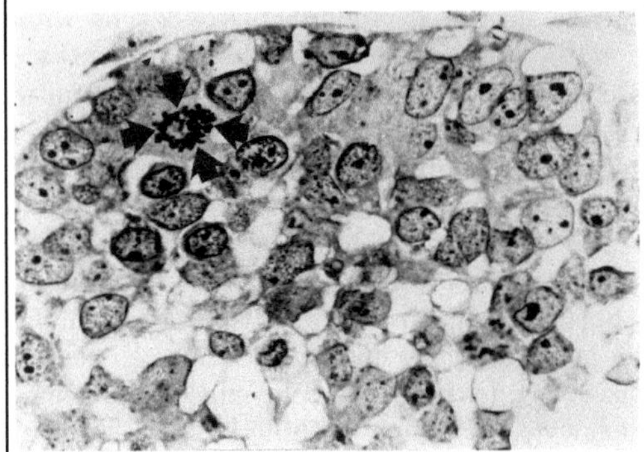

Figure 3. Dog fetal thymus (118 mm body length, 49 days of gestation). Thymic cortex under the light microscope. Methacrylate embedding; 3 μm thin section. Schaffer fixation. Hematoxylin-eosin staining. Magnification 500x.

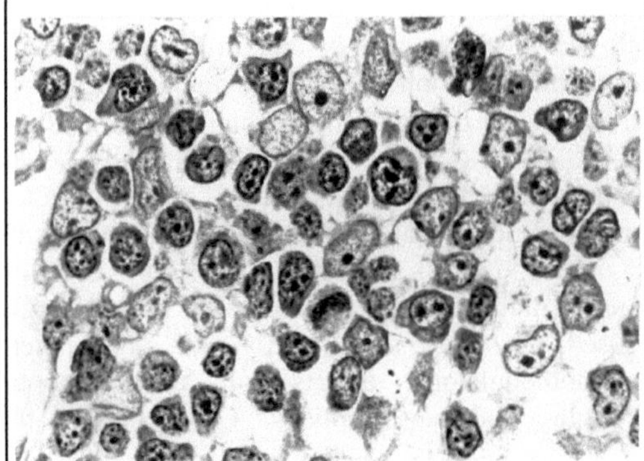

Figure 4. Dog fetal thymus (91 mm body length, 46 days of gestation). Characteristic organization of thymic medullar stromal microenvironment. Methacrylate embedding; 3 μm thin section. Schaffer fixation. Hematoxylin-eosin staining. Magnification 630x.

spectacular increase in this enzyme's activity in calves, chicken, rat, and mice (206, 208, 209). In normal children, Janossy and co-authors (46) investigated the presence of the TdT with immunoperoxidase technique. In the cortical areas, 60 to 85% of the thymocytes showed nuclear TdT. Very few number of cells had TdT localized in their cytoplasm. These seemed to be dividing lymphatic elements. In contrary, only 3% of the medullary thymocytes expressed nuclear TdT. A note: the abundant non-lymphatic cell populations were TdT negative. In contrast to thymocytes, all of the lymphocytes in the control human tonsils were TdT negative and these results were also reflected in biomedical assays (57 units/10^8 and 0.1 units/10^8 resp.). Further, these TdT positive thymocytes were analyzed in smears (in suspension) with a monoclonal antibody (NA 1/34) to HTA-1, a membrane antigen, typical for the thymocytes in cortex. Both, membrane HTA-1 and nuclear TdT were expressed in more than 65% of the thymocytes (TdT$^+$, HTA-1$^+$ cells). Only 1 to 3% were TdT$^+$ and HTA$^-$1$^-$ thymocytes; frequently these were larger cells with an oval-shaped nucleus and one nucleolus. These cells stained with Giemsa, had a dark blue cytoplasm, which is in contrast to the pale cytoplasmic staining of the small thymocytes. A somewhat larger proportion (5 to 10%) of the thymocytes were TdT$^-$, HTA-1$^+$ and 10 to 20% of the population (the bulk of the medullary thymocytes) was TdT$^-$, HTA-1$^-$. Our conclusion is that the majority of the cortical thymocytes in the child thymus are TdT$^+$, HTA-1$^+$, whereas most medullary thymocytes are TdT$^-$, HTA-1$^-$. Monoclonal antibodies to HLA-A, B, C antigens (W 6/32 and 34/28) labeled a minor proportion (10 to 17.5%) of thymocytes, which did not stain for nuclear TdT. In addition, the Ia-like antigens were undetectable on human thymocytes and that HLA-A, B, C antigens were detectable on medullary, but not on cortical ones. Janossy (46) in his observations on the ontogeny of the human thymus had analyzed fetal thymuses (at 14, 17, 21 and 22 weeks of gestation) by immunohistology. The tissue architecture of the thymus lobe earliest studied in 14 and 17 week old fetuses already showed the characteristic demarcation between cortex and medulla. In these cases more than 50% of fetal cortical thymocytes expressed HTA-1 antigen (but the staining intensity was lower than on infant ones) and was a characteristic for small thymocytes. The medullary thymocytes were always HTA-1$^-$. Nevertheless, in the biochemical essays very low TdT values were measured (0.0 to 1.1 units/10(8) cells). Similarly, the thymic tissue demonstrated no presence of TdT in 14 week old fetus. At the prenatal age of 17 weeks more TdT$^+$ cells were detected in the thymus (10-15% were small thymocytes). Certainly, these cell suspensions from fetal thymic tissue contained a higher percentage (15-40%) of large thymoblasts, than the samples from child thymus (1-3%). These blasts demonstrated dark blue cytoplasmic staining after Giemsa and reacted with anti-HuTLA serum. On the other hand, the large thymoblasts had an unremarkable phenotype, lacked detectable TdT, failed to stain with anti-Ia-like sera and the anti-ALL antiserum, and showed no or only weak reactivity with monoclonal antibodies to HTA-1 antigen and to a human leukocyte antigen (HLe-1) (210, 211). It is also interesting that considerable numbers (about 5%) of TdT$^+$, HTA-l$^+$ cells could be detected in the medulla. These represent thymocytes of cortical origin, migrated in the medulla during their maturation, and leave the thymus through the postcapillary venules, present in the cortico-medullary boundary as has been suggested in mice (212, 213). Cytotoxic depletion studies with anti-Lyt-1, anti-Lyt-2, and anti-Lyt-3 antisera in mouse, demonstrated that the functional mouse T cell subpopulations are different in their expression of these surface antigens (214).

Mouse helper T cells were reactive to MoAB anti-Lyt-1 (human with MoAB anti-Leu 3/a + 3/b) whereas mouse suppressor and cytotoxic T cells could be depleted only with MoAB anti-Lyt-2 or anti-Lyt-3 (197,198) (human with MoAB anti- Leu 2/a). Thus, the helper T lymphocytes were classified as Lyt-1$^+$, Lyt-2$^-$, Lyt-3$^-$ and the suppressor and cytotoxic T cells as Lyt-1$^-$, Lyt-2$^+$, Lyt-3$^+$. In contrast immature T lymphocytes were sensitive to depletion by all three antisera, and were therefore classified as Lyt-1$^+$, Lyt-2$^+$, Lyt-3$^+$. These latter cells were present in 50% of the T cells in the adult spleen, but almost 100% in spleens from seven days old mice (197, 198). With quantitative of these antigens on lymphocyte subpopulations by flourosence-activated cell sorter (FACS) in adult mice, showed that Lyt-1 is present on all lymphocytes and surprisingly showed that Lyt-1 is also found on essentially all peripheral T cells. Therefore, it established Lyt-1 as a quantitative rather than a qualitative marker for distinguishing functional and maturational T cell subpopulations. Lyt-2, in contrast, was shown to be present on fewer peripheral T lymphocytes and was demonstrated on most thymocytes (215). Thus, it remains as a qualitative marker for a T cell subpopulation in the periphery and because its higher frequency in the thymus, as a marker for distinguishing immature from mature T cell populations during organogenesis. Lyt-3, which is located on the same surface macromolecule as Lyt-2 (216), constitutes a second determinant, marking the same populations of the thymocytes.

The adult thymus contains a medullary subpopulation, resistant to hydrocortisone depletion, which is believed to be mature and to include cells that will emigrate and contribute to the peripheral T cell pool (212). On average, this population has less surface Thy-1 and more surface Lyt-1 than the thymocytes in cortex (217). In the thymus, the thymocytes can also be divided in optimal conditions into functional subclasses of T lymphocytes having a wide repertory of antigen specificity (197, 198). The serological analysis has revealed that this process involves a number of marked changes in surface antigens except for the so-called TL system and are quantitative rather than qualitative features (218, 219).

Kasai and co-workers (220) reported that the majority of the cells from the mouse fetal thymus (from 13th day of gestation) were brightly stained with fluorescein isothiocyanate (FITC)-conjugated Dolichos biflorus agglutinin (DBA) lectin and the proportion of such cells drastically decreased with the increase in gestation time (13th day-92%; 14th day-76%; 16th day-41%; 18th day-7%). Unlike other T cell antigens, the DBA receptor is not expressed on cells of adult lymphoid tissues. Therefore this antigen was provisionally called "fetal thymus antigen" (FT-1). A small percentage of FT-1 positive cells still remained in the thymus of neonatal mice, but is almost completely absent from that of adult mice (7-8 weeks old). The molecular weight of this new surface antigen is 130 kD and chemically is a glycoprotein, examined by two-dimensional gel electrophoresis.

Bodey (221) investigated the differentiation and homing properties of pre-thymocytes and thymocytes in human fetuses during the prenatal ontogenesis. In the early stages of the organogenesis, we can call all the presenting cells as stem cells, which as early as during the 8th fetal week migrate into the thymus. Through immunofluorescent and immunocytogenetic methods with monoclonal antibody libraries, we clearly proved the surface antigen formation of T-10$^+$ cells, the so-called pan thymic cells and only such cells were present between the 8th and the 12th weeks of the prenatal development. During the fast fetal changes of the third lunar month, the formation of the thymocyte subpopulations already began. The definite tendency and pathway of T lymphocyte maturation is directed from the cortex to the medulla. In the thymic cortex the following T lymphocyte subpopulations are mainly present, demonstrating the heterogeneity of surface antigen distribution: T-helper (with the typical CD4$^+$, CD8$^-$), T-suppressor (CD8$^+$, CD56$^-$) and T-cytotoxic cells (with the selective CD8$^+$, CD11/b$^-$), investigated in detail by Reinherz and Schlossman (222) and "dissected" by Lanier (181, 182), using four-color FACS analysis and different fluorochrome conjugated Becton-Dickinson MoABs.

It has long been known that malnutrition often results in zinc deficiency and that the biologically active form of thymulin, formerly called "facteaur thymique serique" or FTS, a thymic hormone requires presence of zinc. Zinc was identified in secretory granules associated with cytoplasmic vacuoles in thymic RE cells. Zinc is an essential component of over 200 enzymes, it is required for all DNA- and RNA-polymerases, and Zn-metalloenzymes all being participants in replication or repair of DNA. Mitotic activity in the thymocytes is increased when either calcium or magnesium is elevated in the cellular microenvironment. Parathyroidectomy results in the abolition of fluctuations in plasma level of calcium and cell mitosis, causing severe thymic atrophy. Enhanced epithelial growth and an increase in the number of macrophages was detected in thymic cultures with the addition of glutarine (a parathyroid hormone) to the culture medium. Recently calcitrol (biologically the most active metabolite of vitamin D), has been investigated to inhibit IL-1 and IL-2 dependent thymocyte proliferation. The IL-2R bearing thymocytes are the intrathymic stem cells.

Numerous mesenchymal cells around the primary (early) thymic anlagen and the connective tissue cells that form the primitive capsule and septae within the organ were demonstrated to originate from a well developed neural crest (223-227). Since this special, embryonal type mesenchyme derives from ectoderm, it is referred as ectomesenchyme in the International Histologic Nomenclature (IHN). The neural crest, through formation of the thymic ectomesenchyme, and through production of local, *in situ* acting growth factor(s), probably before the secretion of an 11 kD chemotactic factor (thymotaxin), monitors the maturation and differentiation of epithelial and connective tissue components of the early thymic microenvironment (31-33, 54, 70, 71, 228-235). Interestingly, later the *in vivo* route of entry for pre-T stem cells is *via* the highly vascularized cortico-medullary junction, using special homing receptors, adhesion molecules for better contact to the endothelial cells and integrin molecules like addressin or ICAM-1 (LFA-1), L-Selectin, also s.E- and s.P-Selectin and VCAM-1 (VLA-4) related structures. In the second step, the hematopoietic cells migrate to the subcapsular thymic microenvironment. The integrins $\alpha_4\beta_1$ and α_5 and β_1 are able to induce a dependent proliferation (236). Major portion of double positive ($CD4^+$ and $CD8^+$) human thymocytes express $\alpha_4\beta_1$ integrin in active form. These integrin molecules are also involved in cellular traffic and positive intrathymic selection of T lymphocytes.

As was mentioned earlier the tissue composition of the "first" large blood vessels of the thymus, which appear at the embryonic length of 20-30 mm except for the endothelial cell layer, are also of ectomesenchymal, i.e. neural crest origin.

5. THE THYMUS AS THE KEY ORGAN OF IMMUNO-NEUROENDOCRINE REGULATION

Complex endocrine relationships exist between the thymus, hypothalamus, pituitary and ovaries. Experimental or spontaneous, neural crest ablation early during the ontogenesis, always resulted in abnormal thymic organogenesis, and also abnormality within the heart, the position of great vessels, thyroid, and parathyreoid histogenesis (237). An interesting association of immune-neuroendocrine regulatory is seen in the mouse mutation "staggerer" in which defective cerebellar ontogenesis results in defective thymic histogenesis (238). The neuropeptides oxytocin (OT) and vasopressin (VP) are under the regulatory control of two independent genes expressed in the magnocellular neurons (MN) of the hypothalamus and they are produced in the neurosecretory cell granules of the paraventricular and supraoptic region nuclei (55, 228-231, 239-244). Earlier thymic studies defined a subcapsular $A_2B_5^+$ and $Thy\text{-}1^+$ endocrine RE cell or nurse cell subpopulation within the cortical RE network. Thymic stromal elements possess an anterior pituitary hormone-stimulating activity (245). In our immunocytochemical studies, human postnatal thymic RE cells, located under the capsule (outer cortex) and in the medulla also demonstrated presence of UJ13/A, UJ127.11, UJ167.11 and UJ181.4 early neuroectodermal (adhesion) antigens. In recent study, Nagano and Kelly (246) observed the tissue distribution the short and long forms of rat prolactin receptors. The thymic RE cells expressed both types of prolactin receptor and the expression was clearly regulated by the hormonal environment associated with the oestrus cycle. IL-1 immunoreactivity, presence of OT-like peptides and neurohypophysial activity in the RE stromal cells were further discovered. The co-existence of OT and VP secretion in the same RE cells in the thymus is in sharp contrast to the organization of hypothalamo-neurohypophysial system, where VP- and OT-containing neurons are separate. IL-1 may regulate the release of OT and VP like peptides. The physiologic action of OT- and VK-like peptides within the thymus remains to be defined. Thus, OT and VP are capable of replacing the IL-2 requirement for interferon γ (IFN-γ) production *in vitro* (37, 67, 79, 247-249).

The cytokine, IL-1 was detected as an endogenous pyrogen, a macrophage product that stimulates thymocyte proliferation, activates B lymphocytes, release of soluble ICAM-1 adhesion molecule and mediates a wide range of inflammatory effects. Within the CNS IL-1 also exerts a number of effects including modulation of slow wave sleep, food intake regulation, analgesia, corticotropin-releasing factor production and stimulation of growth in astroglial cells.

There is also experimental evidence, demonstrating that the gonadotropin releasing hormone (GnRH) is involved in modulating the cellular immune system functions (250). Absolute levels of intrathymic T lymphocytes, $CD8^+$ cells, CD4 and CD8 expressing immature cells and other immature T cell subsets, differentiating toward CD4 were significantly reduced two weeks after agonist treatment in slow release microcapsule formulation. Single i.m. injection of agonist decreased both absolute and relative thymic weights and the absolute thymocyte counts (250).

Recently a new thymus-neuroendocrine-liver pathway is proposed for maintaining the body homeostasis (251). Thymectomy in adult male rats resulted in a decrease of liver microsomal cytochrome P-450 and aminopyrine-N-demethylase activities. There was a decrease in hypothalamic luteinizing hormone-releasing hormone (LHRH), plasma luteinizing hormone (LH) and testosterone levels. Supplementation of testosterone propionate restored the normal levels of the above mentioned. In female thymectomized rats increase of liver malondialdehyde (MDA), accompanied with decrease of liver superoxide dismutase (SOD) and glutathione (GSH). There was also a decline of fluidity and calcium ion uptake in liver microsomal and mitochondrial membranes, in the levels of hypothalamic LHRH and the plasma estradiol. Administration of estradiol benzoate eliminated all symptoms.

6. IMMUNO-ONTOGENESIS LYMPHATIC STEM CELL MIGRATION TO THE THYMUS

In humans, pluripotent stem cells, capable of colonizing the thymus, emigrate from the yolk sac (until day 60), from the fetal liver (between days 50 and 150) or from the bone marrow (after day 79 of prenatal development). The RE cells at this stage also produce two or more additional thymic peptides: thymosin α-1 and thymosin β-4 (9, 37, 252). Thymosin α-1 (Tα-1) is the first biologically active peptide to be isolated from thymosin fraction 5. Tα-1 is an acidic peptide, which contains 28 amino acid residues, with a molecular weight of 3108 daltons. It has been demonstrated that this thymic hormone increases the mitogenic responses of thymocytes and the production of interferons, lymphotoxin and tumor necrosis factor (TNF), stimulates antibody production, and augments the development of a Thy-1$^+$, Lyt-1,2,3$^+$ cell clone. Tα-1 also induces the differentiation of CD4 (helper/inducer) lymphocytes.

Thymosins - f.e. thymosin β-4, with a molecular weight of 4963 daltons, has been demonstrated to induce expression of terminal deoxynucleotidyl transferase in TdT-prothymocytes and stem cells in the bone marrow both *in vivo* and *in vitro*.

Thymopoietin, an another thymic polypeptide was isolated by its neuromuscular effects, in electromyographic assays. Even in small quantities (4 ng/mouse), this polypeptide showed its potent action. It has later been shown that the biologic activity of thymopoietin resides in the pentapeptide Arg-Lys-Asp-Val-Tyr, corresponding to residues 32-36 (TP 32-36) and was named thymopentin or TP-5. TP-5 has been shown to enhance several T lymphocyte functions *in vivo*:

1. rejection of 3LL carcinoma cells;
2. generation of CD8$^+$, cytotoxic T lymphocytes; and
3. elimination of autoreactive T cells.

Thymulin, is a well-defined nonapeptide thymic hormone, initially isolated from porcine serum and later from human and calf. The thymic humoral factor (THF) is an another polypeptide, with molecular weight below 3000 daltons. Its administration to humans with immunodeficiencies has been shown to restore cellular immune reactions. It has been defined that during its action, THF involves a mandatory rise in cellular cAMP levels.

7. CELLULAR NETWORK COMBINATION INFLUENCES DETERMINE THE THYMIC MICROENVIRONMENT

Normal T lymphocyte maturation requires an extremely heterogeneous cellular, rich, functional, humoral and molecular biologically well coordinated intrathymic microenvironment (11-13, 25, 38, 39, 46, 62, 63, 68, 69, 87, 89, 114, 118, 135-137, 140, 249, 253-271). The microenvironment includes all types of thymic cells: RE cells, macrophages (extrinsic/intrinsic), fibroblasts, dendritic cells (DC), Langerhans cells (LC), interdigitating cells (ID), and various caliber vascular endothelial cells, involved in complex cell to cell biochemical and immuno-humoral interaction with one another. Thymic MHC restricted, CD4$^+$ dendritic cells population was detected employing immunofluorescense and flow cytometric techniques (272). These dendritic cells were efficient stimulators of allogenic T cells.

As was already mentioned, the RE network is composed of heterogeneous in function and surface IP profile cells. The genetically determined changes of the thymic microenvironment appear during the embryonal, very early stages of human ontogenesis. Haynes (257) described an A2B5$^+$, Thy-1$^+$, UJ13/A$^+$, UJ 127.11$^+$ subcapsullarly located RE cell subpopulation, termed endocrine RE cells. Several research groups such as Haynes (62, 63, 257-259), van Ewijk (265, 273) and van Vliet (274, 275) separately developed libraries of mouse anti-human monoclonal antibodies (MoABs), which reacted positively with the non-lymphatic, RE cells of the thymic microenvironment: from the laboratory of Haynes: **TE3** for cortical RE cells and macrophages; **TE4** for detection a subcapsular and medullar subpopulation of RE cells; **TE7** against mesodermal derived fibrous connective tissue elements; **TE8** for the cells forming the outer cell layer of HBs and detecting the same antigen on cells located in the granulosa cell layer of epidermis; **TE15** detecting cells from the inner cell layer in HBs; and **TE19** stained the antigen on cells surrounding the HBs. The MoABs also reacted positively with numerous cells of astrocytomas and primitive neuroectodermal brain tumors (medulloblastomas-PNETs) during the systematic cell surface and cytoplasmic antigenic distribution screening of primary childhood brain tumors (25). A shared antigen expression between the thymic RE cells and primary neuroectodermal brain tumor cells is an embryologically and tumor biologically important possibility. During the same experiments, postnatal thymic tissue was used as negative tissue.

A study and review by Kampinga and co-workers (276) defined the presence of two distinct IP for differently located rat RE cells in the cortex and in the subcapsular region/medulla judged by the binding patterns of three MoABs (HIS 37, 38, and 39) and introduced a nomenclature for the main cell types of the thymic microenvironment. The designation Clusters of Thymic Epithelium Staining (CTES) included the following groups of MoABs:

- Group 1) PAN - these MoABs label all RE cells;
- Group 2) these MoABs bind to subcapsular/perivascular and medullary located RE cells, including the HBs. The subcapsular cell clone can be defined with thick or thin immunocytochemical staining;
- Group 3) PAN CORT.-these MoABs label the whole cortical RE network. This group was subdivided into: 3A - cortical RE cells only; 3B - cortical and small subset of medullar RE cells; 3C - cortical RE cells and subset of leukocytes, which was further subdivided into: 3C1 - cortical RE cells and macrophages or IDC; 3C2 - cortical RE cells and thymocytes;
- Group 4) PAN MED. - these MoABs label all medullary located RE cells and the HBs;
- Group 5) these MoABs predominantly label the Hassall bodies. This group is subdivided into: 5A - PAN HB - only HBs are labeled; 5B - HBs and myeloid cells (probably equivalent of CD15); 5C - HBs and associated medullar RE cells; 5D - HBs and associated medullar RE cells and a subset of leukocytes; and 5E - HBs and associated medullar and subcapsular RE cells.

RE cell surface IP characterization with mouse anti-human, anti-keratin MoABs, developed against keratin, a cytoskeletal protein present predominantly in epithelial tissues: Colic and co-workers (277) subdivided the thymic RE cells using double immunostaining with a panel of 13 MoABs, raised against thymic RE cells and cells of the mesodermal stroma. RE cells were identified with anti-human keratin MoABs. The non-lymphatic cellular microenvironment was subdivided into five distinct cell groups:

1. Pan RE cells (positive reaction with MoAB R-MC 2,3 and 8);
2. Cortically located RE cells (R-MC 13-17 positive cells);
3. MoABs detecting subcapsular and subtrabecular RE cells and the majority of medullary RE cells (positive staining with MoAB R-MC 18-20);
4. IP specific for HBs and a small subpopulation of medullary RE cells (positive staining with MoAB R-MC 22); and
5. IP typical for the mesodermal stroma cells (positive and selective stromal cell staining with MoAB R-MC 23).

Recent observation employing flow cytometry determined that shared or common antigens exist between the thymic RE cells and the dendritic cell population (132). The greater sensitivity of flow cytometric observation showed the great functional complexity of the thymic RE network involved in T lymphocyte maturation.

Another experimental study defined a unique 60 kD surface antigen on the thymic RE cells (278). The immunologically potent molecule participate in the cellular interactions between the cells of RE network and the immature cortical thymocytes. Two rat monoclonal antibodies AS19 and AS32 were directed against the 60 kD protein, but to different epitopes. The same protein was also determined on fibroblasts and vascular endothelial cells.

The development of the thymic microenvironment in chicks was observed employing lectin histochemistry (279) Wheat germ agglutinins (WGA), concavalin-A (Con A), and ricinus agglutinin (RCA-I) were assayed. Con A lectin expression detected several cell clusters of stromal cells and thymocytes. These clusters probably represented lymphostromal cell complexes, with which the Con A receptors are associated in relation to cell to cell adhesion.

8. ONTOGENESIS OF THE T CELL RECEPTOR (TCR)

During the T lymphocyte development the T cell receptor (TCR) α, β, γ, and δ genes are rearranged and expressed. TCR rearrangement requires the coordinating activity of two recombinase activating genes Rag-1 and Rag-2 (280). There are two periods of high Rag-1 and Rag-2 mRNA expression during the ontogenetical development of the T lymphocytes:

1. in the early stage with $CD25^+$ $CD4^-$ $CD8^-$ and $CD3^-$ IP; and
2. later at $CD4^+$ $CD8^+$ stage.

The results suggest an active mechanism regulating the expression of these two recombinases during the early thymic development. The TCRs are disulfide linked heterodimer polypeptides and recognize peptides bound either to class I or class II major histocompatibility complex (1, 17, 21, 22, 32, 50, 51, 58, 74, 84, 89, 100, 103, 112, 113, 133, 138, 281-287). Another important feature of intrathymic T lymphocyte differentiation is its association with DNA rearrangement in the T-cell receptor loci. The thymus is the principal locus of rearrangement of the TCR α, β, γ and, probably, δ genes. In early ontogenesis, a very high level of TCR γ, δ chain rearrangement was detected. γ, δ T cells generally do not express CD4 or CD8. In normal mammals, extrathymic rearrangement also occurs in skin and bone marrow. A basic question is whether rearrangement of TCR γ and δ genes is obligatory for the rearrangement of TCR and α genes and expression of a TCR-α, β receptor, or whether these two types of receptors represent totally different cell lineages. The C-V-D-J recombination brings the constant (C), variable (V), diversity (D) and joining (J) segments together to form a complete V gene. The Vα region for example consists of >20 families each of which may contain from 1 to 8 individual genes. The Constant region (Cα) consists of only one gene having the identical sequence in humans. The J α region consists of 50 short segments separated by long stretches of random sequences of DNA. The β chain is arranged similarly to the α chain. The Vβ region consists of approximately 70 individual genes, which belong to 20 families, each with one to seven members. Rearrangement is a decisive step in T lymphocyte differentiation and thus serves as a genetic marker of commitment to T lymphocyte lineage. There are five different TCR V region epitopes: βV 12, βV 5.2/3, βV 6.7, βV 8, and α V 2.3. The βV 6.7 expression was skewed to CD4 cells and βV 5.2/3 to CD8. DNA analyses revealed D-J, but no V-D-J β gene rearrangements, leading to the hypothesis that the early γ-δ expressing lymphocytes secrete a factor(s) that is required for α-β expressing cell lineage. This differentiating factor may be a new lymphokine. Nothing is known about the mechanism of α rearrangement, whether it involves secretion of a special lymphokine at that early time of development or whether cell to cell interaction plays the leading role is not clear yet. This mechanism is a central question in immuno-ontogenesis. The three basic types of TCR-expressing cells are:

1. double negative TCR-γ, δ;
2. double negative TCR-α , β; and
3. TCR-α , β expressing either CD4 or CD8 or both develop from the triple negative precursors.

9. INTRATHYMIC DIFFERENTIATION PATHWAY OF T LYMPHOCYTES

Intrathymic differentiation produces mature peripheral T lymphocytes which possess the cellular ability of developed effector cells, able to carry out several cell mediated effector functions, such as specific target cell lysis and lymphokine secretion (14, 16, 29, 33, 34, 37, 38, 47, 48, 54, 56, 60, 62, 63, 68, 70, 82, 83, 88, 91, 94, 101, 103, 107, 114, 133, 139, 260, 261, 267, 271, 282, 288-309).

The maintenance of homeostasis in normal mammalian tissues reflects a balance between cell proliferation and cell death. The immunomorphologic phenomenon for the life span of immature cortical thymocytes is termed programmed cell death (PCD), or apoptosis or negative selection (57, 64, 310, 311) leading to a series of genetic steps that ultimately result in cell death (312, 313). Researchers are uncertain as to why and how this event occurs, however, it is believed to represent a biological phenomenon during ontogenesis. Ronald E. Ellis and his co-workers at MIT (Cambridge) have found the gene named ced-9 that regulates apoptosis. Incidentally, the ced-9 gene parallels bcl-2, the first known proto-oncogene protein to block cell death or an oncogene that controls the suicidal activities of T lymphocytes in humans (314). In adults, bcl-2 presence is restricted to progenitor and long-lived cells but is much more widespread during the ontogenesis (315). Bcl-2 was detected in contact with mitochondria, endoplasmic reticulum and the nuclear membranes, sites of reactive oxygen species generation. Bcl-2 prevents oxidative damages to cellular constituents including lipid membranes. A family of bcl-2 related genes is emerging that includes Bax, a conserved homologue that heterodimerizes within the body with bcl-2. Recently, a pre-set ratio of bcl/Bax appears to regulate the survival or death of various cell types after an apoptotic stimulus. Moore and co-workers (316) defined that thymocytes susceptible to apoptosis on TCR ligation express bcl-2 α mRNA suggesting that changing levels of bcl-2 expression cannot be the only determinants regulating susceptibility to apoptosis. Thymic nurse cells (TNC) are thymic subcapsular and cortical RE cells that envelop in specialized microenvironment the nonfunctional, nonselected, apoptotic cortical thymocytes (317). Debatin and co-workers (318) described expression of APO-1 on immature thymocytes. The result suggests that the APO-1 pathway (typical for peripheral lymphocytes) may be involved in negative selection of at least a subpopulation of thymocytes after intrathymic cell activation. Further understanding of the apoptotic mechanism, and further advances in gene transfer methodology will provide a way to induce therapeutic apoptosis in certain cell populations.

Thymocytes, immunophenotypically (IP), are remarkably heterogeneous with respect to the expression of different combinations of five major groups of cell surface markers. Their function to perform certain specialized tasks is, at least partially, dependent upon the expression of various cell surface antigens:

1. the T cell receptor (TCR) as an antigen;
2. growth factor receptors;
3. the recognition glycoproteins CD1, CD3, and CD7 which are thought to participate in interactions with MHC Class I and II antigen molecules (53, 319);
4. homing receptor structures;
5. a 16kD thymocyte cell surface glycoprotein and
6. Glycosaminoglycan (GAG)-binding molecules on the surface of lymphocytes.

Thymocytes express at least 12 distinct GAG-binding molecules with molecular weights (MW) ranging from 10,000 to 250,000 daltons. It has been defined that these molecules can not be correlated with the entry of lymphatic cells into a particular lymphatic tissue. The 16 kD thymocyte membrane glycoprotein was defined by Amarante-Mendes and co-workers (320) and was shown that this protein is in interaction with medullary RE cells trough the gp23/45 epithelial adhesion molecule.

The very earliest stages of T lymphocyte maturation, that is the stages of differentiation from a pluripotent stem cell to the stage of T lineage committed, intrathymic progenitor cells have remained unsolved (24, 27, 89, 105, 321, 322). Most recent studies concur that the most immature intrathymic thymoblasts, alone capable of repopulating the thymus *in vivo*, resides in the TdT$^+$ and TdT$^-$, CD3$^-$, CD4$^-$, and CD8$^-$ (triple negative IP) (2, 21, 28-31, 37, 41, 61, 64, 65, 68, 86, 133, 139, 233, 254, 323, 324), but CD7$^+$, CD2$^+$, CD1$^+$ and IL$^-$2R$^+$ cell subset. The CD3$^-$, CD4$^-$, CD8$^-$ subpopulation is itself very heterogeneous by other cell surface antigens, and can be divided into at least 11 distinct subsets (8, 71, 325). Among the earliest thymocytes to enter the CD4/CD8 developmental pathway are CD4$^-$CD810 precursor cells that differentiate into CD4$^+$ CD8$^+$ thymocytes (123). This differentiation pathway requires at least one cell division and that the progression of the early thymocytes through a cell cycle is specifically dependent to a cell-cell contact with RE cells expressing TGF-β proteins. TGF-β type II receptors were detected on the subcapsular and cortical thymic RE cells during the embryonal development. Thus, the TGF-β type II receptor expressing thymic RE cells can regulate the rate at which double positive (CD4$^+$ CD8$^+$) thymocytes are generated from the CD4$^-$ CD810 lymphopoietic precursor cells.

The major population (70-80% of all thymic cells) of immature, non-functional, non-immunocompetent (CD4$^+$8$^+$), "common" thymocytes also express the thymic antigen CD1. Therefore, mature thymocytes can be isolated from human thymus by treatment with anti-CD1 MoAB, supplemented with complement. CD6 is a cortical thymocyte specific antigen. As expected, about 15-20% of thymocytes are transitional forms (TdT$^+$/$^-$), between cortical (TdT$^+$) and medullary (TdT$^-$) thymocytes (326). The proliferation of triple-negative and common thymocytes may be activated by interactions of CD2 with LFA-3 on the RE cells, as well as by IL-2 and IL-2R. Thymic RE cells secrete IL-1, granulocyte colony-stimulating factor, and myeloid colony-stimulating factor. Antibodies against the Thy-1 antigen induce proliferation of early thymocytes in the presence of IL-1. Two-color immunofluorescence studies of thymic cell subset reveal that CD1- thymocytes comprise at least three distinct subpopulations:

1. immunophenotypically mature, "single positive" T lymphocytes (CD3$^+$4$^+$8$^-$ or CD3$^+$4$^-$8$^+$);
2. a subset of CD3$^+$ "double negative" thymocytes (CD3$^+$4$^-$8$^-$); and
3. a remaining minor subpopulation of CD3$^-$4$^-$8$^-$ (the well-known triple negative IP) (16, 39, 65, 83, 91). Interleukin-12, produced by the thymic stromal cells in combination with the stem cell factor induced significant proliferation of these triple negative pro-T lymphocytes (327). IL-12 already has been implicated in the immunomaturation and activation of peripheral T lymphocytes and natural killer (NK) cells. In

more mature cells, stimulation through CD3 or TCR (probably both), and also through CD28, may become essential for thymocyte proliferation. Double negative thymocytes express CD5 or T1 molecules, CD38 structures and transferrin receptors (T9), but most cells (85-99%) in this compartment express CD2 and MHC Class I molecule (28-30, 68, 233).

It was determined that the expression of three isoforms of CD45, (CD45RA, CD45RC and CD45RO) seem to be in coordination with the intrathymic T lymphocyte maturation and with the stage of their activation (328). Anti-CD45RA MoAB reacted only with a subset of medullary located thymocytes, whereas the anti-CD45RC was expressed on the majority of medullary thymocytes. Thymocytes that bear CD45RA also express CD45RC, but do not CD45RO. A subset of CD45 RA^- RO^+ thymocytes was also defined. CD45RC was selectively present on mature and already medullary located T lymphocytes during the intrathymic maturation pathway of thymocytes.

Recent experimental studies employing knockout mice helped to determine the roles of some molecules in the sequence of stages during the intrathymic T lymphocyte maturation. It is accepted that:

1. the protein-tyrosine kinase (PTK) p56lck is required for signal transduction through CD4/CD8, while
2. the tyrosine phosphatase CD45 (exon 6 of the encoding gene) is required for dephosphorylation of regulatory tyrosines at the C-terminal of PTK molecules (91, 328). The experiments showed that p56lck knockout animals have few immunocompetent intrathymic T lymphocytes and only half of the normal level of double negative $CD4^-CD8^-$ thymocytes, suggesting that p56lck is important for the intrathymic proliferation and selection of thymocytes.

10. MAJOR HISTOCOMPATIBILITY COMPLEX (MHC) RESTRICTION (POSITIVE INTRATHYMIC SELECTION OF T LYMPHOCYTES

The primary role of the major histocompatibility complex (MHC), the class I and class II antigens, is the presentation of foreign peptides derived from foreign antigens to the effector cells of the immune system (3, 32, 65, 80, 122, 133, 138, 285, 321, 323, 324, 329, 330). In the absence of foreign proteins the peptides bound to MHC molecules will be produced from the proteins of the host itself (331). The most novel and widely accepted concept of intrathymic thymocyte selection is the affinity-avidity model (332).

The hypothesis for tolerance induction to "self" and to MHC restriction during T cell differentiation is as follows: Thymocytes learn to tolerate self (the body) during the close intracellular and cell to cell contacts with the MHC class I antigens positive cortical RE cells, necessary for physiologic T cell maturation (3, 32, 66, 80, 89, 282, 285, 333-338). The selected thymocytes also receive a protective signal, which rescues them from programmed cell death (112, 131, 132, 270, 271, 285, 308, 309, 336-340). Whether positive selection (self restriction) requires cell to cell contact for MHC molecules and presentation of foreign peptides. Whether the cell to cell contact is made with the MHC molecules complexed to various self peptides is unclear (335). Thymocytes also acquire a restriction to MHC class II antigens being in "examination contact" with the most dominant accessory cells of the cortico-medullary junction, the interdigitating (ID) cells. Other dendritic cells produce a thymic stroma derived T cell growth factor, distinct from all interleukins. Thymic stromal elements also contain an anterior pituitary hormone stimulating (releasing) activity, named thymic neuroendocrine-releasing factor (TNRF) (245). TNRF stimulates prolactin (PRL) release, potentiates thyrotropin-releasing hormone (TRH) and is additive to the physiologic role of growth hormone-releasing hormone (GHRH) on GH release. Anterior pituitary cells directly respond to TNRF with immediate increases in hormone release. Other thymic peptides, such as the thymic hormones, may regulate pituitary gland function as well. The role of thymic stroma cells in T lymphocyte tolerance induction and maturation is not yet clear. MHC restriction regulates an intrathymic negative and positive selection for the lymphatic cells as described below.

10.1 Negative Selection

A basic factor in shaping the TCR repertoire during the intrathymic maturation process of thymocytes is the susceptibility of $CD4^+$ $CD8^+$ (double positive intermediary) cells to induction of apoptosis when the TCR is engaged by self antigens (316). The intrathymic negative selection during the T lymphocyte immuno-maturation pathway involves selective induction of apoptosis in thymocytes.

Negative selection takes place in the thymic cortex as soon as the maturing thymocytes are expressed to a fully mature, functionally capable α/β positive TCR-CD3 (T3) (T cell receptor complex) or γ/δ TCR-CD3 complex and are exposed to cells bearing major histocompatibility complex (MHC) class I and II antigenic glycoproteins (95, 325) (reticulo-epithelial (RE) cells, ectomesenchymal elements, dendritic cells, interdigitating cells, fibroblasts, macrophages). The expression of TCR-δ occurs before the appearance of TCR α/β cell surface heterodimer surface molecules.

The role of class I and class II MHC molecules is to bind peptides of intracellularly or extracellularly processed antigenic proteins, respectively, so that they can be recognized as foreign antigens by α/β TCR-Ti complexes, which exhibit dual specificity: for the antigenic peptide as well as for the peptide binding polymorphic domains of MHC molecules. After this differentiation step, the lymphatic cells interact with the interdigitating (ID) cells at the cortico-medullary junction level. Those thymocytes, which happen to react strongly with "self" MHC antigens will be tolerated (negative selection). The majority of these thymocytes have the "mixed" cell surface antigenic expression: $CD7^+$, $CD2^+$, $CD3^+$, $CD4^+$, $CD1^-$, $CD8^-$, and MHC Class I^+.

The critical event in the tolerance modification is the strength of interaction between the recently assembled TCR complex and the abundant MHC class II antigens on the ID cells, in the absence of foreign antigens. The ID cells express the highest quantities of MHC class II antigens in the human body. The MHC class II antigens are cell surface heterodimeric glycoproteins composed of 34 kD α, and 29 kD β chains and play a crucial role in the development of self tolerance, MHC restriction, and antigen presentation (15). Each of these processes involves the interaction of MHC with TCR α, β, γ or δ receptors, antigen molecules and accessory molecules such as CD4. After this interaction, "untolerized," "virgin" T cells may leave the thymus, possessing some residual affinity to MHC (the RE cell influence) while remaining unreactive to unmodified MHC antigens until foreign antigens are be presented in the context of self MHC products (341). The antigen induced activation of the T cells can be inhibited by antigen analogs that have been called TCR peptide antagonists (342). The first peptide antagonists were identified for the cytochrome c-specific T cell clone AD10. The antagonist appears to act by inducing the formation of nonstimulatory chemical complexes between the TCRs and the MHC molecules presenting the necessary peptides. Though unable to mediate mature T cell activation, the presence of TCR peptide antagonists resulted in a negative intrathymic selection because of the deletion of $CD4^+$ $CD8^+$ thymocyte subpopulation (342). Although considerable data exist for the molecular basis of mature T lymphocyte signal transduction, the enzymes that participate in the TCR selection processes have remained a mystery. The CD45RO protein thyrosine phosphatase, a cellular enzyme is involved in the regulation of the intrathymic thymocyte apoptosis and in the TCR selection mechanisms during the immunological maturation of $CD4^+CD8^+$ thymocyte population (343). Thus, augmented thymic presence of the CD45RO protein thyrosine phosphatase increased the efficacy of TCR-mediated apoptosis and MHC restricted negative selection of intrathymic thymocytes.

Recent *in vitro* experiments determined a potential novel regulatory molecule associated with the extracellular matrix (ECM) of thymic stromal cells and stimulated a thymic T cell line of immature IP (111, 344). The stimulatory activity of the ECM was resistant to high salt extraction, acetic acid and strong denaturants but showed sensitivity to treatment with trypsin and heat over 80 $^{\circ}$C. Purified ECM proteins (lamilin, collagen, fibronectin, and vitronectin) were either ineffective or showed only partial stimulatory effects. The ECM produced by other cells such as fibroblast cell lines was nonstimulatory.

Cultured thymic RE and stromal cells produced the steroids pregnenolone and deoxycorticosterone and immunocytochemical observations demonstrated presence of steroidogenic enzymes in radioresistant thymic RE cells (11, 345). The locally produced glucocorticoids, because of their antagonist role to the TCR-mediated regulation for apoptosis may be a key mechanism of antigen-specific T lymphocyte selection. The relationship between these interactions and the biology of the development of tolerance are, however, poorly understood.

The extreme stimulatory effect of the so-called superantigen (staphylococcal enterotoxin B, streptococcus pyogenes, representing 27 kD exotoxins, and superantigens produced by some retroviruses and Mycoplasma arthridis) on the T cells of several mammalian species is also well defined (346). These stimulatory molecules also provide information on tolerance mechanisms induced by apoptosis.

10.2 Positive Selection

It is certain that the initial negative selection realized on host cortical $CD40^+$, HLA^-ABC^+, HLA^-DR^+, also expressing homing associated cell-adhesion molecule (H-CAM), intercellular adhesion molecule-1 (ICAM-1), leukocyte function associated antigen 3 (LFA-3) and β 1 subfamily integrins RE cells (347-349). During a second ontogenetic step, the above mentioned thymocyte elimination is followed by a positive selection mechanism on the level of the cortico-medullary junction on the host interdigitating (ID) cells (347). In experimental, pre-clinical animal models (usually mouse and rat), treatment with Cyclosporin A, resulted in serious depletion of the medullary located ID cell population. The suppression of ID cells was manifested in a loss of tolerance to "self," resulting in the development of T helper lymphocyte deficiency, followed by autologous graft versus host disease. To determine the developmental stages at which the mechanisms of positive selection produce immunocompetent T cells early thymocytes of large dividing and small nondividing types were transferred into the thymus (350). In contrast to earlier studies, both types were able to produce mature thymocytes with $CD4^+ CD8^-$ and $CD4^- CD8^+$ immunophenotype. Thus, the developmental window for positive selection also includes the population of small cortical thymocytes, certainly after the initiating multi-step activation. Further experiments are required with human thymocyte populations to understand the interim steps of tolerance induction and MHC (self) restriction (285, 351).

11. FIGURES

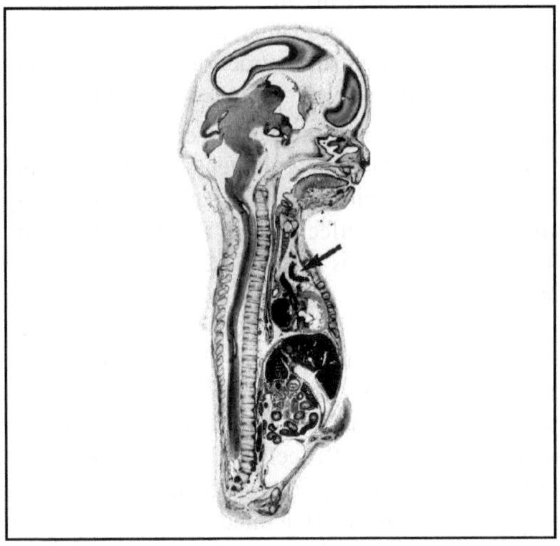

Figure 1. Serial section of 5 cm long human fetus, embedded in wax. Black arrow demonstrate the thymus. 5 μm thick section. Staining after the method of Giemsa. Magnification 40x.

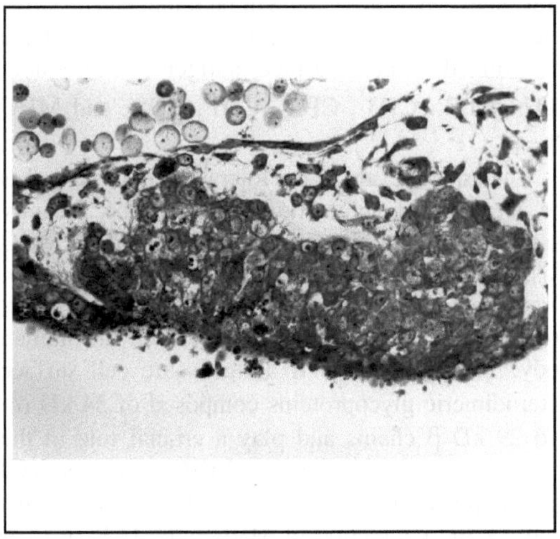

Figure 2. Epithelial stage of thymic histogenesis. Dog embryo in 23rd day of gestation. Proliferation of the primitive epithelial cells in direction of the ectomesenchymal cells. Primitive hematopoietic stem cells are already present. Methacrylate embedding. 3 μm thin section. Modified staining after Giemsa. Magnification 400x.

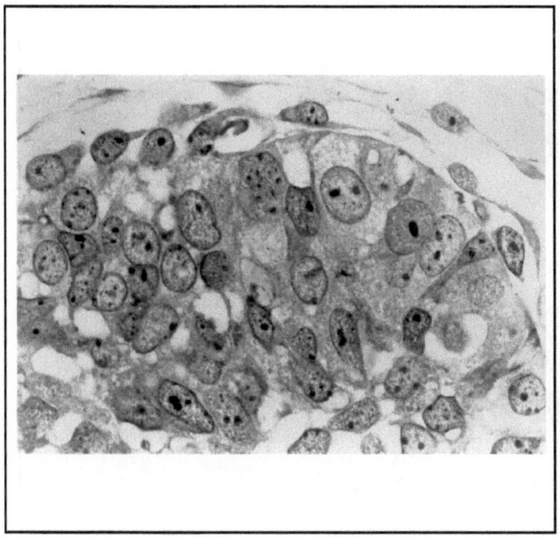

Figure 3. Developing stage of thymic cortex. Dog thymus. Some already committed lymphatic stem cell immigration is detectable. The close cell to cell character of the epithelium is over. The primitive epithelial cells are forming a pseudo-reticular tissue, adopting in the microenvironment also some ectomesenchymal cells. Methacrylate embedding. 3 µm thin section. Modified Giemsa staining. Magnification 400x.

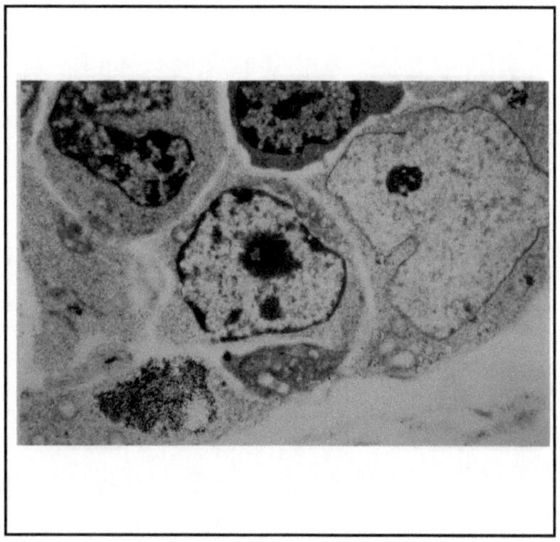

Figure 5. Developing thymic cortex. Dog fetus in 34th day of gestation.Typical for the stage cellular thymic microenvironment. Epon embedding. TEM picture. Magnification 10000x.

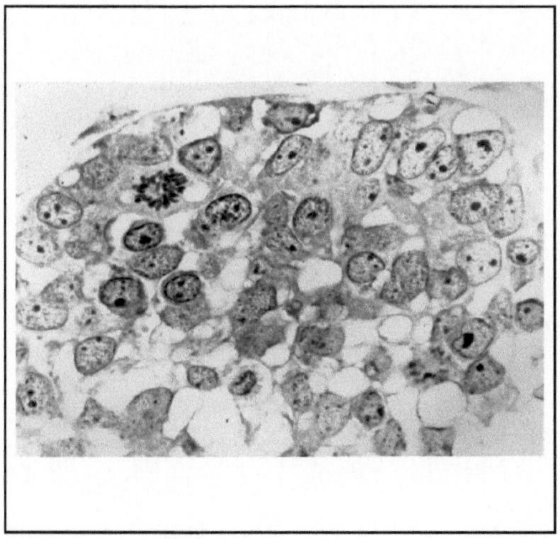

Figure 4. Thymic cortex in an early stage of lymphopoiesis. Dog fetus in 33rd day of gestation. Subcapsular, secretory reticulo-epithelial (RE) cell is in mitosis (black arrows). Other RE cells are still remaining in groups. This close cell to cell epithelial distribution is typical for this early stage of T lymphocyte producing thymus. The first lymphatic blast cells are already present. Methacrylate embedding. 3 µm thin section. Hematoxylin-eosin staining. Magnification 630x.

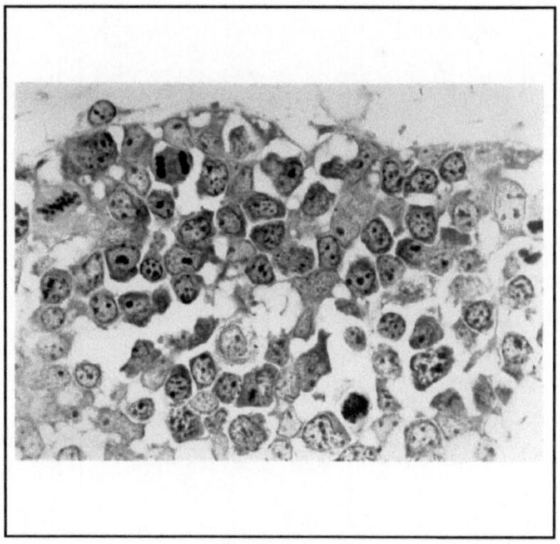

Figure 6. Thymic cortex. Dog fetus in 36th day of gestation. Well developed lymphopoietic thymus, with numerous lymphatical and RE cell mitoses. Well developed RE cell processes. Methacrylate embedding. 3 µm thin section. Modified Giemsa staining. Magnification 500x.

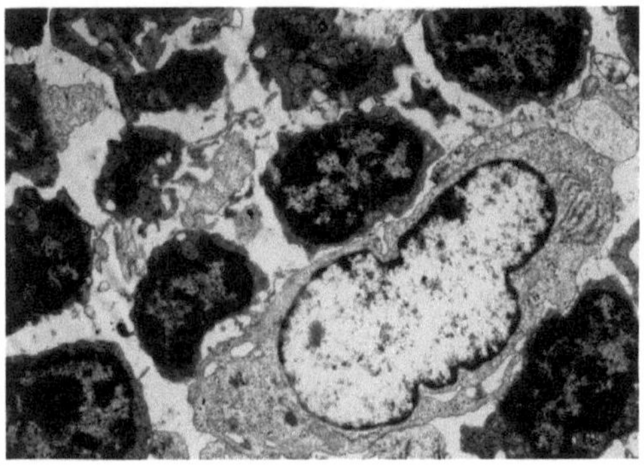

Figure 7. Thymic cortex. Human fetus of 21 cm body length (early 5th lunar month). RE cell surrounded with thymocytes showing several nuclear envelope related cell-cell contacts. The type of cellular contact is not desmosomal. Epon embedding. TEM picture. Magnification 10500x.

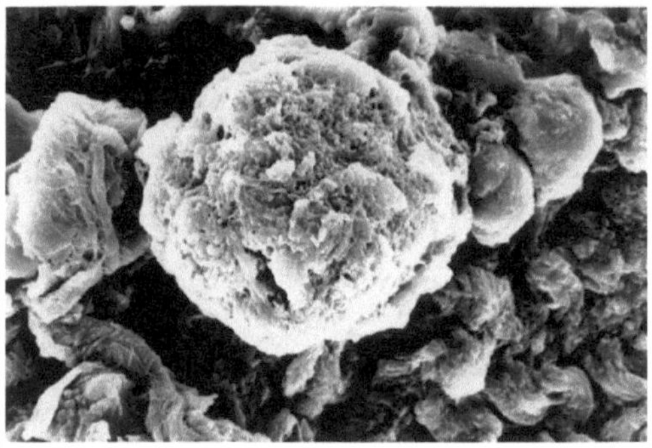

Figure 8. Thymic medulla. Three month old mouse. The most characteristic morphological structure, of the mammalian thymic medulla, the body of Hassall's appears only when the thymic histogenesis is completed. SEM picture. Magnification 20000x.

REFERENCES

1. Clevers H, Alarcon B, Willeman T, Terhorst C: The T cell receptor/ CD3 complex: a dynamic protein ensemble. Annu Rev Immunol *6:* 629-662, 1988.
2. Cross M, Dexter TM: Growth factors in development, transformation, and tumorigenesis. Cell *64:* 271-280, 1991.
3. 3. Dellabona P, Peccoud J, Kappler J, Marrack P, Benoist C, Mathis D: Superantigens interact with MHC Class II molecules outside of the antigen groove. Cell *62:* 1115-1121, 1990.
4. Dembic Z, Haas W, Weiss S, McCubrey J, Kiefer H, von Boehmer H, Steinmetz M: Transfer of specificity by murine α and β T cell receptor genes. Nature (London) *320:* 232-238, 1986.
5. Dembic Z, Haas W, Zamoyska R, Parnes J, Steinmetz M, von Boehmer H: Transfection of the CD8 gene enhances T-cell recognition. Nature (London) *326:* 510-511, 1986.
6. Denning SM, Tuck DT, Singer KH, Haynes BF: Human thymic epithelial cells function as accessory cells for autologous mature thymocyte activation. J Immunol *138:* 680-686, 1987.
7. Doolittle RF, Hunkapiller MW, Hood LE, Devare SG, Robbins KC, Aaronson SA, Antoniades HN: Simian sarcoma virus onc gene, v-sis, is derived from the gene (or genes) encoding a platelet-derived growth factor. Science *221:* 275-277, 1983.
8. Engel I, Ottenhoff THM, Klausner RD: High-efficiency expression and solubilization of functional T cell antigen receptor heterodimers. Science *256:* 1318-1321, 1992.
9. Fabien N, Auger C, Monier JC: Immunolocalization of thymosin α-1, thymopoietin and thymulin in mouse thymic epithelial cells at different stages of culture: a light and electron microscopic study. Immunology *63:* 721-727, 1988.
10. Bertagnolli M, Herrmann S: IL-7 supports the generation of cytotoxic T lymphocytes from thymocytes. Multiple lymphokines required for proliferation and cytotoxicity. J Immunol *145:* 1706-1712, 1990.
11. Bodey B: Histomorphology and histochemistry of the human thymus during the prenatal development. pp. 1-360, Ph. D. Thesis Sofia (590 references), 1977.
12. Bodey B, Calvo W, Prummer O, Fliedner TM, Borysenko M: Development and histogenesis of the thymus in dog. A light and electron microscopical study. Developmental Compar Immunol *11:* 227-238, 1987.
13. Brelinska R: Thymic nurse cells: division of thymocytes within complexes. Cell Tissue Res *258:* 637-643, 1989.
14. Brelinska R, Warchol JB, Boniver J, Houben-Defresne MP: Distribution of Lyt antigens on the surface of thymocytes associated with thymic macrophages and dendritic cells. Histochemistry *88:* 553-556, 1988.
15. Crispe NI, Moore M, Hussmann L, Smith L, Bevan M, Shimonkevitz R: Differentiation potential of subsets of CD4-8- thymocytes. Nature (London) *329:* 336-339, 1987.
16. Dalloul AH, Mossalayi MD, Dellagi K, Bertho JM, Debre P: Factor requirements for activation and proliferation steps of human $CD2^+CD3^-CD4^-CD8^-$ early thymocytes. Eur J Immunol *19:* 1985-1990, 1989.
17. de la Hera A, Muller U, Olsson C, Isaaz S, Tunnacliffe A: Structure of the T cell antigen receptor (TCR): two CD3 β subunits in a functional TCR/CD3 complex. J Exp Med *173:* 7-17, 1991.
18. Defresne MP, Humblet Ch, Rongy AM, Greimers R, Boniver J: Effect of interferon-γ and tumor necrosis factor-α on lymphoepithelial interactions within thymic nurse cells. Eur J Immunol *20:* 429-432, 1990.
19. Eshel I, Savion N, Shoham J: Analysis of thymic stromal cell subpopulations grown in vitro on extracellular matrix in defined medium. I. Growth conditions and morphology of murine thymic epithelial and mesenchymal cells. J Immunol *144:* 1554-1562, 1990.
20. Eshel I, Savion N, Shoham J: Analysis of thymic stromal cell subpopulations grown in vitro on extracellular matrix in defined medium. II. Cytokine activities in murine thymic epithelial and mesenchymal cell culture supernatants. J Immunol *144:* 1563-1570, 1990.
21. Chien Y-H, Iwashima M, Kaplan KB, Elliot JF, Davis MM: A new T cell receptor gene located within the α locus and expressed early in T cell differentiation. Nature (London) *327:* 677-682, 1987.
22. Chien Y-H, Iwashima M, Wettstein DA, Kaplan KB, Elliot JF, Born W, Davis MM: T-cell receptor δ gene rearrangements in early thymocytes. Nature (London), *330:* 722-727, 1987.
23. Davila DG, Minoo P, Estervig DN, Kasperbauer JL, Tzen C-Y, Scott RE: Linkages in control of differentiation and proliferation in murine mesenchymal stem cells and human keratinocyte progenitor cells: the effects of carcinogenesis. In: Mechanisms of differentiation, PB. Fisher (ed), CRC Press, Boca Raton Ann Arbor Boston, pp. 2-13, 1990.
24. Auerbach R: Genetic control of thymus lymphoid differentiation. Proc Natl Acad Sci U.S.A. *47:* 1175-1181, 1961.
25. Beard J: The source of leukocytes and the true function of the thymus. Anat Anz *18:* 550-573, 1900.
26. Brown KM, Spirito S, Basch RS: Thymic stromal cells in culture. I. Establishment and characterization of a line which is cytotoxic for normal thymocytes and produces hematopoietic growth factor(s). Cell Immunol *134:* 442-457, 1991.
27. Burgess A, Nicola N: Growth factors and stem cells. Academic Press, New York, pp. 1-38, 1983.
28. Fink PJ, Weissman IL, Kaplan HS, Kyewski BA: The immunocompetence of murine stromal cell associated thymocytes. J Immunol *132:* 2266- 2272, 1984.
29. Fowlkes BJ, Pardoll DM: Molecular and cellular events of T-cell development. Adv Immunol *44:* 207-264, 1989.
30. Fujiwara H: Thymic stroma-derived T cell growth factor: its role in the growth of immature thymocytes and T cell repertoire selection. Int Symp Princess Takamatsu Cancer Res Fund *19:* 115-126, 1988.
31. Gallatin M, St. John TP, Siegelman M, Reichert R, Butcher EC, Weissman IL: Lymphocyte homing receptors. Cell *44:* 673-680, 1986.
32. Goldberg AL, Rock KL: Proteolysis, proteasomes and antigen presentation. Nature (London) *357:* 375-379, 1992.
33. Grabstein KH, Namen AE, Shanebeck K, Voice RF, Reed SG, Widmer MB: Regulation of T cell proliferation by IL-7. J Immunol *144:* 3015-3020, 1990.
34. Greenbaum LA, Horowitz JB, Woods A, Pasqualini T, Reich E-P, Bottomly K: Autocrine growth of $CD4^+$ T cells. J Immunol *140:* 1555-1560, 1988.
35. Gutierrez JC, Palacios R: Heterogeneity of thymic epithelial cells in promoting T-lymphocyte differentiation in vivo. Proc Natl Acad Sci USA *88:* 5685-5689, 1991.

36. Hadden JW, Galy A, Chen H, Hadden EM: A pituitary factor induces thymic epithelial cell proliferation in vitro. Brain Behav Immunol *3:* 149-159, 1989.
37. Hall NR, McGillis JP, Spangelo BL, Goldstein AL: Evidence that thymosin and other biologic response modifiers can function as neuroactive transmitters. J Immunol *135:* 802s-805s, 1985.
38. Haynes BF, Denning SM, Singer KH, Kurtzberg J: Ontogeny of T-cell precursors: a model for the initial stages of human T-cell development. Immunol Today *2:* 87-89, 1989.
39. Haynes BF, Singer KH, Denning SM, Martin ME: Analysis of expression of CD2, CD3, and T cell antigen receptor molecule during early human fetal thymic development. J Immunol *141:* 3776-3784, 1988.
40. Imhof BA, Ruiz P, Hesse B, Palacios R, Dunon D: EA-1, a novel adhesion molecule involved in the homing of progenitor T lymphocytes to the thymus. J Cell Biol *114:* 1069-1078, 1991.
41. Janossy G, Campana D, Akbar A: Kinetics of T lymphocyte development. Curr Top Pathol *79:* 59-99, 1989.
42. Janossy G, Greaves MF: Lymphocyte activation. I. Response of T and B lymphocytes to phytomitogens. Clin Exp Immunol *9:* 483-498, 1971.
43. Janossy G, Pizzolo G: Lymphocytes. 2. Differentiation. Differentiation of lymphoid precursor cells. J Clin Pathol (suppl.) *13:* 48-58, 1979.
44. Janossy G, Bofill M, Poulter LW, Rawlings E, Burford GD, Navarrete C, Ziegler A, Kelemen E: Separate ontogeny of two macrophage-like accessory cell populations in the human fetus. J Immunol *136:* 4354-4361, 1986.
45. Janossy G, Tidman N, Papageorgiou ES, Kung PC, Goldstein G: Distribution of T lymphocyte subsets in the human bone marrow and thymus: An analysis with monoclonal antibodies. J Immunol *126:* 1608-1613, 1983.
46. Janossy G, Thomas JA, Bollum FJ, Granger S, Pizzolo G, Bradstock KF, Wong L, McMichael A, Ganeshaguru K, Hoffbrand AV: The human thymic microenvironment: an immunohistologic study. J Immunol *125:* 202-212, 1980.
47. Jotereau FV, Le Douarin NM: Demonstration of a cyclic renewal of the lymphocyte precursor cells in the quail thymus during embryonic and perinatal life. J Immunol *129:* 1869-1877, 1982.
48. Kadish JL, Basch RS: Hematopoietic thymocyte precursors. III. A population of thymocytes with the capacity to return ("home") to the thymus. Cell Immunol *30:* 12-24, 1977.
49. Wiles MV, Keller G: Multiple hematopoietic lineages develop from embryonic stem (ES) cells in culture. Development *112:* 605-614, 1991.
50. Cavender DE: Organ-specific and non-organ-specific lymphocyte receptors for vascular endothelium. J Invest Dermatol *94:* 41S-48S, 1990.
51. Chervenak R, Moorhead JW, Cohen JJ: Prethymic T cell precursors express receptors for antigen. J Immunol *134:* 695-698, 1985.
52. Clark MJ, Gagnon J, Williams AF, Barclay AN: MRC OX-2 antigen: A lymphoid/neuronal membrane glycoprotein with a structure like a single immunoglobulin light chain. EMBO J *4:* 113-118, 1985.
53. Auerbach R: Morphogenetic interactions in the development of the mouse thymus gland. Devel Biol *2:* 271-276, 1960.
54. Auerbach R: Experimental analysis of the origin of cell types in the development of mouse thymus. Develop Biol *3:* 336-354, 1961.
55. Bernton EW, Beach JE, Holaday JW, Smallridge RC, Fein HG: Release of multiple hormones by a direct action of interleukin-1 on pituitary cells. Science *238:* 519-521, 1987.
56. Borgulya P, Kishi H, Muller U, Kirberg J. von Boehmer H: Development of the CD4 and CD8 lineage of T cells: instruction versus selection. EMBO J *10:* 913-918, 1991.
57. Brune B, Hartzell P, Nicotera P, Orrenius S: Spermine prevents endonuclease activation and apoptosis in thymocytes. Exp Cell Res *195:* 323-329, 1991.
58. Campana D: The developmental stages of the human T cell receptors: A review. Thymus *13:* 3-18, 1989.
59. Couture C, Patel PC, Potworowski EF: A novel thymic epithelial adhesion molecule. Eur J Immunol *20:* 2769-2773, 1990.
60. Dargemont C, Dunon D, Deugnier MA, Denoyelle M, Girault JM, Lederer F, Ho Diep Le K, Godeau F, Thiery JP, Imhof BA: Thymotaxin, a chemotactic protein, is identical to β2-microglobulin. Science *246:* 803-806, 1989.
61. Granstein RD, Staszewski R, Knisely TL, Zelinsky DM: Aqueos humor contains transforming growth factor-β and a small (<3500 daltons) inhibitor of thymocyte proliferation. J Immunol *144:* 3021-3027, 1990.
62. Haynes BF: Human T lymphocyte antigens as defined by monoclonal antibodies. Immunol Rev *57:* 127-161, 1981.
63. Haynes BF: Human thymic epithelium and T cell development: Current issues and future directions. Thymus *16:* 143-157, 1990.
64. Hiramine C, Hojo K, Koseto M, Nakagawa A, Mukasa A: Establishment of a murine thymic epithelial cell line capable of inducing both thymic nurse cell formation and thymocyte apoptosis. Lab. Invest., *62:* 41-54, 1990.
65. Inaba K, Steinman RM: Accessory cell - T lymphocyte interactions. Antigen-dependent and -independent clustering. J Exp Med *163:* 247-261, 1986.
66. Ingold AL, Landel C, Knall C, Evans GA, Potter TA: Co-engagement of CD8 with the T cell receptor is required for negative selection. Nature (London) *352:* 721-723, 1991.
67. Farr A.G., Hosier S., Braddy SC, Anderson SK, Eisenhardt DJ, Robles CP: Medullary epithelial cell lines from murine thymus constitutively secrete IL-1 and hematopoietic growth factors and express class II antigens in response to recombinant interferon-γ. Cell Immunol *119:* 427-444, 1989.
68. Fujiwara H, Ogata M, Mizushima Y, Tatsumi Y, Takai Y, Hamaoka T: Proliferation and differentiation of immature thymocytes induced by a thymic stromal cell clone. Thymus *16:* 159-172, 1990.
69. Fujiwara S, Fisher RJ, Bhat NK, Diaz de la Espina SM, Papas TS: A short-lived nuclear phosphoprotein encoded by the human ets-2-protooncogene is stabilized by activation of protein kinase "C". Mol Cell Biol *8:* 4700-4706, 1988.
70. Fukuzawa M, Sharrow SOP, Shearer GM: Effect of cyclosporin-A on T cell immunity. II. Defective thymic education of CD4 T helper cell function in cyclosporin-A-treated mice. Eur J Immunol *19:* 1147-1152, 1989.
71. Galy AH, Dinarello CA, Kupper TS, Kameda A, Hadden JW: Effects of cytokines on human thymic epithelial cells in culture. II. Recombinant IL1 stimulates thymic epithelial cells to produce IL6 and GM-CSF. Cellular Immunol *129:* 161-175, 1990.
72. Ham RG, Veomett MJ: In: Mechanisms of development, RG, Ham MJ, Veomett CV, Mosby (eds), St. Louis-Toronto-London, pp. 3-109; 165-282; 317-471, 1980.

73. Raulet DH: The structure, function and molecular genetics of the γ/δ T cell receptor. Annu Rev Immunol *7:* 175-207, 1989.
74. Raulet R, Garman R, Saito H, Tonegawa S: Developmental regulation of T-cell gene expression. Nature (London) *314:* 103-107, 1985.
75. Snodgrass HR, Kisielow P, Kiefer M, Steinmetz M, von Boehmer H: Ontogeny of the T-cell antigen receptor within the thymus. Nature (London) *313:* 592-595, 1985.
76. Strominger JL: Developmental biology of T cell receptors. Science *244:* 944-950, 1989.
77. Weiss A, Imboden J, Hardy K, Manger B, Terhorst C, Stobo J: The role of the T3/antigen receptor complex in T-cell activation. Annu Rev Immunol *4:* 593-619, 1986.
78. Weiss A, Imboden J, Hardy K, Stobo J: The role of the antigen receptor/T3 complex in T cell activation. Adv Exp Med Biol *213:* 45-49, 1987.
79. Deschaux P, Rouabhia M: The thymus: key organ between endocrinologic and immunologic systems. Ann NY Acad Sci *496:* 49-55, 1987.
80. Gray D: Understanding germinal centers. Res Immunol *142:* 237-242, 1991.
81. Hara T, Fu SM, Hansen JA: Human T-cell activation 2. A new activation pathway used by a major T-cell population via a disulfide-bonded dimer of a 44 Kd polypeptide 9.3 antigen. J Exp Med *161:* 1513-1524, 1985.
82. Havran WL, Allison J: Developmentally ordered appearance of thymocytes expressing different T-cell antigen receptors. Nature (London) *335:* 443-445, 1988.
83. Janossy G, Bofill M, Trejdosiewicz LK, Willcox HN, Chilosi M: Cellular differentiation of lymphoid subpopulations and their microenvironments in the human thymus. Curr Top Pathol *75:* 89-125, 1986.
84. Jenkinson EJ, Kingston R, Owen JJT: Importance of IL-2 receptors in intrathymic generation of cells expressing T-cell receptors. Nature (London) *329:* 160-162, 1987.
85. Jiraskova Z, Necas E, Holub M, Ludvik J: The dysgenetic thymus in nu/nu mice is a potent source of colony stimulating factor (CSF). Folia Biol (Praha) *36:* 231-235, 1990.
86. Jung LKL, Haynes BF, Nakamura S, Pahwa S, Fu SM: Expression of early activation antigen (CD69) during human thymic development. Clin exp Immunol *81:* 466-474, 1990.
87. Kaneshima H, Itoh M, Asai J, Taguchi O, Hiai H: Reorganization of thymic microenvironment during development and lymphomagenesis. Jpn J Cancer Res *78:* 799-806, 1987.
88. Kisielow P, von Boehmer H: Negative and positive selection of immature thymocytes: timing and the role of the ligand for α, β T cell receptor. Sem Immunol *2:* 35-44, 1990.
89. Kisielow P, von Boehmer H: Development of T cells in T-cell-receptor transgenic mice: direct evidence for positive and negative selection in the thymus. Thymus Update *3:* 14-27, 1990.
90. Kosaka H, Ogata M, Hikita I, Maruo S, Sugihara S, Matsubara H, Takai Y, Hamaoka T. Fujiwara H: Model for clonal elimination in the thymus. Proc Natl Acad Sci USA *86:* 3773-3777, 1989.
91. Kosaka H, Ogata M, Sano H, Sakuma S, Mizushima Y, Nakamura E, Teraoka H, Hamaoka T, Fujiwara H: Thymic stroma-derived T cell growth factor (TSTGF): III. Its ability to promote T cell proliferation without stimulating interleukin 2 or 4-dependent autocrine mechanism. J Leukocyte Biol *47:* 121-128, 1990.
92. MacDonald HR, Nabholz M: T-cell activation. Annu Rev Cell Biol *2:* 231-253, 1986.
93. Mackay CR: T-cell memory: the connection between function, phenotype and migration pathways. Immunol Today *12:* 189-192, 1991.
94. Mackay CR: Lymphocyte homing. Skin-seeking memory T cells. Nature (London) *349:* 737-738, 1991.
95. Marrack P, Parker DC: T-cell selection. A little of what you fancy... Nature (London) *368:* 397-398, 1994.
96. Martorell J, Rojo I, Vilella R, Martinez-Caceres E, Vives J: CD27 induction on thymocytes. *145:* 1356-1363, 1990.
97. Metcalf D: The nature and regulation of lymphopoiesis in normal and neoplastic thymus. In: The Thymus: Experimental and Clinical Studies, pp. 242-263, CIBA Foundation Symposium, Editors: G.W.E. Wolstenholme & R. Porter, Churchill, London, 1966.
98. Moulder K, Roberts K, Shevach EM, Coligan JE: The mouse vitronectin receptor is a T cell activation antigen. J Exp Med *173:* 343-347, 1991.
99. Nossal GJ: The double cascade of lymphoid proliferation: current chalanges and problems areas. Amer J Anat *170:* 253-259, 1984.
100. Oettgen H, Terhorst C: The T cell receptor-T3 complex and T lymphocyte activation. Human Immunol *18:* 187-204, 1987.
101. Ostergaard HL, Clark WR: Evidence for multiple lytic pathways used by cytotoxic T lymphocytes. J. Immunol *143:* 2120-2126, 1989.
102. Ostergaard HL, Trowbridge IS: Coclustering CD45 with CD4 or CD8 alters the phosphorylation and kinase activity of p56 lck. J Exp Med *172:* 347-350, 1990.
103. Palacios R, Pelkonen J: Prethymic and intrathymic mouse T-cell progenitors. Growth requirements and analysis of the expression of genes encoding TCR/T3 components and other T-cell-specific molecules. Immunol Rev *104:* 8-27, 1988.
104. Palacios R, Samaridis J: Thymus colonization in the developing mouse embryo. Eur J Immunol *21:* 109-113, 1991.
105. Palacios R, Sideras P, von Boehmer H: Recombinant mouse interleukin 4/BSF-1 promotes growth and differentiation of intrathymic T cell precursors from fetal mice in vitro. EMBO J *6:* 91-95, 1987.
106. Pardoll DM, Fowlkes BJ, Lechler RI, Germain RN, Schwartz RH: Early genetic events in T cell development analyzed by in situ hybridization. J Exp Med *165:* 1624-1638, 1987.
107. Penit C: In vivo proliferation and differentiation of prothymocytes in the thymus. Immunol Res *6:* 271-278, 1987.
108. Penninger J: Phenotypic and functional analysis of thymic nurse cell (TNC)-lymphocytes. Wien Klin Wochenschr *103:* 45-50, 1991.
109. Turka LA, Ledbetter JA, Lee K, June CH, Thompson CB: CD28 is an inducible T cell surface antigen that transduces a proliferative signal in $CD3^+$ mature thymocytes. J Immunol *144:* 1646-1653, 1990.
110. Uckun FM, Tuel-ahlgren L, Obuz V, Smith R, Dibirdik I, Hanson M, Langlie MC, Ledbetter JA: Interleukin 7 receptor engagement stimulates thyrosine phosphorylation, inositol phospholipid turnover, proliferation, and selective differentiation to the CD4 lineage by human fetal thymocytes. Proc Natl Acad Sci USA *88:* 6323-6327, 1991.
111. Utsumi K, Sawada M, Narumiya S, Nagamine J, Sakata T, Iwagami S, Kita Y, Teraoka H, Hirano H, Ogata M, Hamaoka T, Fujiwara H: Adhesion of immature thymocytes to thymic stromal cells through fibronectin

molecules and its significance for the induction of thymocyte differentiation. Proc Natl Acad Sci USA *88:* 5685-5689, 1991.

112. von Boehmer H: The developmental biology of T lymphocytes. Annu Rev Immunol *6:* 309-326, 1988.
113. Wesselborg S, Janssen O, Pechhold K, and Kabelitz D: Selective activation of δ γ+ T cell clones by single anti-CD2 antibodies. J Exp Med *173:* 297-304, 1991.
114. Zlotnik A, Ranson J, Frank G, Fischer M, Howard M: Interleukin 4 is a growth factor for activated thymocytes: possible role in T cell ontogeny. Proc Natl Acad Sci USA *84:* 3856-3860, 1987.
115. Muller R: Proto-oncogenes and differentiation. Trends Biochem Sci *11:* 129-132, 1986.
116. Pfeifer-Ohlsson S, Rydnert J, Scott Goustin A, Larsson E, Betsholtz C, Ohlsson R: Cell-type-specific pattern of c-myc protooncogene expression in developing human embryos. Proc Natl Acad Sci USA *82:* 5050-5054, 1985.
117. Greenberg ME, Ziff EB: Stimulation of 3T3 cells induces transcription of the c-fos proto-oncogene. Nature (London) *311:* 433-437, 1984.
118. Heldin C-H, Westermark B: Growth factors as transforming proteins. Eur J Biochem *184:* 487-496, 1989.
119. Heldin C-H, Westermark B: Signal transduction by the receptors for platelet derived growth factor. J Cell Sci *96:* 193-196, 1990.
120. Carding S, Reem GH: C-myc gene expression and activation of human thymocytes. Thymus *10:* 219-229, 1987.
121. Seckinger P, Milili M, Schiff C, Fougereau M: Interleukin-7 regulates c-myc expression in murine T cells and thymocytes: a role for tyrosine kinase(s) and calcium mobilization. Eur J Immunol *24:* 716-722, 1994.
122. Ignotz RA, Heino J, Massague J: Regulation of cell adhesion receptors by transforming growth factor β. J Biol Chem *264:* 389-392, 1989.
123. Takahama Y, Letterio JJ, Suzuki H, Farr AG, Singer A: Early progression of thymocytes along the CD4/CD8 developmental pathway is regulated by a subset of thymic epithelial cells expressing transforming growth factor β. J Exp Med *179:* 1495-1506, 1994.
124. Colic M, Matanovic D, Hegedis L, Dujic A: Immunohistochemical characterization of rat thymic non-lymphoid cells. I. Epithelial and mesenchymal components defined by monoclonal antibodies. Immunology *65:* 277-284, 1988.
125. Ruther U, Muller W, Sumida T, Tokuhisa T, Rajewsky K, Wagner EF: c-fos expression interferes with thymus development in transgenic mice. Cell *53:* 847-856, 1988.
126. Kinashi T, Tashiro K, Lee KH, Inaba K, Toyama K, Palacios R, Honjo T: Growth factors involved in lymphocyte differentiation. Phil Trans R Soc (London) *B327:* 117-125, 1990.
127. Spooncer E, Boettiger D, Dexter TM: Continuous in vitro generation of multipotential stem cell clones from src-infected cultures. Nature (London) *310:* 228-230, 1984.
128. Kurtzberg J, Denning SM, Nycum LM, Singer KH, Haynes BF: Human immature thymocytes can be driven to differentiate into non-lymphoid lineages by thymic epithelial-derived cytokines. Proc Natl Acad Sci USA *86:* 7575-7579, 1989.
129. Kroczek RA, Gunter KC, Germain RN, Shevach EM: Thy-1 functions as a signal transduction molecule in T lymphocytes and transfected B lymphocytes. Nature (London) *322:* 181-184, 1986.
130. van Ewijk W, Kisielow P, von Boehmer H: Immunohistology of T cell differentiation in the thymus of H-Y-specific T cell receptor α /β transgenic mice. Eur J Immunol *20:* 129-137, 1990.
131. von Boehmer H: Learning by the immune system studied in T-cell receptor transgenic mice. Thymus *13:* 19-26, 1989.
132. von Boehmer H., Kirberg J, Rocha B: An unusual lineage of α /β T cells that contains autoreactive cells. J Exp Med *174:* 1001-1008, 1991.
133. Benoist C, Mathis D: Positive and negative selection of the T cell repertoire in MHC class II transgenic mice. Semin Immunol *1:* 117-124, 1989.
134. Agus DB, Surh CD, Sprent J: Reentry of T cells to the adult thymus is restricted to activated T cells. J Exp Med *173:* 1039-1046, 1991.
135. Andrews P, Boyd RL: The murine thymic nurse cell: An isolated thymic microenvironment. Eur J Immunol *15:* 36-42, 1985.
136. Beard J: The origin and histogenesis of the thymus in Raja batis. Zool Jahrbucher, Abt Ontogenie Tiere *17:* 403-480, 1902.
137. Bell ET: The development of the thymus. Amer J Anat *5:* 29-62, 1905.
138. Brown JH, Jardetzky T, Saper MA, Samraoui B, Bjorkman PJ, Wiley DC: A hypothetical model of the foreign antigen binding site of class II histocompatibility molecules. Nature (London) *332:* 845-850, 1988.
139. Fink PJ, Bevan MJ, Weissman IL: Thymic cytotoxic T lymphocytes are primed in vivo to minor histocompatibility antigens. J Exp Med *159:* 436-451, 1984.
140. Fujiwara S, Kobayashi H, Awaya A: The distribution and structure of FTS immunoreactive cells in the thymus of the mouse. Arch Histol Jpn *51:* 467-472, 1988.
141. Cantor H, Weissman I: Development and function of subpopulations of thymocytes and T lymphocytes. Prog Allergy *20:* 1-64, 1976.
142. Stutman O: Intrathymic and extrathymic T cell maturation. Immunol Rev *42:* 138-184, 1978.
143. Cohen A, Lee JW, Dosch HM, Gelfand EW: The expression of deoxyguanosine toxicity in T lymphocytes at different stages of maturation. J Immunol *125:* 1578-1582, 1980
144. Shortman K, Egerton M, Spangrude GJ, Scollay R: The generation and fate of thymocytes. Semin Immunol *2:* 3-12, 1990.
145. Egerton M, Shortman K, Scollay R: The kinetics of immature murine thymocyte development in vivo. Int Immunol *2:* 501-507, 1990.
146. Hugo P, Waanders GA, Scollay R, Shortman K, Boyd RL: Ontogeny of a novel $CD4^{+}CD8^{-}CD3^{-}$ thymocyte subpopulation: a comparison with $CD4^{-}$ $CD8^{+}$ $CD3^{-}$ thymocytes. Int Immunol *2:* 209-218, 1990.
147. Tourneux F, Herrmann G: Sur l'evolition histologique du thymus chez l'embryon humain et chez les mammifères. C R Soc Biol Paris X *4*, 1887.
148. Maurer F: Schilddrüse, Thymus und sonstige Schlundspalten-derivate bei Schidna und ihre Beziehungen zu den gleichen organen bei anderen Wirbeltieren. Semons Forschunsreisen *3*, 1899.
149. Prenant A, Contribution à l'étude organique et histologique du thymus de la glande thyroide et de la glande carotidienne. Cellule *10*, 1894.
150. Stöhr P: Über die Natur der Thymuselemente. Annt N *31:* 409-457, 1906.
151. Sainte-Marie G, Leblond CP: Cytologic features and cellular migration in the cortex and medulla of thymus in the young adult *rat.* Blood *23:* 275-299, 1964.

152. Fichtelius KE: The mammalian equivalent to bursa Fabricii of birds. Exp Cell Res *46:* 231-234, 1967.
153. Yoffey JN : The lymphomyeloid complex. (CIBA Found Symph on Haemaopoiesis. Cell production and its regulation. Eds. Wolstenholme, GEW and O'Connor NC, Churchill, London, 1960), 1-42.
154. Yoffey JN: The role of the lymphocyte in haemopoiesis. (Abstr 8th Congress Eur Soc Haemat, Wein, 1961; Karger, Basel-New York, 1962), I,1.
155. Tachibana T, Suzuki M: Proceedings: Case study of T3-thyrotoxicosis associated with periodic paralysis and clinical observation of 17 cases of thyrotoxic paralysis encountered in the past. Nippon Naibunpi Gakkai Zasshi. *50:* 396, 1974.
156. Tachibana F, Imai Y, Kojima M: Development and regeneration of the thymus: the epithelial origin of the lymphocytes in the thymus of the mouse and chick. J Reticuloendothel Soc *15:* 475-496, 1974.
157. Moore MA, Owen JJ: Experimental studies on the development of the thymus. J Exp Med *126:* 715-726, 1967.
158. Moore MA, Owen JJ: Chromosome marker studies in the irradiated chick embryo. Nature *215:* 1081-1082, 1967.
159. Moore MA, Owen JJ: Chromosome marker studies on the development of the haemopoietic system in the chick embryo. Nature *208:* 956, 1965.
160. Owen JJ, Ritter MA: Tissue interaction in the development of thymus lymphocytes. J Exp Med *129:* 431-442, 1969.
161. Le Douarin N, Barq G: Use of Japanese quail cells as "biological markers" in experimental embryology. C R Acad Sci Hebd Seances Acad Sci D *269:* 1543-1546, 1969.
162. Le Douarin N: Details of the interphase nucleus in Japanese quail (Coturnix coturnix japonica). Bull Biol Fr Belg *103:* 435-452, 1969.
163. Le Douarin N, Teillet MA: Localization, by the method of interspecific grafts of the neural area from which adrenal cells arise in the bird embryo. C R Acad Sci Hebd Seances Acad Sci D *272:* 481-484, 1971.
164. Le Douarin N: Comparative ultrastructural study of the interphasic nucleus in the quail (Coturnix coturnix japonica) and the chicken (Gallus gallus) by the regressive EDTA staining method. C R Acad Sci Hebd Seances Acad Sci D *272:* 2334-2337, 1971.
165. Le Douarin N, Jotereau F: Embryologic origin of thymus lymphocytes in bird embryos. C R Acad Sci Hebd Seances Acad Sci D *276:* 629-632, 1973.
166. Deparis P, Jaylet A: Thymic lymphocyte origin in the newt Pleurodeles waltlii studied by embryonic grafts between diploid and tetraploid embryos. Ann Immunol (Paris), 1976
167. Jaylet A, Deparis P: An investigation into the origin of the thymocytes of the newt Pleurodeles waltlii using grafts between normal animals and animals with a marker chromosome. Dev Comp Immunol *3:* 175-180, 1979.
168. Nagata S: Electron microscopic study on the early histogenesis of thymus in the toad, Xenopus laevis. Cell Tissue Res *179:* 87-96, 1977.
169. Turpen JB, Volpe EP, Cohen N: Stem cell influx following the heterotopic transplantation of the thymic primordia between frog embryos. Dev Comp Immunol *1:* 255-262, 1977.
170. Volpe EP, Turpen JB: Lymphocyte differentiation and allograft reactivity. Experimental studies on the origin of thymic lymphocytes. Transplant Proc *9:* 785-788, 1977.
171. Volpe EP, Tompkins R, Reinschmidt D: Clarification of studies on the origin of thymic lymphocytes. J Exp Zool *208:* 57-66, 1979.
172. Tompkins R, Reinschmidt D, Volpe EP: Colonization of the thymic primordium by migratory lymphocytoblasts in amphibian embryos. Dev Comp Immunol *3:* 635-642, 1979.
173. Borum K: Pattern of cell production and cell migration in mouse thymus studied by autoradiography. Scand J Haematol *5:* 339-52, 1968.
174. Hemmingsson EJ, Alm GV: Migration of haemopoietic cells from the yolk sac to the thymus and the bursa of Fabricius in the chick embryo. Acta Pathol Microbiol Scand [A]. *81:* 79-84, 1973.
175. Metcalf D, Moore MA: Factors modifying stem cell proliferation of myelomonocytic leukemic cells in vitro and in vivo. J Natl Cancer Inst *44:* 801-808, 1970.
176. Bach JF, Dardenne M, Goldstein AL, Guha A, White A: Appearance of T-cell markers in bone marrow rosette-forming cells after incubation with thymosin, a thymic hormone. Proc Natl Acad Sci U S A *68:* 2734-2738, 1971.
177. Komuro K, Boyse EA: Induction of T lymphocytes from precursor cells in vitro by a product of the thymus. J Exp Med *138:* 479-482, 1973.
178. Komuro K, Boyse EA: In-vitro demonstration of thymic hormone in the mouse by conversion of precursor cells into lymphocytes. Lancet *1:* 740-743, 1973.
179. Komuro K, Boyse EA, Old LJ: Production of TL antibody by mice immunized with TL- cell populations. A possible assay for thymic hormone. J Exp Med. *137:* 533-536, 1973.
180. Reinherz EL, Hussey RE, Schlossman SF: A monoclonal antibody blocking human T cell function. Eur J Immunol *10:* 758-762, 1980.
181. Lanier LL, Cwirla S, Phillips JH: Genomic organization of T cell γ genes in human peripheral blood natural killer cells. J Immunol *137:* 3375-3377, 1986.
182. Lanier LL, Weiss A: Presence of Ti (WT31) negative T lymphocytes in normal blood and thymus. Nature *324:* 268-270, 1986.
183. Galili U, Seeley J, Svedmyr E, Klein E, Klein G, Weiland O: Blood lymphocytes in infectious mononucleosis share the following characteristics with activated T cells: natural attachment, stable E rosetting and glucocorticoid sensitivity. J Clin Lab Immunol *3:* 153-158, 1980.
184. Galili N, Galili U, Klein E, Rosenthal L, Nordenskjold B: Human T lymphocytes become glucocorticoid-sensitive upon immune activation. Cell Immunol *50:* 440-444, 1980.
185. Galili U, Klein E, Klein G, Singh Bal I: Activated T lymphocytes in infiltrates and draining lymph nodes of nasopharyngeal carcinoma. Int J Cancer *25:* 85-89, 1980.
186. Casali AM, Grossi CE, Manzoli Guidotti L, Manzoli FA: Cytophotometric characterization of cortical and medullary lymphocytes in the chicken bursa and thymus. Beitr Pathol *160:* 231-244, 1977.
187. Grossi CE, Casali AM, Bartoli S, Governa M, Manzoli FA: Separation and characterization of cortical and medullary bursal lymphocytes. Eur J Immunol *4:* 150-152, 1974.
188. Murray RG, Murray AS, Pizzo A: The fine structure of the thymocytes of young rats. Anat Rec *151:* 17-39, 1965.
189. Weiss L: Electron microscopic observations on the vascular barrier in the cortex of the thymus of the mouse. Anat Rec *145:* 413-438, 1963.
190. Gad P, Clark SL Jr: Involution and regeneration of the thymus in mice, induced by bacterial endotoxin and

studied by quantitative histology and electron microscopy. Am J Anat *122:* 573-605, 1968.
191. Raitsina SS, Kalinina II: Ultrastructure and antigens in differentiation of thymus lymphocytes in human embryogenesis. Biull Eksp Biol Med *88:* 470-473, 1979.
192. Haar JL: Light- and electron microscopy of *human* fetal thymus. Anat Rec *179:* 463-476, 1974.
193. Hwang WS, Ho TY, Luk SC, Simon GT: Ultrastructure of the rat thymus. A transmission, scanning electron microscope, and morphometric study. Lab Invest *31:* 473-487, 1974.
194. Raviola E, Karnovsky MJ: Evidence for a blood-thymus barrier using electron-opaque tracers. J Exp Med *136:* 466-498, 1972.
195. Bhalla DK, Karnovsky MJ: Surface morphology of mouse and rat thymic lymphocytes: an in situ scanning electron microscopic study. Anat Rec *191:* 203-220, 1978.
196. Stinson R, Glick B: Scanning electron microscopy of chicken lymphocytes: a comparative study of thymic, bursal, and splenic lymphocytes. Dev Comp Immunol *2:* 311-318, 1978.
197. Cantor H, Boyse EA: Functional subclasses of T lymphocytes bearing different Ly antigens. II. Cooperation between subclasses of Ly^+ cells in the generation of killer activity. J Exp Med *141:* 1390-1399, 1975.
198. Cantor H, Boyse EA: Functional subclasses of T-lymphocytes bearing different Ly antigens. I. The generation of functionally distinct T-cell subclasses is a differentiative process independent of antigen. J Exp Med *141:* 1376-89, 1975.
199. Katz DH, Dorf ME, Benacerraf B: Control of t-lymphocyte and B-lymphocyte activation by two complementing Ir-GLphi immune response genes. J Exp Med *143:* 906-918, 1976.
200. Germain RN, Benacerraf B: The involvement of suppressor T cells in Ir gene regulation of secondary antibody responses of primed (responder X nonresponder)F1 mice to macrophage-bound L-glutamic acid60-L-alanine30-L-tyrosine. J Exp Med *148:* 1324-1337, 1978.
201. Benacerraf B: A hypothesis to relate the specificity of T lymphocytes and the activity of I region-specific Ir genes in macrophages and B lymphocytes. J Immunol. 1978 *120:* 1809-1812, 1978.
202. Benacerraf B, Germain RN: The immune response genes of the major histocompatibility complex. Immunol Rev *38:* 70-119, 1978.
203. Cantor H, Weissman I: Development and function of subpopulations of thymocytes and T lymphocytes. Prog Allergy *20:* 1-64, 1976.
204. Blomgren H, Andersson B: Evidence for a small pool of immunocompetent cells in the mouse thymus. Exp Cell Res *57:* 185-192, 1969.
205. Aoki T, Hammerling U, De Harven E, Boyse EA, Old LJ: Antigenic structure of cell surfaces. An immunoferritin study of the occurrence and topography of H-2' theta, and TL alloantigens on mouse cells. J Exp Med *130:* 979-1001, 1969.
206. Grégoire KE, Goldschneider I, Barton RW, Bollum FJ: Ontogeny of terminal deoxynucleotidyl transferase-positive cells in lymphohemopoietic tissues of rat and mouse. J Immunol *123:* 1347-1352, 1979.
207. Bollum FJ: Terminal deoxynucleotidyl transferase as a hematopoietic cell marker. Blood *54:* 1203-1215, 1979.
208. Chang LM, Bollum FJ: Doxynucleotide-polymerizing enzymes of calf thymus gland. IV. Inhibition of terminal deoxynucleotidyl transferase by metal ligands. Proc Natl Acad Sci U S A *65:* 1041-1048, 1970.
209. Chang LM: Development of terminal deoxynucleotidyl transferase activity in embryonic calf thymus gland. Biochem Biophys Res Commun *44:* 124-131, 1971.
210. Jenkinson EJ, Van Ewijk W, Owen JJ: Major histocompatibility complex antigen expression on the epithelium of the developing thymus in normal and nude mice. J Exp Med *153:* 280-292, 1981.
211. Owen JJ, Jenkinson EJ: Embryology of the lymphoid system. Prog Allergy *29:* 1-34, 1981.
212. Weissman IL: Thymus cell maturation. Studies on the origin of cortisone-resistant thymic lymphocytes. J Exp Med *137:* 504-510, 1973.
213. Weissman IL, Levy R: In vitro cortisone sensitivity of in vivo cortisone-resistant thymocytes. Isr J Med Sci *11:* 884-888, 1975.
214. Haaijman JJ, Micklem HS, Ledbetter JA, Dangl JL, Herzenberg LA, Herzenberg LA: T cell ontogeny. Organ location of maturing populations as defined by surface antigen markers is similar in neonates and adults. J Exp Med *153:* 605-614, 1981.
215. Ledbetter JA, Rouse RV, Micklem HS, Herzenberg LA: T cell subsets defined by expression of Lyt-1,2,3 and Thy-1 antigens. Two-parameter immunofluorescence and cytotoxicity analysis with monoclonal antibodies modifies current views. J Exp Med *152:* 280-95, 1980.
216. Ledbetter JA, Herzenberg LA: Xenogeneic monoclonal antibodies to mouse lymphoid differentiation antigens. Immunol Rev *47:* 63-90, 1979.
217. Micklem HS: B lymphocytes, T lymphocytes and lymphopoiesis. Clin Haematol *8:* 395-419, 1979.
218. Kamarck ME, Gottlieb PD: Expression of thymocyte surface alloantigens in the fetal mouse thymus in vivo and in organ culture. J Immunol *119:* 407-415, 1977
219. Owen JJ, Raff MC: Studies on the differentiation of thymus-derived lymphocytes. J Exp Med *132:* 1216-1232, 1970.
220. Kasai H, Nishimura S: Hydroxylation of the C-8 position of deoxyguanosine by reducing agents in the presence of oxygen. Nucleic Acids Symp Ser *12:* 165-167, 1983.
221. Bodey B: Differentiation and homing of T lymphocytes in the human thymus in the course of prenatal development. Experimental Hematology *14:* 163, 1983.
222. Reinherz EL, Schlossman SF: Current concepts in immunology: Regulation of the immune response--inducer and suppressor T-lymphocyte subsets in human beings. N Engl J Med *303:* 370-373, 1980.
223. Kirby ML, Bockman DE: Neural crest and normal development: A new perspective. Anat Rec *209:* 1-6, 1984.
224. Le Douarin NM: The neural crest. Cambridge Univ. Press, London, pp. 1-259, 1982.
225. Le Douarin N, Jotereau FV: Tracing of cells of the avian thymus through embryonic life in interspecific chimeras. J Exp Med *142:* 17-40, 1975.
226. Le Lievre CS, Le Douarin NM: Mesenchymal derivatives of the neural crest: Analysis of chimeric quail and chick embryos. J Embryol Exp Morphol *34:* 125-154, 1975.
227. Weston JA: The migration and differentiation of neural crest cells. Adv.Kabelitz D (1991) Selective activation of $\delta/\gamma+$ T cell clones by single anti-CD2 antibodies. J Exp Med *173:* 297-304, 1970.
228. Elands J, Resink A, de Kloet ER: Oxytocin receptors in the rat thymic gland. Eur J Pharmacol *151:* 345-346, 1988.
229. Geenen V, Defresne M-P, Robert F, Legros J-J, Franchimont P, Boniver J: The neurohormonal thymic microenvironment: immunocytochemical evidence that

thymic nurse cells are neuroendocrine cells. Neuroendocrinology *47:* 365-368, 1988.
230. Geenen V, Legros J-J, Franchimont P, Baudrihaye M, Defresne MP, Boniver J: The neuroendocrine thymus: coexistence of oxytocin and neurophysin in the human thymus. Science *232:* 308-310, 1986.
231. Geenen V, Legros J-J, Franchimont P, Defresne MP, Boniver J, Ivell R, Richter D: The thymus as a neuroendocrine organ: Synthesis of vasopressin and oxytocin in human thymic epithelium. Ann NY Acad Sci *496:* 56-66, 1987.
232. Gillitzer R, Pilarski LM: In situ localization of CD45 isoforms in the human thymus indicates a medullary location for the thymic generative lineage. J Immunol *144:* 66-74, 1990.
233. Giunta M, Favre A, Ramarlio D, Grossi CE, Corte G: A novel integrin involved in thymocyte-thymic epithelial cell interactions. J Exp Med *173:* 1537-1548, 1991.
234. Godfrey DI, Izon DJ, Tucek CL, Wilson TJ, Boyd RL: The phenotypic heterogeneity of mouse thymic stromal cells. Immunology *70:* 66-74, 1990.
235. Gold DP, Clevers H, Alarcon B, Dunlap S, Novotny J, Williams AF, Terhorst C: Evolutionary relationship between the T3 chains of the T-cell receptor complex and the immunoglobulin supergene family. Proc Natl Acad Sci USA *84:* 7649-7653, 1987.
236. Salomon DR, Mojcik CF, Chang AC, Wadsworth S, Adams DH, Coligan JE, Shevach EM: Constitutive activation of integrin α 4 β 1 defines an unique stage of human thymocyte development. J Exp Med *179:* 1573-1584, 1994.
237. Fabris N, Mocchegiani E, Mariotti S, Pacini F, Pinchera A: Thyroid function modulates thymic endocrine activity. J Clin Endocrinol Metab *62:* 474-478, 1986.
238. Trenkner E, Hoffmann MK: Defective development of the thymus and immunological abnormalities in the neurological mouse mutation "straggerer". J Neurosci *6:* 1733-1737, 1986.
239. Argiolas A, Melis MR, Stancampiano R, Mauri A, Gessa GL: Hypothalamic modulation of immunoreactive oxytocin in the rat thymus. Peptides *11:* 539-543, 1990.
240. Cross RJ, Campbell JL, Markesbery WR, Roszman TL: Transplantation of pituitary grafts fail to restore immune function and to reconstitute the thymus glands of aged mice. Mechanisms Ageing Develop *56:* 11-22, 1990.
241. Elands J, Resink A, de Kloet ER: Neurohypophyseal hormone receptors in the rat thymus, spleen and lymphocytes. Endocrinology *126:* 2703-2710, 1990.
242. Fabris N, Mocchegiani E, Muzzioli M, Provinciali M: Interazioni immuno-neuroendocrine nello sviluppo e nell'invecchiamento. Gerontologia *70:* 679-686, 1989.
243. Moll VM, Lane BL, Robert F, Geenen V, Legros JJ: The neuroendocrine thymus. Abundant occurrence of oxitocin-, vasopressin-, and neurophysin-like peptides in epithelial cells. Histochemistry *89:* 385-390, 1988.
244. Robert F, Geenen V, Schoenen J, Burgeon E, de Groote D, Defresne MP, Legros JJ, Franchimont P: Colocalization of immunoreactive oxytocin, vasopressin and interleukin-1 in human thymic epithelial neuroendocrine cells. Brain Behavior Immunity *5:* 102-115, 1991.
245. Spangelo BL, Ross PC, Judd AM, McLeod RM: Thymic stromal elements contain an anterior pituitary hormone-stimulating activity. J Neuroimmunology *25:* 37-46, 1989.
246. Nagano M, Kelly PA: Tissue distribution and regulation of rat prolactin receptene expression. Quantitative analysis by polymerase chain reaction. J Biol Chem *269:* 13337-13345, 1994.
247. Su Y-L, Ho KL, Dalakas MC, Mutchnick MG: Localization of immunoreactive thymosin α 1 in astrocytes of normal human brain. Annu Neurol *26:* 277-280, 1989.
248. Wekerle H, Linington C, Lassmann H, Meyermann R: Cellular immune reactivity within the CNS. TINS *9:* 271-277, 1986.
249. Weller GL: Development of the thyroid, parathyroid and thymus glands in man. Contrib Embryol Carnegie Inst *24:* 95-138, 1933.
250. Rao LV, Cleveland RP, Ataya KM: Alterations in thymic and bone marrow lymphocyte subpopulations in GnRH agonist treated prepubertal female mice. J Rep Immunol *25:* 167-184, 1993.
251. Li L, Zhou JH, Xing ST, Lau BH: Thymus-neuroendocrine-liver pathway. Med. Hypotheses *41:* 470-472, 1993.
252. Roson E, Garcia-Caballero T, Heimer EP, Felix AM, Dominguez F: Cellular distribution of prothymosin α and parathymosin in rat thymus and spleen. J Histochem Cytochem *38:* 1889-1894, 1990.
253. Andrews P., Boyd R.L, Shortman K: The limited immunocompetence of thymocytes within murine thymic nurse cells. Eur J Immunol *15:* 1043-1048, 1985.
254. Farr A: Epithelial heterogeneity in the murine thymus: a cell surface glycoprotein expressed by subcapsular and medullary epithelium. J Histochem Cytochem *39:* 645-653, 1991.
255. Hammar JA: Zur Histogenese und Involution der Thymusdruse. Anat Anz *27:* 23-30 and 41-89, 1905.
256. Hammar JA: Zur Frage der Histogenese der Thymusdruse. Zentrbl Pathol *33:* 505-513, 1923.
257. Haynes BF: The human thymic microenvironment. Adv Immunol *36:* 81-142, 1984.
258. Haynes BF: Use of monoclonal antibodies to identify antigens of human endocrine thymic epithelium. In: Thymic hormones and lymfokines: biochemical chemistry and clinical applications. AL, Goldstein (ed), Plenum Press, pp. 43-56, 1984.
259. Haynes BF, Crummett B, Scearce RM, Wolf LS, Tuck D, Singer KG: Summary of reactivity patterns of the thymic epithelial panel of antibodies in human skin and thymus. In: Leucocyte Typing III., AJ, McMichael (ed), Oxford University Press, Oxford, pp. 262-269, 1987.
260. Haynes BF, Denning SM, Le PT, Singer KH: Human intrathymic T cell differentiation. Semin Immunol *2:* 67-77, 1990.
261. Kyewski BA, Kaplan HS: Lymphoepithelial interactions in the mouse thymus: phenotypic and kinetic studies on thymic nurse cells. J Immunol *128:* 2287-2292, 1982.
262. Landry D, Lafontaine M, Barthelemy H, Paquette N, Chartrand C, Pelletier M, Montplaisir S: Human thymic dendritic cell-thymocyte association: Ultrastructural cell phenotype analysis. Eur J Immunol *19:* 1855-1860, 1989.
263. Lobach DF, Scearce RM, Haynes BF: The human thymic microenvironment: Phenotypic characterization of Hassall's bodies with the use of monoclonal antibodies. J Immunol *134:* 250-257, 1985.
264. Takacs L, Marinova Ts: The ontogeny of human thymic epithelium specific antigens as defined by monoclonal antibodies. Thymus *15:* 147-152, 1990.
265. van Ewijk W: Cell surface topography of thymic microenvironment. Lab Invest *59:* 579-590, 1988.
266. Vollger LW, Tuck DT, Springer TA, Haynes BF, Singer KH: Thymocyte binding to human thymic epithelial cells is inhibited by monoclonal antibodies to CD-2 and LFA-3 antigens. J Immunol *138:* 358-363, 1987.

267. Weissman IL: Thymus cell maturation. Studies on the origin of cortisone resistant thymic lymphocytes. J Exp Med *137:* 504-510, 1973.
268. Wekerle H, Ketelsen U-P, Ernst M: Thymic nurse cells: Lymphoepithelial cell complexes in murine thymuses: Morphological and serological characterization. J Exp Med *151:* 925-944, 1980.
269. Yu E, Lee I: Reticular network of the human thymus. J Korean Med. Science *8:* 431-436, 1993.
270. Zinkernagel RM: Thymus and lymphohemopoietic cells: their role in T cell maturation, in selection of T cells, H-2 restriction specificity and H-2 linked Ir gene control. Immunol Rev *42:* 224-270, 1978.
271. Zinkernagel RM: The thymus: its influence on recognition of self major histocompatibility antigens by T-cells and consequences for reconstitution of immunodeficiency. In: Cooper MD, Lawton AR, Milscher PA, Mueller-Eberhard HJ (eds) Immunodeficiency. Springer, Berlin Heidelberg New York pp. 171-181, 1979.
272. Sotzik F, Rosenberg Y, Boyd AW, Honeyman M, Metcalf D, Scollay R, Wu L, Shortman K: Assessment of CD4 expression by early T precursor cells and by dendritic cells in the human thymus. J Immunol *152:* 3370-3377, 1994.
273. van Ewijk W: Immunohistology of lymphoid organs. Curr Opinion Immunol *1:* 954-965, 1989.
274. van Vliet E, Melis M, van Ewijk W: Monoclonal antibodies to stromal cell types of the mouse thymus. Eur J Immunol *14:* 524-529, 1984.
275. van Vliet E, Melis M, van Ewijk W: Immunohistology of thymic nurse cells. Cell Immunol *87:* 101-109, 1984.
276. Kampinga J, Berges S, Boyd RL, Brekelmans P, Colic M, van Ewijk W, Kendall MD, Ladyman H, Nieuwenhuis P, Ritter MA, Schuurman H-J, Tournefier A: Thymic epithelial antibodies: Immunohistological analysis and introduction of nomenclature. Thymus *13:* 165-173, 1989.
277. Colic M, Jovanovic S, Mitrovic S, Dujic A: Immunohistochemical identification of six cytokeratin-defined subsets of the rat thymic epithelial-cells. Thymus *13:* 175-185, 1989.
278. Takeuchi T, Kuzuhara H, Tamura H, Hiramine C, Hojo K, Yamamoto H: A 60-kDa thymic epithelial cell surface protein as a potent molecule mediating the cellular interaction with immature T cells. Cellular Immunol *146:* 324-334, 1993.
279. Fernandez JG, Sanchez AJ, Melcon C, Chamorro CA, Garcia C, Paz P: Development of the chick thymus microenvironment: a study by lectin histochemistry. J Anatomy *184:* 137-145, 1994.
280. Wilson A, Held W, MacDonald HR: Two waves of recombinase gene expression in developing thymocytes. J Exp Med *179:* 1355-1360, 1994.
281. Bjorkmann P, Saper MA, Samaoui B, Bennett WS, Strominger JL, Wiley DC: Structure of the human class I histocompatibility antigen, HLA-A2. Nature (London) *329:* 506-512, 1987.
282. Claverie J-M, Kourilsky P: The peptidic self model: A reassessment of the role of the major histocompatibility complex molecules in the restriction of the T-cell response. Annu Immunol *137:* 425-442, 1987.
283. Kronenberg M, Siu G, Hood L, Shastri N: The molecular genetics of the T-cell antigen receptor and T-cell antigen recognition. Annu Rev Immunol *4:* 529-591, 1986.
284. Littman DR: The structure of the CD4 and CD8 genes. Annu Rev Immunol *5:* 561-584, 1987.
285. Sprent J, Schaefer M: Antigen-presenting cells for unprimed T cells. Immunology Today *10:* 17-19, 1989.
286. Toribio ML, de la Hera A, Regueiro JR, Marquez C, Marcos MAR, Bragado R., Arnaiz-Villena A, Martinez AC: α /β heterodimeric T-cell receptor expression early in thymocyte differentiation. J Mol Cell Immunol *3:* 347-362, 1988.
287. Zelechowska MG, Lecomte J, Potworowski EF: Immunocytochemical localization of a thymic medullary epithelial glycoprotein. Scand J Immunol *23:* 561-565, 1986.
288. Aspinall R, Kampinga J: A novel lymphocyte surface antigen recognized by the monoclonal antibody HIS45. Thymus *13:* 245-252, 1989.
289. MacDonald HR, Budd RC, Howe RC: A CD3- subset of $CD4^-CD8^+$ thymocytes: a rapidly cycling intermediate in the generation of $CD4^+CD8^+$ cells. Eur J Immunol *18:* 519-523, 1988.
290. Peault B, Chen C-LH, Cooper MD, Barbu M, Lipinski M, Le Douarin NM: Phylogenetically conserved antigen on nerve cells and lymphocytes resembles myelin-associated glycoprotein. Proc Natl Acad Sci USA *84:* 814-818, 1987.
291. Raedler A, Raedler E, Becker WM, Arndt R, Thiele HG: Subcapsular thymic lymphoblast expose receptors for soybean lectin. Immunology *46:* 321-328, 1982.
292. Reinherz EL, Schlossman SF: The differentiation and function of human T lymphocytes: a review. Cell *19:* 821-827, 1980.
293. Ruiz P, Dunon D, Hesse B, Imhof BA: T-lymphocyte precursors adhere to thymic endothelium. In: Lymphatic tissues and in vivo immune responses. Imhof BA, et al. (eds) Marcel Dekker, New York, pp. 963-965, 1991.
294. Saalmuller A, Hirt W, Reddehase MJ: Phenotypic discrimination between thymic and extrathymic $CD4^-CD8^-$ and $CD4^+CD8^+$ porcine T lymphocytes. Eur J Immunol *19:* 2011-2016, 1989.
295. Scollay R, Andrews P, Boyd R, Shortman K: The role of the thymic cortex and medulla in T cell differentiation. Adv Exp Med Biol *186:* 229-234, 1985.
296. Scollay R, Chen W-F, Shortman K: The functional capabilities of cells leaving the thymus. J Immunol *132:* 25-30, 1984.
297. Scollay R, Shortman K: Thymocyte subpopulations: an experimental review, including flow cytometric cross-correlations between the major murine thymocyte markers. Thymus *5:* 245-295, 1983.
298. Scollay R, Smith J, Stauffer V: Dynamics of early T cells: Prothymocyte migration and proliferation in the adult mouse thymus. Immunol Rev *91:* 129-157, 1986.
299. Scollay R, Weissman I: T cell maturation: thymocyte and thymus migrant subpopulations defined with monoclonal antibodies to the antigens Lyt-1, Lyt-2 and ThB. J Immunol *124:* 2841-2844, 1980.
300. Scollay R, Wilson A, D'Amico A, Kelly K, Egerton M, Pearse M, Wu L, Shortman K: Developmental status and reconstitution potential of subpopulations of murine thymocytes. Immunol Rev *104:* 81-120, 1988.
301. Serra HM, Ledbetter JA, Krowka JF, and Pilarski LM: Loss of CD45 (Lp 220) represents a post-thymic T cell differentiation event. J Immunol *140:* 1435-1441, 1988.
302. Snow PM, van de Rijn M, Terhorst C: Association between the human thymic differentiation antigens T6 and T8. Eur J Immunol *15:* 529-532, 1985.
303. Stutman O: Ontogeny of T cells. Clin Immunol Allergy *5:* 191-234, 1985.
304. Swat W, Ignatowicz L, von Boehmer H, Kisielow P: Clonal deletion of immature $CD4^+8^+$ thymocytes in suspension culture by extrathymic antigen presenting cells. Nature (London) *351:* 150-153, 1991.

305. Toribio ML, Martinez AC, Marcos MAR, Marquez C, Cabrero E, de la Hera A: A role for T3+4-6-8- transitional thymocytes in the differentiation of mature and functional T-cells from human prothymocytes. Proc Natl Acad Sci USA *83:* 6985-6988, 1986.
306. van Dongen JJM, Hooijkaas H, Comans-Bitter M, Hahlen K, de Klein A, van Zanen G.E. van't Veer MB, Abels J, Benner R: Human bone marrow cells positive for terminal deoxynucleotidyl transferase (TdT), HLA-DR and a T cell marker may represent prothymocytes. J Immunol *135:* 3144-3150, 1985.
307. Weiss A, Manger B, Imboden J: Synergy between the T3/antigen receptor complex and T44 in the activation of human T cells. J Immunol *137:* 819-825, 1986.
308. Zinkernagel RM, Callahan GN, Althage A, Cooper S, Streilein JW, Klein J: On the thymus in the differentiation of H-2 self-recognition by T cells: evidence for dual recognition? J Exp Med *147:* 882-896, 1978.
309. Zinkernagel RM, Callahan GN, Klein J, Dennert G: Cytotoxic T cells learn specificity for self H-2 during differentiation in the thymus. Nature (London) *271:* 251-253, 1978.
310. Rothenberg EV: Death and transfiguration of cortical thymocytes: a reconsideration. Immunol Today *11:* 116-119, 1990.
311. Smith CA, Williams GT, Kinston R, Jenkinson EJ, Owen JJT: Antibodies to CD3/T cell receptor complex induce death by apoptosis in immature T cells in thymic cultures. Nature (London) *337:* 181-184, 1989.
312. Kizaki H, Tadakuma T: Thymocyte apoptosis. Microbiol Immunol *37:* 917-925, 1993.
313. Pellicciari C,Manfredi AA, Bottone MG, Schaack V, Barni S: A single-step staining procedure for the detection and sorting of unfixed apoptotic thymocytes. Eur J Histochem *37:* 381-390, 1993.\
314. Fujii Y, Okumura M, Takeuchi Y, Inada K, Nakahara K, Matsuda H, Tsujimoto Y: Bcl-2 expression in the thymus and periphery. Cellular Immunol *155:* 335-344, 1994.
315. Korsmeyer SJ, Shutter JR, Veis DJ, Merry DE, Oltvai ZN: Bcl-2/Bax: a rheostat that regulates an anti-oxidant pathway and cell death. Sem Cancer Biol *4:* 327-332, 1993.
316. Moore NC, Anderson G, Williams GT, Owen JJ, Jenkinson EJ: Developmental regulation of bcl-2 expression in the thymus. Immunology *81:* 115-119, 1994.
317. Aguillar LK, Aguillar-Cordova E, Cartwright J, Jr, Belmont JW: Thymic nurse cells are sites of thymocyte apoptosis. J Immunol *152:* 2645-2651, 1994.
318. Debatin KM, Suss D, Krammer PH: Differential expression of APO-1 on human thymocytes: implications for negative selection? Eur J Immunol *24:* 753-758, 1994.
319. Bockman DE, Kirby ML: Dependence of thymus development on derivates of neural crest. Science *223:* 498-500, 1984.
320. Amarante-Mendes JG, Couture C, Potworowski EF: Identification of a 16 kDa thymocyte membrane glycoprotein involved in the thymocyte/thymic medullary epithelial cell interaction. Immunol Letters *37:* 47-52, 1993.
321. Grey HM, Sette A, Buus S: How T cells see antigen? Scientific Amer Sept Issue, 38-46, 1989.
322. Peters JP, Borst J, Oorschot V, Fukuda M, Krahenbulh O, Tschopp J, Slot JW, Geuze HJ: Cytotoxic T lymphocyte granules are secretory lysosomes, containing both perforin and granzymes. J Exp Med *173:* 1099-1109, 1991.
323. Doyle C, Strominger JC: Interaction between CD4 and class II MHC molecules mediates cell adhesion. Nature (London) *330:* 256-259, 1987.
324. Dunon D, Kaufman J, Salomonsen J, Skjoedt K, Vainio O, Thiery JP, Imhof BA: T cell precursor migration towards β 2-microglobulin is involved in thymus colonization of chicken embryos. EMBO J *9:* 3315-3322, 1990.
325. Eichmann K, Jonsson JI, Falk I, Emmrich F: Effective activation of resting mouse T lymphocytes by cross-linking submitogenic concentration of the T cell antigen receptor with either Lyt-2 or L3T4. Eur J Immunol *17:* 643-650, 1987.
326. Colic M, Matanovic D, Hegedis L, Dujic A: Heterogeneity of rat thymic epithelium defined by monoclonal anti-keratin antibodies. Thymus *12:* 123-130, 1988/89.
327. Godfrey DI, Kennedy J, Gately MK, Hakimi J, Hubbard BR, Zlotnik A: IL-12 influences intrathymic T cell development. J Immunol *152:* 2729-2735, 1994.
328. Zapata JM, Pulido R, Acevedo A, Sanchez-Madrid F, de Landazuri MO: Human CD45RC specificity. A novel marker for T cells at different maturation and activation stages. J Immunol *152:* 3852-3861, 1994.
329. Edelman GM: Cell adhesion molecules in the regulation of animal form and tissue pattern. Annu Rev Cell Biol *2:* 81-116, 1986.
330. Pilarski LM, Turley EA, Shaw AR, Gallatin WM, Laderoute MP, Gillitzer R, Beckman IG, Zola H: FCM46, a cell protrusion-associated leukocyte adhesion molecule-1 epitope on human lymphocytes and thymocytes. J Immunol *147:* 136-143, 1991.
331. Kozono H, White J, Clements J, Marrack P, Kappler J: Production of soluble MHC class II proteins with covalently bound single peptides. Nature (London) *369:* 151-154, 1994.
332. Sebzda E, Wallace VA, Mayer J, Yeung RS, Mak TW, Ohashi PS: Positive and negative thymocyte selection induced by different concentrations of a single peptide. Science *263:* 1615-1618, 1994.
333. Hunkapiller T, Hood L: Growing immunoglobulin gene superfamily. Nature (London) *323:* 15-16, 1986.
334. Kyewski BA, Fathman CG, Rouse RV: Intrathymic presentation of circulating non-MHC antigens by medullary dendritic cells. An antigen-dependent microenvironment for T cell differentiation. J Exp Med *163:* 231-246, 1986.
335. Pullen AM, Bill J, Kubo RT, Marrack P, Kappler JW: Analysis of the interaction site for the self superantigen Mls-1(a) on T cell receptor Vβ. J Exp Med *173:* 1183-1192, 1991.
336. von Boehmer H, Haas W, Jerne NK: Major histocompatibility complex-linked immune-responsiveness is aquired by lymphocytes of low-responder mice differentiating in thymus of high responder mice. Proc Natl Acad Sci USA *75:* 2439-2442, 1978.
337. von Boehmer H, Hafen K: Minor but not major histocompatibility antigens of thymus epithelium tolerize precursors of cytolytic T cells. Nature (London) *320:* 626-628, 1986.
338. von Boehmer H, Schubiger K: Thymocytes appear to ignore class I major histocompatibility complex antigens expressed on thymus epithelial cells. Eur J Immunol *14:* 1048-1052, 1984.
339. Sprent J, Gao E-K, Kanagawa O, Webb SR: T-cell selection in the thymus. In: Immune system and cancer. Hamaoka T. et al. (eds), Japan Sci Soc Press, Taylor & Francis LTD., Tokyo-London, pp. 127-136, 1989.
340. von Boehmer H: Self recognition by the immune system. Eur J Biochem *194:* 693-698, 1990.

341. Dalloul AH, Arock M, Fourcade C, Hatzfeld A, Bertho JM, Debre P, Mossalayi MD: Human thymic epithelial cells produce interleukin-3. Blood *77:* 69-74, 1991.
342. Page DM, Alexander J, Snoke K, Appella E, Sette A, Hedrick SM, Grey HM: Negative selection of $CD4^+$ $CD8^+$ thymocytes by T-cell receptor peptide antagonists. Proc Natl Acad Sciences USA *91:* 4057-4061, 1994.
343. Ong CJ, Chui D, Teh HS, Marth JD: Thymic CD45 tyrosine phosphatase regulates apoptosis and MHC-restricted negative selection. J Immunol *152:* 3793-3805, 1994.
344. Pinto VB, Rock KL: Extracellular matrix-induced stimulation of a CD4-8- thymic T cell line. Cellular Immunol *155:* 144-155, 1994.
345. Vacchio MS, Papadopoulos V, Ashwell JD: Steroid production in the thymus: implications for thymocyte selection. J Exp Med *179:* 1835-1846, 1994.
346. Fleischer B: Superantigens. APMIS *102:* 3-12, 1994.
347. Anderson G, Owen JJ, Moore NC, Jenkinson EJ: Thymic epithelial cells provide unique signals for positive selection of $CD4^+$ $CD8^+$ thymocytes in vitro. J Exp Med *179:* 2027-2031, 1994.
348. Fernandez E, Vicente A, Zapata A, Brera B, Lozano JJ, Martinez C, Toribio ML: Establishment and characterization of cloned human thymic epithelial cell lines. Analysis of adhesion molecule expression and cytokine production. Blood *83:* 3245-3254, 1994.
349. Nakafusa Y, Goss JA, Flye MV: Viable antigen-presenting cells are required for induction of intrathymic tolerance. Transpl Proceed *26:* 845-846, 1994.
350. Lundberg K, Shortman K: Small cortical thymocytes are subject to positive selection. J Exp Med *179:* 1475-1483, 1994.
351. Pullen AM, Marrack P, Kappler JW: The T-cell repertoire is heavily influenced by tolerance to polymorphic self-antigens. Nature *335:* 796-801, 1988.

Chapter 5

The Hassall's Bodies

Abstract: During the thymic ontogenesis, the HBs appear when lymphopoiesis is already established and the cortex, medulla and the cortico-medullary junction are capable of conducting the positive and negative selection of T lymphocytes undergoing progressive maturation. The HBs are structurally organized from RE cells, which usually undergo hypertrophy prior to their inclusion in the outer cell layer of the corpuscles. In our observations the greatest developmental progression and main cell-tissue organization of the HBs was observed between 45 and 54 days of gestation in dogs and between 6 and 10 lunar months in humans. The cellular microenvironment of the thymic medulla is composed of networks of cell types, of a variety of origins, and all of them may participate in the construction of growing, progressive HBs. Histochemically, we detected a rich content of basic non-histone proteins, PAS positive substance (glycogen) and acid mucopolysaccharides within the bodies. Employing the histological stain of Pasini and immunocytochemical methods with monoclonal antibodies (MoABs) AE2 and AE3, high molecular weight (56.5 to 67 kD) basic keratins were defined in human HBs. Employing a panel of MoABs developed against thymic RE cell surface antigens, we observed immunoreactivity localized to the outer cell layer of the HBs with MoABs TE8, TE16 and TE19, while the centrally located cells reacted positively with TE15 and TE19. Immunoreactivity in human skin, employing the TE8, TE16 and TE19 MoABs was also observed in the epidermal granulosa cell layer, while TE15 reacted with cells of the stratum corneum. The presence of endocrine, peptide secreting RE cells within the HBs was defined with the use of MoAB A_2B_5, which binds to the GQ ganglioside. The hypertrophied, physiologically active RE cells of the peripheral cell layer of the HBs reacted positively with medium to strong intensity when stained with MoABs UJ127.11, J1153, A_2B_5, 215.D11, and 275.G7. We also observed the expression of transforming growth factor-β type II receptors in HBs. The recently detected expression of the homeobox gene products B3, B4, and C6, transcription factors involved in developmental processes related to hematopoiesis within HBs provides further evidence that HBs are important functional components of the RE network of the thymus which provide developing thymocytes with paracrine and juxtacrine signals to ensure their proper functional maturation during the intrathymic lymphopoiesis. Our transmission electron microscopical (TEM) studies on HBs determined the existence of groups of RE cells connected to one another by desmosomes. We went on to further observe long cytoplasmic processes originating from medullary RE cells and directly contacting thymic T lymphocytes and accessory antigen presenting cells (macrophages, dendritic cells, interdigitating cells, Langerhans cells, etc.) by the use of scanning electron microscopy (SEM). Thus, our results indicate that the HBs are unique, antigenically distinct, functionally active, multicellular components of the nonlymphocytic, cellular microenvironment of the thymic medulla, and participate in the physiological activities of the prenatal and adult thymus. Future immunohistochemical and genetic studies will clarify the exact origin of the various non-lymphatic thymic cells participating in the determination of the particular physiological activities, progressive growth, and the terminal cell differentiation within the HBs.

Key words: Thymic Ontogenesis in Vertebrates; Origin of the Hassall's Bodies (HBs) - Hypotheses; Reticulo-Epithelial Cell Origin of HBs; Capillary Involvement; Morphological Observations; Scanning Electron microscope (SEM) Study; Transmission Electron microscope (TEM) Observation; Immunocytochemistry; Streptavidin-biotin antigen detection technique; Mouse Anti-Human Monoclonal Antibody (MoAB); Immunophenotype (IP); Thymic Organ and Tissue Cultures; Total Body and Local Thymic Irradiation; Thymic involution.

1. HISTORICAL OVERVIEW

The Hassall's bodies (HBs) are the most characteristic, organ-specific structures of the mammalian thymus. However, thymic research, as demonstrated in the literature, has increasingly moved towards a better understanding of T lymphocyte maturation and differentiation as a main function of the thymus (1-4). Lymphopoiesis is critically related to the function of accessory cells and the epithelial cell network of the thymus, organized in so-called "*lympho-epithelial symbiosis*" (5-7). Therefore, the development, histogenesis and function of HBs are of great importance. The literature, however, has yet to be enriched with a thorough review article on this subject (our present attempt is unique in this regard).

Phylogenetic observations have concluded that HBs are not found in vertebrates of the lower orders, e.g. in cyclostomes they are completely absent. In addition, a detailed ultrastructural study has determined that mammalian like HBs are also absent in reptiles (8). On the other hand, HBs are always present in the thymic medulla of mammals.

Arthur Hill Hassall, in his *The Microscopic Anatomy of the Human Body, in Health and Disease* (9), first described the HBs as large cellular structures of the thymic medulla, and named them *"cytoblasts or mother cells"* in 1849. His account mirrored his teacher's (Simon) opinion:

> "in early life there exists in the thymus gland no trace whatever of complete cells, that is only in later life that nucleated cells are formed, and that these are developed out of the granular corpuscles, which are alone present in the gland in the first years of its existence. The same statements are applied to the thyroid body....the opinions entertained by Mr. Simon, in his "Essay on the Thymus" have not been overstated...the dotted corpuscles are undoubtedly quite similar to those which we have recognized as becoming the nuclei of cells in the thyroid body, and in other organs."

Hassall finished the chapter concerning organized fluids with the words: *"the corpuscles of the thymus are mixed up with those corpuscles ... in every way similar to the white corpuscles of the blood, but very distinct from the true cell corpuscles of the gland"*. To the present day, these bodies bear his name and the expression *"the concentric corpuscles of Hassall"* was introduced into histological nomenclature by Friedrich Henle in 1849 (10). These characteristic thymic structures were also described by Rudolf Virchow, and thus they are sometimes referred to as Hassall-Virchow corpuscles (11, 12).

2. HYPOTHESES CONCERNING THE ORIGIN OF HBS

During the past century, a great number of researchers have observed and speculated on the development, histogenesis, and possible functions of HBs in various vertebrates. The origin of HBs has been the subject of considerable controversy and the question still remains without a definitive answer.

In this historical view, we must first point out that the great majority of published hypotheses are speculative, far from scientific reality. Secondly, all significant cellular changes occurring in the thymic microenvironment during the ontogenesis of this primary lymphatic organ will be considered in our discussion.

Early investigators of thymic histogenesis, such as Bruch (13), Friedleben (14), Maurer (15), Magni (16), Stöhr (17, 18) and Fulci (19) proposed a possible thymocytic cell origin of HBs. According to them, HBs originate from thymocytes or *"small thymic cells"* located in the thymic epithelial microenvironment: the precursor, lymphoblastic cells of future thymocytes, which have migrated into the thymus during development. Ghika (20) and Goldner (21, 22) also shared this view, that the thymocytes were, in some ways, connected to the origin and development of HBs.

The contention that HBs arose from the endothelial cells of degenerated small blood vessels in the thymus was proposed by Cornil and Ranvier (23) and supported in a number of articles written by Afanassiew (24), Nusbaum and Machowski (25), Jordan and Horsley (26), and Juba and Mihalik (27). Jordan and Looper (28) observed the ontogenesis of the thymus in box-turtles and criticized this hypothesis.

Since 1860, the thymic literature has emerged from studies concerning the normal pre- and postnatal histogenesis of the thymic microenvironment in a variety of vertebrates by a number of authors: Kölliker (29, 30), His (31, 32), Toldt (33), Krause (34), Stieda (35), Capobianco (36), Pierson (37), von Ebner (38), Prymak (39), Wallish (40), Goodall (41), Klose & Vogt (42), Rabl (43), Hart (44), Marine (45), Scammon (46), Woollard (47), Webster (48), and others. This research converged on the proposal that the HBs originate exclusively from the cells of the primordial, primary and pure epithelial thymic anlage, and that, later on, the HBs represent remnants of these omnipotent embryonal epithelial cells. This was also a speculative hypothesis, not taking into account the dynamic nature of postnatal thymic histogenesis, the regularly occurring new formation of HBs, and the significant differentiation processes taking place within the complex thymic cellular microenvironment.

Schambacher (49), Marine (45) and even Shier (50) considered that the HBs are parts of cellular substances enveloping the primary thymic epithelium in the early thymic duct or that the HBs differentiate secondarily from different tubular structures. Hammar (51) criticized both hypotheses based upon his embryological observations: "Als ein Gewinn der Forschung in der Periode vor 1910 ist zu verzeichnen, daß die Auffaßung von den Hassallschen Körpern als Überreste epithelialen Thymusanlage oder überhaupt als starre unveränderliche Organbestandteile als aus der Diskussion abgeführt betrachtet werden kann, was natürlich nicht besagt, daß einzelne Rückfälle in altere Auffaßungen nicht auch in der hier zu berücksichtigenden Periode Vorkommen." Juba and Mihalik (27) expressed the importance of the primary hypertrophy of the RE cells of the HB central core:

> "Die Bildung scheint damit anzufangen, daß eine Zelle - seltener ein paar nebeneinander liegende Zellen - des Retikulums bedeutend an Größe gewinnen und eine mehr sphärische Gestalt annehmen. Indem sie bei dieser Vergrößerung die Nachbarzellen erreichen, werden diese durch den Wachstumsdruck seitwärts verschoben und lagern sich der zentralen Zelle schalenförmig an. Auch die peripheren Zellen werden bald hypertrophisch, wodurch das Gebilde eiter wachst und neue Zellen an seine Peripherie angefügt werden. Bei dem so fortschreitenden Wachstum können zwei oder mehrere Körperchen einander erreichen, sich einander an fügen und nun als eine einheitliche Bildung weiterwachsen. Es entstehen also zusammengesetzte Hassallsche Körper."

Jordan and Looper (28) were the first to report the presence of *"unicellular bodies of Hassall"*, identified as individual hypertrophied RE cells with late fibrotic changes in their cytoplasm in 1928.

To study the cellular origin, Kostowiecki (52) reconstructed the HBs of 5, 6, 7, and 9 months old human fetuses. He supported both hypotheses concerning HB genesis. His contribution lies in the discovery of additional thymic microscopic bodies which were *"mixed in origin"*, and have concentric lamellate structure. These bodies were organized around a degenerated blood vessel, surrounded by hypertrophied RE cells. In Weller's words (53):

> "Reconstructions of the corpuscles thus demonstrated numbers of irregularly shaped structures scattered throughout the thymus, most of them, however, at a distance from the blood-vessels. These findings agree well with the conception of Hassall's corpuscles as epithelial pearls."

The so-called *true* HBs usually contain one or two enlarged and partially destroyed central RE cells, and, occasionally, such RE cells also constitute their peripheral core (54).

An unusual interpretation is found in the works of Norris (55), who observed human thymic ontogenesis in serial total body tissue sections and reported the migration of ectodermally derived epithelial cells from the cervical sinus into the thymic medulla. He suggested that the HBs of the human thymus are formed only by the migration and proliferation of these ectodermal epithelial cells. The possible ectodermal derivation of HBs was also discussed by Zotterman (56) in pigs and later by Massart (57, 58), who described typical HBs in both the ectodermal and endodermal parts of the thymus of *Vesperrugo pipistrellus*.

Winiwarter (59) concluded that the HBs mostly originate from the cells of the RE meshwork, but that there was no denying the participation of connective tissue elements following some kind of thymic involution.

Dustin and Baillez (60) argued the myoepithelial derivation of the thymic bodies:

> "Les cellules myo-epitheloides, en general, isolees ou groupees en corps hassaliens, ne derivent pas de l'ebauche epitheliale primitive du thymus; elles sont d'origine mesodermique; ce sont des elements de la lignee conjonctive penetrant a l'interieure du thymus et y subissant des metaplasies particulieres a cet organe. Ces cellules peuvent, ou bien etre eparses dans la parenchyme, ou bien se grouper en corpuscules, ou bien, tres frequemment appartenir a des formations peritheliales et se metaplasier lors de l'involution vasculaire."

Other authors have demonstrated two ways of interpreting the origin and histogenesis of HBs. These descriptive studies *concocted* another way of differentiation of the RE cells:

> "The epithelial cells of the primordium of the thymus differentiate into two kinds of cells. One is the reticular cell and the other is myoid cell or striated muscle. These cells or muscles are destined to degenerate. To remove these degenerated cells or muscles, the neighboring reticular cells appear around them. They acquire the phagocytic character to clear the degenerated substances. Due to a large quantity of the degenerated substances, those reticulum cells also get degenerated. Then new phagocytic reticular cells appear again around the degenerated reticular cells. In this way the concentric arrangement of the degenerated reticular cells are formed. As the result the same corpuscle consists of at most about four to five layers. If two or more degenerated myoid cells near close to each other, the polynuclear Hassall's corpuscle comes into existence. The degeneration of the myoid cells in the frog and snake does not form concentric Hassall's corpuscles. The unicellular Hassall's corpuscles of the previous investigators are probably equivalent to the above mentioned rudimental

> myoid cells or Hassall's corpuscles in early stage."

The observations of Imai *et al.* (61) were significant because they defined the presence of another cell type within HBs, and they considered a possible dual cell origin and histogenesis of HBs. According to Imai and co-workers, in humans and other mammals, the kernel of the HB represents a rudiment of a myoid cell. In amphibian (frog), reptilian (snake, if the HBs have developed), and avian (fowl) thymic tissues, the central part of the HB has been found to be composed of some type of striated muscle. As stated previously, the other cellular components of HBs are the RE cells of the thymic medullary microenvironment. More specifically, RE cells surround the degenerated myoid (i.e. also described as "giant") cell(s), forming the concentric structure of the HBs. According to Wassjutotschkin (62) it is difficult to accept the origin of HBs exclusively from myoid cells; this is also our opinion. Under certain conditions, however, it is possible that such cells comprise the central cellular core of HBs. During our detailed observation of the prenatal histogenesis of the human thymus, we defined that the number of myoid cells in HBs is minimal under physiological conditions and, considering the basic principles of human embryology and thymic ontogenesis, the above mentioned histogenetic mechanisms are simply unacceptable.

Hammar, in his review book on the thymus (51), interpreted the humoral experiments of Kliwanskaja-Kroll (63) in which rats were fed thyroid tissue:

> "...bei ihren sorgfältigen, auch numerisch bearbeiteten Versuchen mittels Hyperthyreodisierung von weißen Ratten mit Schilddrüse vom Rind positiven Erfolg verzeichnet. Die Gesamtzahl der Hassall'schen Körper war nach 10 tägiger Verfütterung mit Thyreoideasubstanz 2.5 mal und die Zahl der kleinsten etwa 5mal die des Kontrolltieres mit immer abnehmendem Übergewicht der höheren Körpergruppen...bei den späteren Tieren (nach 30, 60, 120 und 210 Tagen) lagen die Gesamtzahlen niedriger (1.6 resp. 1.1, 1.3 und 1.4 mal der Kontrollwerte) und auch mit die einzelnen Gruppenwerte wären im allgemeinen niedriger und mit der größten Anzahlsteigerung nicht länger in die Gruppe der kleinsten Körper verlegt."

He also held that the connective tissue elements migrating into the thymus could not be the precursor cells of HBs. Hammar (51) also gave an important interpretation of some earlier publications:

> "Dustin (1923) bezeichnet die Konjuctivo-vaskulären Züge des Thymus als einer Metaplasie fähig, durch welche die Hassall'schen Körper und Zilienzysten ihren Ursprung geben. Dies gilt zunächts für die Katze, Abrauch für weiße Mäuse, deren Thymus sehr arm an Hassall'schen Körpern und bisweilen ganz frei von ihren ist. Ähnliche Bilder von einen Ursprung der Körper aus venösen Zügen hat er beim Axolotl, beim Frosch, bei der Katze, bei der weißen Maus und beim Menschen gesehen....Florentin (1929) und Florentin und Weis (1932) findet in einem isoliert liegenden Thymus Krötchen Hassallsche Körper, die hautig um eine kleine Arterie mit sehr deutlichem Endothel zentriert sind. In gewissen Fällen war das Arterienlumen thrombosiert. Es fitt Dustins Aussicht über den vaskularen Ursprung der Körper bei mit dem von Pensa gemachten Zusatz, daß eine oder mehrere Schichten von epithelialen Markzellen hinzukommen können, sodaß der Körper solchenfalls gemischten Ursprungs ist."

Bastenie (64, 65), studying the human thymus gland, found that HBs originated from hypertrophied perithelial (mesodermal) cells. He also reported that during chronic, high grade thymic involution, the so-called *"process of diffuse Hassallisation"* is preceded by the solid infiltration of the thymic reticulum by mesenchymally derived connective tissue elements and that HBs originate from these cells as part of thymic tissue reorganization following the pathobiological condition of involution.

The origin of HBs from the cells of the mesenchymal RE meshwork was proposed in the works of Paulitzky (66), Watney (67, 68), and Tourneux & Verdun (69). These authors described the presence of HBs invaded by mesenchymally derived reticular cells, which could also represent components of the complex cellular architecture of HBs. Another interesting hypothesis was proposed by Grégoire (70) when he expressed that HBs were derived only from the cells of the *"reticular cell"* network of mesodermal origin. Specifically, he stated that they originated and specialized from capillary pericytes: *"tout simplement l'expression de l'action metaplasiante du milieu thymique, action d'une extreme energie qui peut se faire sentir aussi bien sur les elements mesodermiques que sur les elements epitheliaux."* Béclère & Pigache (71) and Pigache & Worms (72) stated that large cells formed unicellular HBs and, in later stages of their histogenesis, the concentric layers of degenerated leukocytes around them formed the well-known, characteristic HBs.

Kingsbury (73-75), it seems, did not agree that the thymic medulla is in a state of constant degeneration, and arguing with himself wrote:

> "If they are endothelial hypertrophies of obliterated vascular channels, the exceedingly great abundance of such degeneration occurring practically throughout the life history of the organ presents so unique a feature that it raises at once an inquiry as to the underlying conditions and the reasons for their occurrence."

Two extrathymic, but cell based hypotheses have also been published. Beard (76) concluded that the HBs were degenerated parts of the parathyroids, taking their places as accessory glands within the thymus. Another embryologically unrealistic and completely false thesis was proposed by Popoff (77, 78) when he stated that the HBs were degenerated rudimentary organs of the genital and follicular cells.

According to Goldner (21, 22), after adrenaline injection in dogs, one normal and one hypertrophied RE cell appeared as unicellular HBs. He described the initial formation of HBs to be most dependent on the appearance of these unicellular bodies and, furthermore, that all RE cells in HBs are degenerative in nature. The same histogenetic mechanism was reported by Goldner following x-ray irradiation (probably around a necrotic mass).

The most acceptable hypothesis remains the one indicating a cellular origin of HBs from the cells of the thymic medullary RE network, and has been recognized following the systematic and detailed observations of Hammar (79-82), the *father of modern thymology* (Fig. 1 and Fig. 2). Others with the same opinion in classical thymic literature include Prenant (83), Bell (84), Dantschakoff (85), Mietens (86), Maximow (87), Pappenheimer (88,89), Jaffe & Plavska (90), and this view is shared by all relatively recent systematic studies by Smith (91), Bodey (92, 93), and Liberti *et al.* (94).

Why do the HBs appear to be round? Hammar (79) considered that from the very beginning of the formation of HBs, the hypertrophy of one or two RE cells, which eventually impinge upon and fuse with other RE cells causes the concentric arrangement of the RE cells in the medulla around these, RE cells undergoing expansion due to hypertrophy and also characterized by slow atrophy. The lamellated structure of the HBs also results from this continued expansion.

Kingsbury (74) proposed a purely mechanistic view of the origin and development of HBs:

> "The epithelial (epitheloid) foundation, however, grows under very peculiar conditions due to 1) the fact that it is no longer a surface tissue and 2) its conversion into a cytoreticulum in correlation with the 'lymphocytic' invasion and proliferation. These conditions are, I think, clearly responsible for the varied forms the growth takes producing the epithelial or epithelioid structures, which have come to be known as corpuscles of Hassall.... the Hassall corpuscles are expressions of 1) growth in a confined space-a result of absence of a free surface; and 2) of a disjunctive growth differentiation due to extreme reticulation that the epithelium has undergone. That they are epithelial in origin is easily determined. The comparison of thymic corpuscles and other epithelial growths in confined space, such as epithelial pearls, is justified. The interpretation presented here requires no assumption of an adaptive significance underlying thymic corpuscle formation." Several years later in his paper concerning HBs, he added: "...in the thymus lobule, the differential growth becomes centripetal with cell conglomerates forming centrally as expressions of a differential growth in a confined space....in the thymus of the teleostome fish where the surface value of the thymic epithelium is retained (placoid thymus) thymic corpuscles are lacking - as would be expected" (95).

3. COMPARATIVE ONTOGENETIC CHRONOLOGY

3.1 Appearance of the very first HB in various vertebrate species

The first HBs appear on the 27th day of incubation in chickens (87, 96) and at various other stages in mammals: in the 19 mm long fetus in rats, the 29 mm ones in guinea pigs, and in the 70 mm ones in cats and calves. In dog fetuses, we detected the first HB on the 38th day of gestation (97). As for humans, Hammar (80, 98) and Bodey (92, 93) have described the initial appearance of HBs in 65-70 mm long (in second to third week of the third intrauterine lunar month) fetuses, while Lobach and co-workers (99) detected them at the beginning of the first week of the 4th lunar month.

3.2 Patho-morphological cell changes in HBs during thymic histogenesis

Hammar (51) described necrotic changes in the centrally located RE cells within the HBs:

> "Es handelt sich dabei erstlich um eine Degeneration den bei verschiedenen Spezies einen verschiedenen Charakter hat, habe ich schon im vorigen Bericht hervogehoben, und ich habe keinen Grund auf diese Seite der Sache zurückzukommen. Auch ist es als leicht verständlich, daß die zentrale Degeneration bei

> Spezies, deren Hassall'sche Körper nie eine etwas beträchtlichere Größe erreichen eine wenig augenfallige Rolle spielt, während sie bei anderen, wie bein Menschen, wo die Körper u.U. bis über 1.0 mm Durchmesser erreichen, durch ihren Umfang und ihre wechselnde Beschaffenheit die Aufnerksamkeit in hohem Grade auf sich zieht....Das besondere Aussehen, welches ein Hassall'sche Körper darbietet, wen sein größter, zentraler Abschnitt aus kernlosen, schuppenartigen oder schollenformigen kernlosen Zellen besteht, und bloß seine Randpartie lebende kernführende Zellen aufueist, hat Veranlassung dazu gegeben, solche Körper als besandere degenerierte Formen aufzufassen und zu bezeichnen.....Wenn ein Körper eine gewisse Größe erreicht, präsentiert er sich als degeneriert."

Hammar (51, 100) morphologically characterized six distinct categories of HBs:

> "1) zystische Körper; 2) leukozyteninvadierte Körper; 3) partiell und 4) total verkalkte Körper; 5) entkalkte Körper nach vorheriger partieller und 6) nach vorheriger totaler Verkalkung. Etwas anders ist die Gruppierung von Ishibashi und Watanabe (1926) ausgefallen. Sie unterscheiden fünf Arten von Hassall'schen Körpern: zellige, hyaline, verkalkte, kolloidale und zusammengesetzte Formen. Diese verschiedenen Arten treten je nach dem Grade der Alters- und Krankheitsveränderungen der Thymus in einen bestimmten prozentuellen Verhältnis auf. Zellige Formen finden sich in allen Lebensaltern, hyaline vorwiegend bei Kindern über 10 Jahren, verkalkte nach dem 2. Lebensjahr, kolloidale meistens im mittleren Alter, nur spärlich im Kindesalter....Die zystischen Körper sind schon durch ihre relativ mäßige Größe und ihren Wandbelag von konzentrisch geschichteten oder einfachen platten Epithelien meistens unschwer von Zysten allerlei Herkunft zu unterscheiden. Selten sind sie leer, häufig enthalten sie Leukozyten. Wenn die Menge dieser häufig zusammengebackenen Zellen, nach Augenmaß beurteilt, einen geringeren Durchmesser hat als die Hälfte des Zystenraumes, wurde das Gebilde noch als zystisch gerechnet. Wenn der leukozytäre Inhalt hin gegen das betreffende Maß übertraf, wurde der Körper als Leukozyten invadiert bezeichnet. Da fast alle Körper Leukozyte enthalten, war eine zwar willkürliche Grenze unumgänglich."

Relying on our systematic observations of over 200 human and 70 dog embryos and fetuses during ontogenesis, with emphasis on the histogenesis and histochemistry of the thymus (92, 93), we have concluded that HBs originate and develop from the medullary RE cells and regularly undergo a variety of morphological changes. The first HBs could be observed on the 38th day of gestation in dogs and at the middle or end of the third lunar month (second or third week) in human fetuses. Thymic organogenesis can be regarded as commencing development from this moment. The HBs appear when lymphopoiesis has already been established and the cortical and medullary areas are recognizable. In our observations, the size of the bodies varied from 10 to 1000 μm. The main development of the HBs occurred between 45 and 54 days of gestation in dogs and between 6 and 10 lunar months of intrauterine life in humans. Histogenesis of new HBs was also observed. This process always began with the nuclear hypertrophy of RE cells (marked changes of the nucleus). This hypertrophy then spread over the cytoplasm as well, resulting in a highly marked acidophilia and gradually the involvement of an increasing number of cells could be observed (in extreme cases up to three-hundred RE cells). Meanwhile, degenerative changes were detected in the central cores of the HBs. These lead to karyorrhexis and eventually to karyolysis while the HBs underwent *circular* development, taking up the smallest space possible (92, 93). During the subsequent development of the bodies, they become infiltrated by granulocytes, thymocytes, and macrophages. Some of these granulocytes are eosinophilic in human fetuses, but not in dogs. During this stage, termed advanced or *"full development"* of HBs by a complex chemotactic mechanism, new cells from the medullary RE meshwork are drawn actively from the HBs' immediate surroundings, which in our judgment, not only aids constructively (progressive growth of HBs), but is also crucial in the situation dependent function of the cellular complex of the HBs (Fig. 3-5). The medulla also takes an active role in this complex building process, which we called a *"stromal or medullar RE tissue reaction."* This reaction could be observed in the medullary RE cells, adjacent to the HBs. Some large bodies have peripheral cell layers, which are attached to one another. The central core of such HBs is filled with cellular debris that represents the product of cell disintegration. During progressive development, this partially necrotic mass is always separated from the outer, viable cell layers.

3.3 Cellular disintegration within the HBs

Kostowiecki (54) described three patterns of *HB regression* in guinea pigs, each ending with their complete disintegration. As the regression continues, the remaining thymocytes, granulocytes, RE cells, and macrophages participate in this process. Small vacuoles are present within the degenerating macrophages. The contents of the HBs gradually shrink, but the regression is by no means uniform. Mature macrophages display great phagocytic activity. Among such cells, small binucleate cells and several thymocytes occur. In the terminal stage, only macrophages and large cells, homogeneously stained, remained in the HBs. When the HBs contained numerous RE cells, we were able to detect a second type of degeneration. The RE cells represent the stroma in the HBs, showing maximum hypertrophy, and are grouped in concentric cell layers around the central core. The nuclei of these cells disappear and their cytoplasm forms hyaline-like laminae. Later, these contents undergo fragmentation and destruction. This detritus, combined with that from degenerating thymocytes, gives the HB a homogeneous, shrunken appearance. The cell detritus is finally phagocitized. In large bodies, the central core might begin to hyalinize (101). When large numbers of thymocytes are present, the third type of HB regression could be present. The HB, while undergoing this kind of regression, rarely shows single RE cells and an intracorpuscular fluid (thymocyte or RE cell secretion), characteristics which are always present otherwise.

4. HISTOCHEMISTRY AND HISTOENZYMOLOGY OF HBS

Histochemically, the central core and the RE cells from the body's periphery stain pink to red and the hypertrophic RE cells give a granular purplish staining with the PAS-reaction. The granular staining was partly reduced following β-amylase digestion, but the central part of the body showed no reduction in its strong stainability (92, 93, 97, 102, 103). There is a rich content of basic non-histone protein located in the central core of the HB and in the circular peripheral RE cell layers, showing various stages of cellular hypertrophy or destruction (92, 93). The extra- and intracellular localization of the neutral lipids and phospholipids as small droplets over the central part of the HBs prove the degenerative process within the bodies. Some of the outer concentric lamellae of the HBs demonstrated a strong reaction for phospholipids employing Baker's (104, 105) method. Gaudecker and Schmale (103) consider that these structures might be identical to the vacuolated cells demonstrated with TEM. In some fetuses in advanced stages of ontogenesis (after the 8th lunar month of intrauterine development), we have been able to demonstrate the accumulation of calcium salts in the central core contents of the HBs, signaling the conclusion of a cell degeneration process.

The existence of a thymic *"enzyme"* was first reported by Jones (106), but it is clear that he was referring to a secreted thymic hormone. Practically all modern histoenzymatic thymic observations have been conducted on vertebrate frozen tissue. Such investigations have been carried out on human fetal and postnatal thymus, obtained after abortions and open heart surgeries (92, 93, 103, 107-112). Other authors conducted their thymic studies in domestic fowls (113-120), as well as other experimental animals such as rabbits (115, 122-124), rats (108, 115, 124, 126, 127), dogs (115, 122), cats (115), guinea pigs (102, 115, 124-126), pigs (121), sheep (115), and oxen (115).

In agreement with other histoenzymatic studies, we observed the strongest enzymatic reactivity in the peripherally located RE cells of the concentrically laminated tissue structure of human thymic HBs. This was especially true for hypertrophic ones in which the enzymatic activity was found to be even greater. Positive activity for α-naphthylacetate-esterase was detected in the peripherally located RE cells within the HBs. In the central core of the HBs, this enzyme was absent. Acid phosphatase activity was also detected within the HBs, and, once again, the enzymatic activity was stronger in the peripherally located, often hypertrophied RE cells. During involution, a weak 5-nucleotidase activity was determined exclusively in HBs, while the surrounding thymic medulla and cortex showed no reactivity. A weak activity was observed in HBs for alkaline phosphatase, but quite strong activity was detected in the adjacent RE cells and capillaries of the surrounding medulla. Because most of the HBs were nearly completely negative for alkaline phosphatase, some authors have assumed that the 5-nucleotidase activity was tissue-specific and not due to other phosphatases (128, 129). Other studies in human thymuses have described the presence of acid phosphatase, esterases, β-glucorinidase and alkaline phosphatase in RE cells comprising the HBs, but the absence of 5'-nucleotidase and ATP-ase activity (130). A direct correlation between the distribution of succinic dehydrogenase (SDH) and of sulphydryl groups was observed, and this demonstrated that the sulphydryl groups are necessary for the enzymatic activity of SDH (124).

D'Anna (120, 131) observed the thymic tissue of *Gallus domesticus*, *Carvia cobaya* and *Ovis* for α-naphthylesterase (ANE) and leucine-aminopeptidase

(LAP) activity. The two enzymatic reactions were positive when the bodies were already concentric and lamellar. LAP was localized in the HB of *Gallus domesticus* from the earliest growth stage, but it is now known that in the initial stages of growth, the bodies require a high level of activity from all enzymes. The process is not the same in different species, but the elements implicated in the genesis of HBs were marked by both ANE and LAP. The differences in species-related patterns may be related to the massive initial enzymatic activity in the formative process, which at least in children does not seem to be destined for fulfillment.

In human thymic tissue observed histoenzymologically, a weak activity was determined in the HBs for the enzymes necessary in the normal metabolism of cells, such as SDH and lactate dehydrogenase (LDH). The activity was relatively stronger in the hypertrophied RE cells (132). The activity of glucoso-6-phosphatdehydrogenase, a key enzyme of the pentose shunt, was observed to be weak. Non-specific esterase, on the other hand, showed a moderate activity in the RE cells of the HBs. The β-D-glucorodinase yielded moderate positive results in the central, degenerating part of the so-called regressive type HBs. These hydrolytic enzymes are of lysosomal nature. The activities of ANE and adenosine-triphosphatase (ATPase) were indicated by dark areas in the peripheral layer of hypertrophied RE cells. The enzymes NADH-diaphorase, NADPH-diaphorase, and NADP-IDH also showed marked activity. The reaction of the intramitochondrial enzyme NAD-IDH was moderate. The DDD-reaction of Barnett & Seligman (133) demonstrated sulphydryl and disulphid groups of the protein in the central lamellae of the HBs after reduction of the SS-groups in 0.5 M-thioglycolic acid (pH 8.0). The peripheral layer of hypertrophied RE cells stained weakly pink, but all of the lymphatic elements were characterized by a uniform, pink stain. The reaction for disulphide groups was very weak. A recent study has identified the presence of acid cysteine proteinase inhibitor (ACPI or cystatin A) in the thymic medullary RE cells and in the HBs (134). The epithelial character of cystatin A positive thymic cells has been established immunocytochemically employing anti-cytokeratin and anti-epithelial membrane antigen (EMA) MoABs. ACPI is a protein, which is also present in the skin and in the dendritic accessory cells of lymphatic tissues.

In conclusion, during our systematic study on 212 human embryos and fetuses we were able to define positive reactivity for all observed enzymes in the peripherally located cells of the HBs, comprised of hypertrophied RE cells (92, 93, 135). Therefore we may conclude that these cells are in a state of very high biological, i.e. metabolic and humoral, activity. On the other hand, as was already mentioned above, in their advanced stage of development, the central core of HBs were often filled with cellular debris, partly necrotic, disintegrated cellular elements, in which the metabolic process had almost ceased. Histochemically we also detected the presence of basic non-histone protein, PAS positive substance (glycogen) in the outer cell layer of HBs (Fig. 22) and acid mucopolysaccharides, which were confined to only the central core. The expression of the histone H1 gene (H1.1) was studied in several human tissues employing *in situ hybridization* (ISH) and *Northern blot analysis* (136). It was revealed that this gene was present in the thymus and in the testes. ISH experiments detected a cell-type-specific expression of the histone H1.1 gene in thymic HBs. Acid mucopolysaccharides were detected by special stainings, which demonstrated their presence with a deep blue color, arranged in the central concentric lamellae. Naturally, non-sulfated mucosubstances were also present because the staining with alcian blue at pH 1.0 was negative. As we already mentioned above the intra- and extracellular, centrally located lipid granules determined the presence of active cell destruction within the HBs.

5. IMMUNOFLUORESCENCE AND IMMUNOHISTOCHEMICAL OBSERVATIONS

The rapid advance in antibody production technology, especially the invention of novel MoABs has opened the gate to a new era of histology on the molecular level. The immunohistochemically detectable antigenic epitopes of certain cell lineage specific macromolecules have aided our understanding of the true origin and tissue distribution of a number of thymic cells. During the past three decades, two-hundred chemical groups of the thymus have been detected in HBs employing immunofluorescent techniques. Indirect immunofluorescence was observed in children and adult thymic tissue after treatment with antisera of active cases of pemphigus vulgaris (137). Fluorescent staining via binding of pemphigus autoantibodies to the intercellular substance (ICS) in HBs, was seen in all investigated thymic tissues. The large deposits of ICS within the HBs produced a fluorescence pattern often detected in the so-called *"epithelial pearls"* in well differentiated squamous cell carcinomas (138). Presence of ICS was never demonstrated in any normal mammalian tissue, except for the squamous epithelium (139). The medullary RE cells not adjacent to the HBs never demonstrated presence of ICS immunofluorescence. These findings represent strong evidence that the HBs are composed of squamous, medullary RE cells. The three well characterized thymic hormones, Thymulin or Facteur Serique Thymique [FTS (140-142)], thymosin-α and thymopoietin (143) have also been detected in HBs. These results have led to the conclusion that HBs possess an active secretory role and thereby may participate in the intrathymic maturation of T-lymphocytes. Furthermore, it is significant that immunocytological evidence for the simultaneous presence of all three thymic hormones in the same RE cells of the human thymic microenvironment has also been defined and published (144).

Drenckhahn and co-workers (145) employed anti-smooth-muscle myosin and anti-actin antibodies for immunofluorescent microscopic observation of the human thymus. The concentrically arranged cells comprising HBs demonstrated strong immunoreactivity with both antibodies, but the intense immunofluorescence also defined antigen expression in some RE cells. Epidermis specific antigens have also been identified in thymic RE cells and in the peripheral layers of HBs (146). Laster and co-workers (147) have established the presence of specific keratins involved in early and advanced stages of epidermal keratinocyte maturation in thymic RE cells. The immunohistochemical characteristics of the human thymic cellular microenvironment have also been observed with extensive panels of poly- and monoclonal antibodies (99, 148-157). Obviously, cell lineage specific immunoreactivity of various antibodies has served well in assessing cell differentiation (maturation) pathways and/or the origin of cell types. Thymic RE cells, both cortical and medullary, are epithelial in origin and nature (e.g. they contain tonofilaments and desmosomes) and the reticular appearance of these cells have not been sufficient to confirm their mesenchymal derivation. The endodermal region of the third pharyngeal pouch, the place of origin for medullary RE cells, is frequently arranged in a duct-like manner and undergoes keratinization and transformation into squamous epithelia (158). Lobach and co-investigators (99) developed four MoABs (anti-TE8, anti-TE15, anti-TE16, and anti-TE19) that selectively demonstrated immunoreactivity with human HBs. TE8 and TE16 reacted with the hypertrophied cells of the outer cell layer of HBs. MoAB TE19 stained the entire cellular content of HBs, as well as the medullary RE cells adjacent to the HBs. More than 90% of the HBs displayed immunoreactivity with MoAB TE15. Three of these antibodies, TE8, TE16, and TE19 also reacted positively with the *stratum granulosum* while TE15 reacted with the *stratum corneum* of human skin. Employing these same antibodies, we also defined the HBs as antigenically distinct parts of the thymic secretory cell microenvironment. It seems that immunoelectron microscopy on thymic thin sections may be the best method to identify stromal cell IPs (159). The employment of MoABs against Mac-1 and Mac-2, which were developed against macrophage antigens, as well as a MoAB against MHC class II antigen determined the IP differences between macrophages and RE cells and bone marrow derived thymic stromal cells. The macrophages were Mac-1$^+$, Mac-2$^+$, and only 50% of them showed Ia positivity. The cortico-medullary interdigitating cells (IDC) were characterized by a Ia$^+$, Mac-1$^+$ and Mac-2$^+$ IP (160, 161), while RE cells displayed cell surface immunoreactivity with Ia, but neither Mac-1 nor Mac-2. These results strongly suggest that macrophages and IDCs are derived from a common precursor stem cell. A recent study has described the use of a novel panel of four MoABs to classify subsets of cells within the thymic microenvironment, two for medullary RE cells and HBs (His-39 and RMC-20) and two for cortical RE cells (His-37 and RMC-17). On the 15th day of gestation, MHC class I and class II molecules can also be identified on thymic RE cells, although they reach their highest density around birth (162).

In our immunocytochemical observations, MoAB UJ 308 has been defined to react positively with cell surface receptors on immature myeloid cells and mature leukocytes. A strong immunoreactivity employing this MoAB was observed in the HBs in frozen tissue sections. Therefore, it is possible that certain leukocytes may contribute to the formation of HBs. The central core of the HBs also demonstrated strong expression of antigens detectable with MoABs UJ 308 and UJ 223.8 and this reactivity was independent of the presence of necrotic cellular changes. The hypertrophied, physiologically active RE cells of the peripheral cell layer of the HBs reacted positively with medium to strong intensity when stained with MoABs UJ127.11, J1153, A_2B_5, 215.D11, and

275.G7. These results further suggest that HBs are not exclusively degenerative structures (163). The employment of MoAB A_2B_5 defined the presence of active endocrine, peptide secreting RE cells associated with or embedded in the structure of the HBs.

In fetal thymic tissue, over 90% of the hypertrophied RE cells that were located in the outer core of the HBs expressed TGF-β type II receptors (TGF-βIIRs). A markedly decreased proportion of RE cells within the HBs of postnatal thymuses demonstrated presence of TGF-βIIR: only 10 to 50% during the early postnatal period which decreased further to between 1 and 10% during the first year of age (164). High density of this growth factor receptor was also established in medullary RE cells adjacent to the HBs.

The homeobox genes are a family of nuclear regulatory genes encoding transcription factors, which possess the capability of both autoregulation as well as activation and inactivation of other genes. The recently detected expression of the homeobox gene products B3, B4, and C6, transcription factors involved in developmental processes related to hematopoiesis within HBs provides further evidence that HBs are important functional components of the RE cellular microenvironment of the thymus which provide developing thymocytes with paracrine and juxtacrine signals to ensure their proper functional maturation during the intrathymic lymphopoiesis (165, 166). Their presence during the postnatal thymic histogenesis (up to 21 years of age) represents immunocytochemical proof that active intrathymic lymphopoiesis and selection remain very much intact (Figs. 19 and 20).

6. TEM AND SEM OBSERVATIONS OF HBS

Ultrastructural observations have led to thorough descriptions of the structure of HBs. Two types of medullary RE cells, keratinized and vacuolated, are present on days 14-15 of intrauterine development in rats. These two cells subsequently differentiate into hypertrophied RE cells and into small HBs (162). Small HBs, comprised of a central cell surrounded by flattened RE cells have been detected in these studies (167-170). Centrally located flattened, concentrically arranged, disintegrated cellular elements without nuclei have been observed on a regular basis in HBs (171-173). Some of these cells were separated by a widened intercellular space lined by short microvilli. These structures showed secondary developmental changes and a cleft-like space around the periphery of several HBs. The presence of vacuolated cells around the central core of the HBs also represents a typical TEM finding. Peripherally, there was a layer of hypertrophied RE cells with large electron lucent nuclei and a *spongy* cytoplasm, containing bundles of filaments. The central lamellae were tightly packed with filaments. The cells in these regions were observed to have a thickened plasma membrane with a thickened inner leaflet. The RE cells in the periphery contained large, pole nuclei with one or two nucleoli (174, 175). Their cytoplasm was filled with free ribosomes, sparsely scattered elements of the rough endoplasmic reticulum, and numerous mitochondria. A Golgi apparatus was rarely observed. Bundles of tonofilaments, β-granules of glycogen, vacuoles of varying sizes, and numerous granules of different electron density were also present. Glycogen deposits appeared more frequently in the most peripheral cells of the HBs. A very interesting finding is that these outer cells had only a few tonofilaments, while in those located centrally, the tonofilaments assembled into tonofibril bundles. Frequently, the cells of the HBs were connected by cytoplasmic processes and desmosomes. Keratohyaline deposits were detected in the more centrally located cells. In some hypertrophied RE cells, membrane-bounded granules about 0.2 μm in diameter were found. They contained a fine granular substance and resembled mucous granules described by Parakkal (176), Hirokawa (177), and Lavker and co-workers (178). Within regressive HBs, vacuoles appeared in central core cells during advanced stages of cellular disintegration, and the cells were more widely separated from each other.

Our TEM observations support the light microscopic and ultrastructural interpretation of other authors, that the HBs are formed from keratinized stratified squamous epithelium (92, 93, 102, 103, 179). The changes detected in the cellular components of the HBs correspond closely to those described in keratinizing epithelia of the epidermis (180, 181). All stages of cell differentiation were observed, ranging from normal RE cells, to RE cells showing characteristics of the cells of the stratum spinosum, and finally, to sheets of keratin-like material. The desmosomal contacts of peripheral hypertrophied RE cells with the central concentric lamellae revealed the same structure as described between the cells of the stratum granulosum and stratum corneum of the epidermis (174). The desmosomal junctions of the central concentric lamellae correspond to the same junctional structures connecting the cells comprising the stratum corneum of the epidermis. Tonofilament synthesis was observed in the hypertrophied RE cells. The small membrane-bounded granules with a lamellar substructure which were detected in some cells of HBs were found to be ultrastructurally identical to the lamellar or membrane coating granules of the spinous cells of various keratinizing epithelia (103, 174, 180, 182, 183), and their presence has been regarded as strong ultrastructural evidence of keratinization. The numerous, large desmosomes, well developed bundles of cytoplasmic tonofibrils, and the large and characteristic keratohyalin granules, described above, lend further ultrastructural evidence to support the hypothesis that the HBs are composed of cells of keratinized squamous stratified epithelium. We also observed long cytoplasmic processes originating from medullary RE cells and directly contacting thymic T lymphocytes and accessory antigen presenting cells (macrophages, dendritic cells, interdigitating cells, Langerhans cells, etc.) by the employment of scanning electron microscopy (SEM).

7. THYMIC OBSERVATIONS UNDER VARIOUS EXPERIMENTAL CONDITIONS

Several authors observed the atrophy of the thymus in malnourished children (79, 184-186). Jonson (187) in his experiments on starving animals suggested the existence of two types of HBs during acute thymic involution, *progressive* and *regressive*:

> "Eine Andeutung davon, daß die einzelligen Formen durch die Degeneration der Zellen verschwanden, habe ich nicht beobachten können. Ich bin daher geneigt anzunehmen, daß ihr relativ schnelles Verschwinden ganz einfach auf einen atrophischen Prozess beruht, wodurch sie den Character gewöhnlicher Retikulumzellen wiedererhalten.....Von den mehrzelligen Körperchen scheinen die kleineren Formen wohl auch teilweise durch Verkleinerung der Zellen und dadurch bedingte Dissoziation eine Auflockerung zu erfahren. Die großeren Formen degenerieren gewöhnlich in ihren zentralen Teilen: die Zellen zerfallen dort und lösen sich auf, während die peripheren Zellen nicht selten sich konzentrisch abplatten und um die entstandene Höhlung herum eine Art epithelialer Bekleidung bilden, in solchen zu Cysten umgewandelten Hassal'schen Körperchen scheinen. Lymphozyten mit einer gewissen Vorliebe sich anzuhäufen."

Hammar (51), in his book written concerning the advances in thymic research during the early part of this century, shared some of his conclusions:

> "...die Grundzüge der regressiven Veränderungen der Hassall'schen Körper.:die Hypertrophie der lebenden Zellen wird rückgängig, und sie nehmen den Character von gewöhnlichen Retikulumzellen wieder an. Die geschädigten oder schon abgestorbenen Zellen werden in verschiedener Weise resorbiert....die kleinen Hassall'schen Körper bei ihrer Rückbildung ins Retikulum gleichsam aufgelöst werden, können die peripheren lebenden Zellen bei größeren Körpern mit mehr oder weniger zahlreichen abgestorbenen Zentralzellen sich bei der Rückbildung verhalten."

He stated that a cystic type HB, by definition, always has a regressive character. A quarter of a century ago, repeated starving experiments defined acute thymic involution in mice (188). The acute involution of the thymus, which may occur as a result of malnutrition or significant stress, was termed *"accidental"* and may be responsible for the consequent immunodeficiency, as well as the absence of HBs (189).

8. CHANGES IN THE CELLS OF THE THYMIC MICROENVIRONMENT AFTER IRRADIATION

Hammar (51) found a very early x-irradiation experiment in the thymic literature:

> "Schon 1909 hatten Aubertin und Bordet bei der Bestrahlung von neugeborenen Katzen und Hunden eine solche Vergrößerung von ungeheuren Dimensionen angegeben. Schon am 3. Tage nach der Bestrahlung sind die Körper 10-15 mal größer, und immer mehr mit Zelltrümmern beladen, wachsen in der Folge weiter, können dabei das siebenhundertfache der Größe der Körper der unbestrahlten Tiere erreichen und seltsame Formen annehmen.."

A rapid, accidental involution of the thymus following total body irradiation (TBI) was reported by Regaud and Crémieu (190) in cats. They found that on the fifth day after TBI the HBs were 4-5 times larger than normal. Between post-irradiation days 8 and 13, the HBs grew giant and represented about half of the involuted thymic parenchyma. As conclusions, the authors stated that the bodies arose from the medullary RE cells, but they did not make any comment on the origin of these cells (i.e. whether they were epithelial or mesenchymal). Malkina (191) provided a detailed summary of the various morphological changes in HBs following x-ray exposure. Jaroslow (192) published evidence obtained during a study of total antigen distribution in the thymus, supporting the blood vessel origin of HBs. Autoradiographic analyses of the thymus after i.v. injections of polymerized flagellin (labeled with ^{125}I) demonstrated that the label was located exclusively to the wall and lumen of the capillaries. In Jaroslow's experiments, new development of HBs around occluded vascular segments was only demonstrated following total body irradiation (TBI). Stimulation of thymic histogenesis and rapid cell differentiation within the thymic microenvironment following a whole head exposure to x-irradiation was demonstrated in rats by Maor & Alexander (193, 194). The possible role of growth hormone in this overall tissue activation has also been discussed.

In our experiments on dogs after TBI employing 400 rad, we observed a reduction in the number of HBs during the first few days (195). Characteristically, the HBs showed a considerable variety of morphological alterations. There were small bodies consisting of one or two basophilic cells. In the cytoplasm of these cells, we frequently detected small basophil granules of uniform size. On the 10th, 20th, and 30th days after TBI, the number of HBs in the thymus fell, both in the zone adjacent to the damage and in the areas remote from it, and this overall alteration of thymic architecture was best observed on the 30th day following TBI. After local irradiation at all stages of repair, the number of HBs and their dimension showed the same changes as after TBI. During the first days after x-ray irradiation, particularly in the zone distant from the damage, large HBs were most commonly encountered and they consisted of swollen cells with large nuclei which quite frequently extruded drops of chromatin. In the cytoplasm of these cells, we observed numerous vacuoles or clumps of basophilic material. Our local, thymus directed, irradiation experiment resulted in the hyperprogression of the HBs observed in a late stage of tissue repair and organ reorganization. In the involuted, cell deficient and reorganized thymic tissue, some large to giant HBs were found (195) (Fig. 7).

8.1 Experimental hormonal treatment - formation of new HBs

Ross and Korenchevsky (196) described new formation of HBs in gonadectomized rats after treatment with estrogens or simultaneous injection of male and female hormones, but not after treatment with male hormones alone. Plagge (197) reported their new formation in normal rats after injection of estrogen for long periods (4-10 months). In his observations, the HBs reached a considerable size with a marked tendency for cyst formation. Ross and Korenchevsky (196) classified their result as a tissue-specific, functional effect of the estrogens, since hyperplasia of numerous epithelial structures occurred after the hormonal treatment. This effect has also been reported earlier in connection with the seminal vesicles (196), the urinary bladder (198), and the uterus (199).

Newly organized HBs have also been located in rats following injection of fatty substances for two months, but this stimulation is most likely non-specific since no RE cell hyperplasia was registered.

8.2 Physiologic role of HBs

The role of the thymus in cellular immunology has been evaluated extensively during the past four decades (4). Naturally, there remain many open questions concerning the significance and physiological role of HBs. Gitlin, Landing, & Whipple (200) demonstrated the presence of homologous albumin and γ-globulin in human thymic HBs. Kouvalainen (201) also noticed the presence of γ-globulin containing cells in the peripherial lamellae of the HBs. He evaluated the significance of cell destruction within the HBs and presented two hypotheses: *"1) Hassall's corpuscles destroy cells that are dangerous to the organism; and 2) Cells are destroyed in Hassall's corpuscles in order to have some key material for the function of the thymus or other lymphoid organ."* The first hypothesis is supported by the clonal selection theory of Burnet (202-205). The γ-globulin containing cells within the HBs may represent abnormal RE cells directing the production of autoreactive clones of T lymphocytes within the thymus. This theory would require more cells undergoing destruction in the HBs of patients suffering from autoimmune diseases. The synthesis or storage of immunoglobulins in HBs has also been reported as one of their functional roles (200, 201, 206-210).

Some experiments have led to the opinion that the HBs' main function might be the phagocytosis and storage of various antigens (177, 206-209, 211, 212). Phagocytic behavior of and cell disintegration within the HBs has been demonstrated in the thymus of guinea pigs following local, thymus directed x-ray irradiation (207). Concurrent local irradiation and i.v. injection of a carbon-containing contrast substance has demonstrated that HBs are able to take up carbon as early as 3 hours after such experimental treatment. The concentration of the carbon reached a maximum on the fourth day after treatment, remained detectable for another 7 to 10 days, and was then eliminated by migrating macrophages simultaneously with the disappearance of a majority of the HBs. New HBs were formed by day 21, restoring normal medullary structure. The phagocytic HBs that remained functional even after treatment were found to be free from carbon after 28 days.

8.3 The thymus in tissue culture

All classic *in vitro* studies, conducted by Wassen (213), Popoff (77, 78), Tschassownikow (214, 215), Murray (216), Mandel & Russell (217), Mandel *et al.* (218), Pyke & Gelfand (219), Papiernik *et al.* (220), Kruisbeek (221, 222), and Ito *et al.* (223) investigated the differentiation of thymic RE cells, their effect on T lymphocyte maturation, and the new formation, histogenesis, and structural progression of HBs under tissue culture conditions. All of these experimental *in vitro* observations have determined that the thymic HBs are derived from medullary RE cells. The cellular heterogeneity of thymic RE cells has been analyzed in murine thymic cultures (224). Several human thymic cultures containing either pure mesodermally derived cells, pure endodermal RE cells or RE cells of ectodermal origin have been studied immunocytochemically to characterize their IP (225). In our *in vitro* studies, we were unable to observe the new formation and histogenesis of HBs, despite the exclusive presence of large, flat and vacuolated RE cells. It is possible that additional cell to cell interactions with cells of different derivations may be necessary for the development of HBs.

Jaffe & Plavska (90) first reported the immediate cell death and cellular reorganization following thymic tissue transplantation in young guinea pigs:

> "...die Rückbildung der alten Hassallschen Körper als die Bildung neuer solcher Körper aus dem übriggebliebenen Epithel. Etwa am 5. Tage begann der letztere Vorgang, während die Läppchen erst am 10. Tage lymphoiden Charakter und um den 14. Markdifferenzierung zeigten."

In our transplantation experiments in C57Bl and DBA2 mice following thymectomy, the thymic graft placed under the kidney capsule first underwent acute (accidental) involution during which the HBs diminished in size and number. After five days, we found numerous newly developed, unicellular HBs.

8.4 HBs in human diseases and patho-biological conditions

Hammar (226) described that in cases of *acute infectious diseases* (diphteria, poliomyelitis anterior acuta, scarlatina, measles, whooping cough, and influenza); the HBs were markedly increased in size and number, and that at the onset of disease they usually disappear. Because of such evidence it has been concluded that the HBs exist for the purpose of neutralizing toxins circulating in the peripheral blood. It might also be possible that during infectious diseases, a more frequent formation of forbidden lymphocyte clones occurs in addition to the development of specific antibodies and cytotoxic T lymphocyte clones against the causative organism. Possible autoantibodies seem to appear in connection with vaccination (227). Kouvalainen (201) also speculated:

> "The infectious changes in Hassall's corpuscles might also be explained, however, according to the second theory mentioned above. It has been shown that some breakdown products of deoxyribonucleic acid enhance antibody production and increase host resistance to various pathogens (Braun et al., 1962). It might be thought that by the destruction of cells in Hassall's corpuscles important nucleic material is given to the antibody-forming cells. The cells under destruction might have in themselves essential information about the pathogens for which antibodies will be produced. It is said that the nucleic acids of lymphocytes will generally be reutilized in the lymphatic organs (Hill, 1961; Krumpf, 1963). The phenomena in the thymus described here might be a form of that general process."

The observation of the existence of a correlation between thymic pathology and the histogenesis of *myasthenia gravis* (MG) commenced half a century ago (228, 229). In MG, most HBs are rich in γ-globulin loaded cells (230, 231). A direct correlation was sought between thymic tissue infection and the development of myasthenia gravis (232). Other papers discussed the significance of thymic extract effects on neuromuscular junctions (233-237). A systematic TEM observation demonstrated the presence of striated muscle cells in the thymus of reptiles and birds (238). During our studies, we also found striated muscle cells in the developing human thymus. A detailed immunohistochemical analysis, carried out by Zoltowska (239) on *myasthenic thymic tissue*, demonstrated that in this pathological condition the HBs originate from medullary cells of a different derivation. One of these cells has monocytogenic characteristics and it belongs to the interdigitating cell (IDC) accessory cell population (160). IDCs have been found to develop and mature from macrophages in both fetal and postnatal rat thymic tissues (240). The common derivation of granulocytes, macrophages, and dendritic cells from an MHC class II-negative precursor cell in bone marrow has also been established (241,242). Desmin and S-100 positive cells, possibly dendritic cells, with elongated forms were also found incorporated within growing HBs (243). Polyclonal keratin and cytokeratin has been detected in large HBs and in the cells of the RE network.

The presence of large HBs, containing vacuolated cell remnants, has been reported in both *thrombotic thrombocytopenic purpura* (244) and in *congenital nephrotic syndrome* (227). The morphological status of the thymus demonstrated the typical changes associated with chronic involution with atypical cell distribution and presence of few, large sometimes cystic HBs full of disintegrated cellular debris (245-249).

The development of *autoimmune hemolytic anemia* following thymectomy and appendectomy in rabbits (250) might be explained as a process of eliminating the forbidden clones previously confined to the HBs.

Changes in HBs during *thyreotoxicosis* (251, 252) may represent signs of an increased rate of metabolism and cell destruction. It is certainly also possible that the rise of forbidden receptors (antigens) leads to the formation of anti-thyroid autoantibodies by the newly differentiated and activated plasma cells.

Marshall and White (253) found an accumulation of circulating antigens in the HBs of guinea pig following *injury* to the thymus. They also described a high concentration of antibody within the HBs, following direct antigen input into the thymus.

In *Swiss type aγglobulinemia*, the thymus demonstrates marked lymphopenia (typical alterations of a chronic involution) and the HBs are decreased in number and size or these concentric structures are completely absent (254, 255). A similar, but stronger disintegration of the typical cell-tissue structure of HBs has also been observed in patients with early and advanced cases of *acquired immunodeficiency syndrome (AIDS)* (256). During the clinical course of AIDS, acute thymic involution progresses to chronic, and the cellular architecture of the thymus is destroyed *in toto*.

9. FIGURES

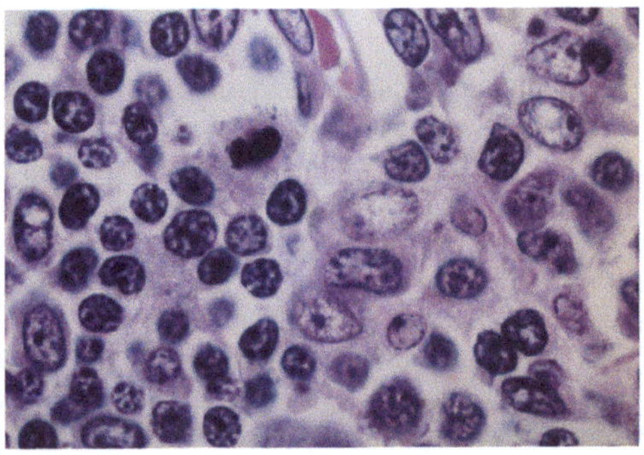

Figure 1. Dog prenatal thymus (49 days of gestation; 118 mm long fetus). Schaffer fixation. Methacrylate embedding. Hematoxylin, eosin staining. Three RE cells are grouped in the thymic medulla. This represents the initial stage of HB formation. Magnification: 1000x.

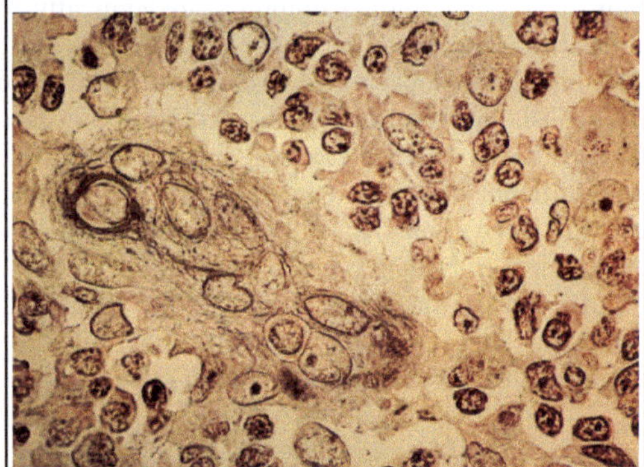

Figure 2. Human prenatal thymus (5th lunar month; 21 cm long fetus). Schaffer fixation. Methacrylate embedding. Gömöri impregnation. A well formed, progressive developing large HB in the thymic medullary microenvironment. The RE cells incorporated within this HB are exhibiting the cellular hypertrophy typical for this stage of development of the HB. New RE cells, adjacent to the concentric body, are involved in the progressive growth of the HB. Magnification: 400x.

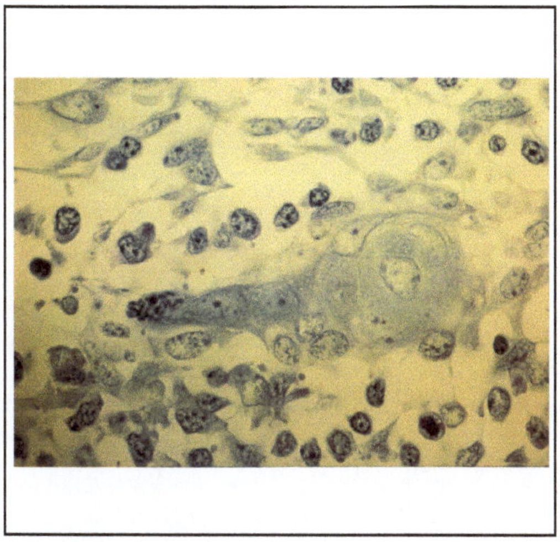

Figure 3. Human prenatal thymic ontogenesis (5th lunar month; 21 cm long fetus). Schaffer fixation. Methacrylate embedding. Giemsa staining. A large, progressively growing HB in the thymic medulla. The involvement of a number of RE cells is well demonstrated. The mitosis of a RE cell at the periphery of the HB identifies the active development and function of thymic bodies. Therefore, the HB in itself cannot be considered a degenerative structure, although cellular destruction may occur within its core. Magnification: 400x.

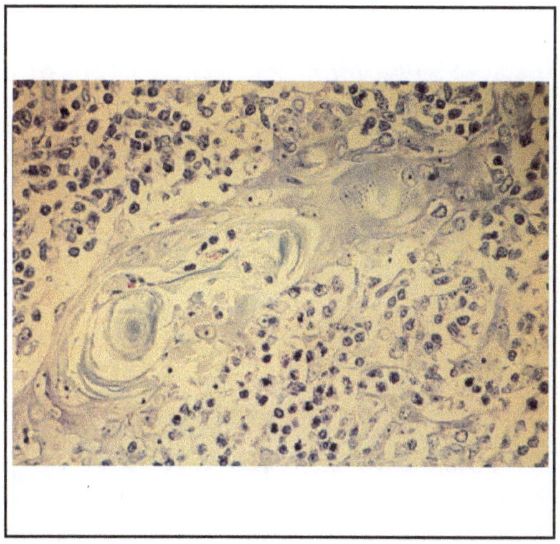

Figure 5. Human prenatal thymus (6th lunar month; 28 cm long fetus). Schaffer fixation. Methacrylate embedding. Giemsa staining. Part of a very large (also called "giant") HB in the thymic medulla. This giant HB was formed via accumulation of a number of other large HBs, a special form of progressive growth. Eosinophilic leukocytes have also been incorporated within the cellular composition of this HB. Magnification: 400x.

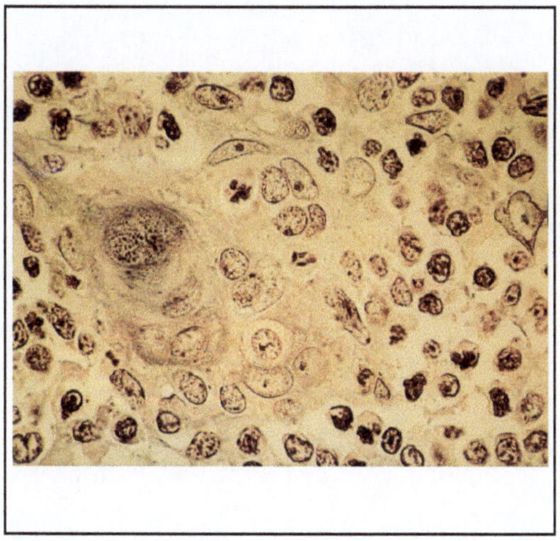

Figure 4. Human prenatal thymus (6th lunar month; 28 cm long fetus). Schaffer fixation. Methacrylate embedding. Gömöri impregnation. A number of neutrophilic granulocytes are involved in the progressive growth of the HB structure. The hypertrophy of some RE cells is also evident. Magnification: 400x.

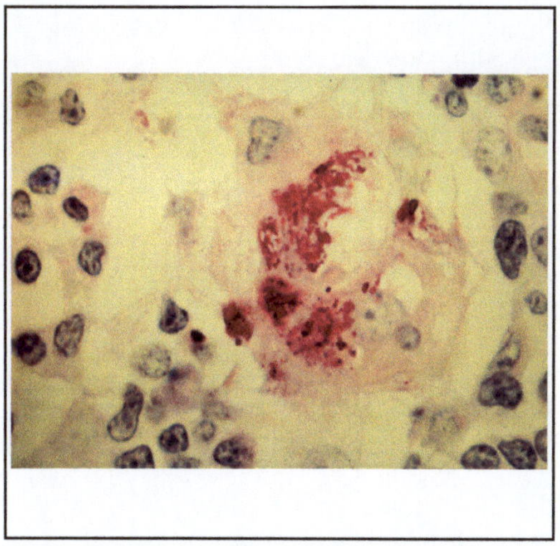

Figure 6. Postnatal dog thymus (4 days after birth; 17.5 cm long). Schaffer fixation. Methacrylate embedding. PAS staining. PAS positive granules in the RE cells and located extracellularly. Magnification: 1000x.

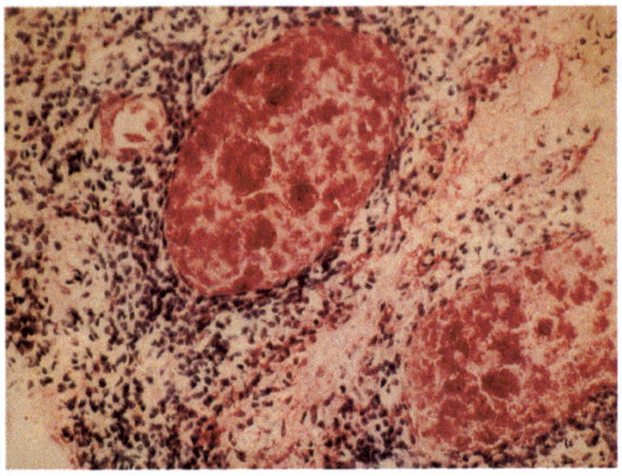

Figure 7. Postnatal human thymus (3 years of age). Frozen section. Alkaline phosphatase conjugated immunostaining employing the mouse, anti-human MoAB AE2. Immunocytochemical demonstration of keratin expression in the cellular microenvironment of the thymic medulla. Two giant HBs are packed with cellular debris, demonstrating a strong presence of keratin. Magnification: 200x.

REFERENCES

1. Miller JFAP: Immunological function of the thymus. Lancet II: 748-749, 1961.
2. Hoshino T, Takeda M, Abe K, Ito T: Early development of thymic lymphocytes in mice studied by light and electron microscopy. Anat Rec 164: 47-65, 1969.
3. Stutman O: Intrathymic and extrathymic T cell maturation. Immunol Rev 42: 138-184, 1978.
4. Bodey B: Development of lymphopoiesis as a function of the thymic microenvironment. Use of $CD8^+$ cytotoxic T lymphocytes for cellular immunotherapy of human cancer. IN VIVO 8: 915-943, 1994.
5. Jolly J: Sur les organes lympho-epitheliaux. Compt Rend Soc Biol 74: 540-543, 1913.
6. Grégoire C: Recherches sur le symbiose lymphoepitheliale au niveau du thymus de mammifere. Arch Biol 46: 717-820, 1935.
7. Grégoire C, Duchateau G: A study on lympho-epithelial symbiosis in thymus. Reactions of the lymphatic tissue to extracts and to implants of epithelial components of thymus. Arch Biol 68: 269-296, 1956.
8. Saad AH, Zapata A: Reptilian thymus gland: an ultrastructural overview. Thymus 20: 135-152, 1992.
9. Hassall AH: The microscopic anatomy of the human body, in health and disease. Vol.1. London, 32 Fleet Street, Samuel Highley, 1849, pp 3-12 & 477-479.
10. Henle J: Ueber Hassall's konzentrische Körperchen des Blutes. Z Radiation Med 7, 1849.
11. Gray E: By candlelight. The life of Dr. Arthur Hill Hassall. London, Robert Hale, 1983.
12. Ortiz-Hidalgo C: Early clinical pathologists 5: The man behind Hassall's corpuscles. J Clin Pathol 45: 99-101, 1992.
13. Bruch: Zeitschrift Radiation. Med 9, 1850.
14. Friedleben A: Die Physiologie der Thymusdrüse in Gesundheit und Krankheit. Frankfurt a/M, 1858, pp 1-336.
15. Maurer F: Die Schilddrüse, Thymus und andere Schlundspaltenderivate bei der Eidesche. Morph Jahrbuch 27: 119-172, 1899.
16. Magni S: Über einige histologische Untersuchungen der normalen Thymusdrüsen eines fünfmonatlichen und eines reifen Fetus. Arch Kinderheilk 38: 14-17, 1903.
17. Stöhr Ph: Über die Natur der Thymuselemente. Anat Hefte 31: 409-457, 1906.
18. Stöhr Ph: Über die Abstammung der kleinen Thymusrindenzellen. Anat Hefte 41: 109-127, 1910.
19. Fulci F: Die Natur der Thymusdrüse nach Untersuchungen über ihre Regenerationsfähigkeit bei Säugetieren. Deutsch Med Wochenschr 39: 1776-1780, 1913.
20. Ghika C: Étude sur le thymus. Paris, Thèse, 1901.
21. Goldner J: Action de l'adrénaline sur le thymus. C R Soc Biol (Paris) 88: 545-548, 1923.
22. Goldner J: Histogénèse du corpuscule de Hassall: unité cytogénique des cellules de charpente, des placards épithéliaux, des corpuscules unicellulaires et des corpuscules hassalliens. C R Soc Biol (Paris) 88: 947-950, 1923.
23. Cornil V, Ranvier L: Manuel d'histologie Pathologique. Paris, 1869, pp133-136.
24. Afanassiew B., Ueber die konzentrischen Körper des Thymus. Arch Mikr Anat 14: 320-342, 1877.
25. Nusbaum J, Machowski J: Die Bildung der concentrischen Körperchen und die phagocytotischen Vorgänge bei der Involution des Amphibienthymus nebst einigen Bemerkungen über die Kiemenreste und Epithelkörperchen der Amphibien. Anat Anz 21: 110-127, 1902.
26. Jordan HE, Horsley GW: The significance of the concentric corpuscles of Hassall. Anat Rec 35: 279-307, 1927.
27. Juba A, Mihàlik P von: Über die Entstchung der Hassallschen Körperchen. Z Anat 90: 278-287, 1927.
28. Jordan HE, Looper JB: The histology of the thymus gland of the box-turtle, Terrapene carolina, with special reference to the concentric corpuscles of Hassall and the eosinophilic granulocytes. Anat Rec 40: 309-337, 1928.
29. Kölliker A von: Entwicklungsgeschichte des Menschen und der höheren Tiere. Leipzig, W Engleman Publ, 1861.
30. Kölliker A von: Entwicklungsgeschichte des Menschen und der höheren Tiere. 2nd Aufl, Leipzig, W Engleman Publ, 1879.
31. His W: Zur Anatomie der menschlichen Thymusdrüse. Z Zool 11: 164, 1862.
32. His W: Anatomie menschlicher Embryonen. Leipzig, 1880.
33. Toldt: Ueber lymphoide Organe der Amphibien. Sitzungsber Akad Wissensch 58 (2), 1868.
34. Krause C: Handbuch des Menschen. Hannover, 1876, p 359.
35. Stieda L: Untersuchungen ueber die Entwicklung der Glandula thymus, Glandula thyreoidea und Glandula carotica. Leipzig, 4: 1-38, 1881.
36. Capobianco F: Della natura dei corpuscoli di Hassall, contribuzione alle conoscenze morphologiche del timo. Boll Soc Natur Napoli 1(4), 1890.
37. Pierson G: Ueber die Entwicklung der embryonalen Schlundspalten und ihre Derivate bei Säugethieren. Zeitschr Zool 47, 1899.
38. Ebner V von: Der Thymus. In: Handbuch der Gewebelehre des Menschen. (Koelliker A, ed) Leipzig, 1899, p 328.
39. Prymak T: Beiträge zur Kenntnis des feineren Baues und der Involution der Thymusdrüse bei den Teleostieren. Anat Anz 21, 1902.
40. Wallish M: Zur Bedeutung der Hassallschen Körperchen. Arch Mikr Anat 63: 274-282, 1903.
41. Goodall A: The postnatal changes in the thymus of guinea-pigs and the effect of castration on thymus structure. J Physiol 32: 191-198, 1905.
42. Klose H, Vogt H: Klinik und Biologie der Thymusdrüse. Beitr Klin. Chir 69: 1-200, 1910.
43. Rabl H: Die Entwicklung der Derivate des Kiemendarmes beim Meerschweinchen. Arch Mikr Anat 82: 79-147.
44. Hart C: Thymusstudien IV. Die Hassallschen Körperchen. Virch Arch Pathol Anat 217: 239-266, 1914.
45. Marine D: The frequency of duct-like spaces in the thymus gland, with remarks on the formation and fate of Hassall's corpuscles. Cleveland Med J 14: 186-195, 1915.
46. Scammon RE: The prenatal growth of the human thymus. Proc Soc Exp Biol Med 24: 906-909, 1927.
47. Woollard HH: The potency of the pharyngeal endoderm. J Anat 66: 242-260, 1932.
48. Webster WD: The development of the thymus bodies in Necturus maculosus. J Morphol 56: 295-323, 1934.
49. Schambacher A: Über die Pericyten von Drüsenkanälen der im Thymus und ihre Beziehung zur Entstehung der Hassalschen Körperchen. Virch Arch 172: 368-394, 1903.
50. Shier KJ: The morphology of the epithelial thymus. Observations on lymphocyte-depleted and fetal thymus. Lab Invest 12: 316-326, 1963.
51. Hammar JA: Die normal-morphologische Thymusforschung im letzten Vierteljahrhundert. Analyse und Synthese. Leipzig, Barth, 1936.
52. Kostowiecki M: Über die Beziehung der Hassallschen Körperchen zu den benachbarten Blutgefässen in dem Thymus menschlicher Phoeten. Bull Int Acad Polon Sci Cracovie Sc Nat Series B: 589-628, 1930.
53. Weller GL: Development of the thyroid, parathyroid and thymus glands in man. Contrib Embryol 24: 95-138, 1933.
54. Kostowiecki M: Development and degeneration of the

second type of Hassall's corpuscles in the thymus of guinea pig. Anat Rec 142: 195-203, 1962.
55. Norris EH: The morphogenesis and histogenesis of the thymus gland in man: in which the origin of the Hassall's corpuscles of the human thymus is discovered. Contrib Embryol 27: 191-208, 1938.
56. Zottermann A: Die Schweinethymus als eine Thymus Ecto-entodermalis. Anat Anz 38: 20-21 and 514-530, 1911.
57. Massart C: Morfologia e sviluppo del sistema tireo-paratireoideo con ricerche originali nei chirotteri (Vesperugo pipistrellus). Arch Ital Anat 44: 22-79, 1940.
58. Massart C: Morfologia e sviluppo del timo con ricerche originali nei chirotteri (Vesperugo pipistrellus). Arch Ital Anat 44: 489-550, 1940.
59. Winiwarter H de: Origine des corps de Hassall du thymus des mammifères (Chien et Chat). C R Soc Biol (Paris) 89: 834-836, 1923.
60. Dustin AP, Baillez G: L' origine des corps de Hassall et des formations kystiques du thymus des mammifères. Bull Soc R SC Med Nat 2: 44-46, 1914.
61. Imai M, Shibata T, Mineda T: A new opinion on the origin of Hassall's corpuscles. Okajimas Fol Anat Jap 45: 51-69, 1968.
62. Wassjutotschkin AM: Untersuchungen über die Histogenese des Thymus. 3. Ueber die myoiden Elemente des Thymus des Menschen. Anat Anz 50: 547-551, 1918.
63. Kliwanskaja-Kroll E: Zur Morphologie des experimentellen Hyperthyreoidismus. Das inkretorische System des im Wachstum begriffenem Organismus bei systematischer Fütterung mit Schilddrüsensubstanz (Schilddrüse und Thymusdrüse). Virch Arch Pathol Anat 272: 430-441, 1929.
64. Bastenie P: La genèse des corps de Hassall dans les thymus humains pathologiques. C R Soc Biol (Paris) 106: 55-56, 1931.
65. Bastenie P: Contribution à l'anatomie pathologique des thymus humains. Arch Int Med Exp 7: 273-345, 1932.
66. Paulitzky A: Disquisitiones de stratis glandulae thymi corpusculis. Halis Habilitationsschr 1863.
67. Watney H: Note on the minute anatomy of the thymus. Proc Roy Soc 27, 1878.
68. Watney H: The minute anatomy of the thymus. Phil Trans Roy Soc 3: 1063-1123, 1882.
69. Tourneux F, Verdun P: Sur les premiers developments de la thyroide, du thymus et des glandules parathyroidennes chez l'homme. J de l'Anat 33: 305-325, 1897.
70. Grégoire C: Action des rayons X sur le thymus au cours de l'histogénèse. C R Assoc Anat (Paris) 27: 328-335, 1932.
71. Béclère H, Pigache R: Action des rayons de Roentgen sur les corpuscles de Hassall. Bull Soc Anat (Paris) 86: 47-51, 1911.
72. Pigache R, Worms G: Le thymus considéré comme glande à secrétion interne. C R Acad Sci (Paris) 154: 234-236, 1912.
73. Kingsbury BF: The development of the human pharynx. I. The pharyngeal derivatives. Am J Anat 18: 329-398, 1915.
74. Kingsbury BF: On the nature and significance of the thymic corpuscles (of Hassall). Anat Rec 39: 141-159, 1928.
75. Kingsbury BF: On the mammalian thymus, particularly thymus IV: the development in the calf. Am J Anat 60: 149-183, 1936.
76. Beard J: The origin and histogenesis of the thymus in Raja batis. Zool Jahrbücher 17: 403-480, 1902.
77. Popoff N: The histogenesis of thymus as shown by tissue culture. Arch J Exptl Zellforsch 4: 395-416, 1926.
78. Popoff NW: The histogenesis of the thymus as shown by tissue cultures, transplantation and regeneration. Proc Soc Exper Biol Med 24: 148-151, 1926.
79. Hammar JA: Zur Histogenese und Involution der Thymusdrüse. Anat Anz 27: 23-30 & 41-89, 1905.
80. Hammar JA: Fünfzig Jahre Thymusforschung. Ergebn Anat Entwicklunggesch 19: 1-274, 1909.
81. Hammar JA: Methode, die Menge der Rinde und des Marks der Thymus, sowie die Anzahl und die Größe der Hassallschen Körper zahlenmäßig festzustellen. Z Angew Anat 1: 311-396, 1914.
82. Hammar JA: Die Menschenthymus in Gesundheit und Krankheit: Ergebnisse der numerischen Analyse von mehr als tausend menschlichen Thymusdrüsen. Teil I. Das normale Organ - zugleich eine kritische Beleuchtung der Lehre des "Status thymicus". Z Mikr Anat Forsch (Leipzig) 6: 1-570, 1926.
83. Prenant A: Contribution a l'étude du développement organique et histologique du thymus, de la glande thyroide et de la glande carotidienne. La Cellule 10: 85-184, 1894.
84. Bell ET: The development of the thymus. Am J Anat 5: 29-62, 1905.
85. Dantschakoff V: Untersuchungen von Blut und Bindegewebe bei Vögeln. Arch Mikr Anat 73: 117-181, 1908.
86. Mietens H: Zur Kenntnis des Thymusretikulum und seiner Beziehungen zu dem der Lymphdrüsen nebst einigen Bemerkungen über die Winterschlafdrüse. Jenaische Zeitschrift Naturwissensch 44: 149, 1908.
87. Maximow AA: Untersuchungen uber Blut und Bindegewebe. II. Uber die Histogenese der Thymus bei Saugetieren. Arch mikr Anat 74: 525-621, 1909.
88. Pappenheimer AM: A contribution to the normal and pathological histology of the thymus gland. J Med Res 22: 1-73, 1910.
89. Pappenheimer AM: Further studies of the histology of the thymus. Am J Anat 14: 299-325, 1913.
90. Jaffe HL, Plavska A: Experimental studies on the formation of Hassall's corpuscles. Proc Soc Exper Biol Med 23: 91-93, 1925.
91. Smith C: The thymus in immunobiology. In: The Thymus in Immunobiology. Structure, function and role in disease. (Good RA, Gabrielsen AE, eds) New York, Harper & Row, Inc., 1964.
92. Bodey B: Histomorphology and histochemistry of the human thymus during its prenatal ontogenesis. Dissertation, Inst Morphol, Bulg Acad Sci, Sofia, Bulgaria, 1977, pp 1-360.
93. Bodey B, Hadjioloff AI: Thymus development, structure and function. Nature (Bulgaria) 26: 11-19, 1977.
94. Liberti EA, Fagundes TP, Perito MA, Matson E, König Junior B: On the size of Hassall's corpuscles in human fetuses. Bull Assoc Anat 78: 15-18, 1994.
95. Kingsbury BF: The interpretation of thymus bodies. Endocrinology 29: 155-160, 1941.
96. Hartmann A: Die Entwicklung der Thymus beim Kaninchen. Arch Mikr Anat 86: 69-192, 1914.
97. Bodey B, Calvo W, Prummer O, Fliedner TM, Borysenko M: Development and histogenesis of the thymus in dog. A light and electronmicroscopical study. Dev Comp Immunol 11: 227-238, 1987.
98. Hammar JA: Über progressive und regressive Formen von Hassallschen Körpern. Z Anat 70: 466-488, 1924.
99. Lobach DF, Scearce RM, Haynes BF: The human thymic microenvironment. Phenotypic characterization of Hassall's bodies with the use of monoclonal antibodies. J Immunol 134: 250-257, 1985.
100. Hammar JA: Zur Frage der Histogenese der Thymusdrüse. Zbl Pathol 33: 505-513, 1923.
101. Caso LV: Histochemical studies of mucosubstances and keratins of thymic corpuscles. Aust J Exp Biol Med Sci 57: 493-496, 1979.
102. Smith C, Parkhurst HT: Studies on the thymus of the

mammal. XI. A comparison of the staining properties of Hassall's corpuscles and of thick skin of the guinea pig. Anat Rec 103: 649-673, 1949.
103. Gaudecker B von, Schmale EM: Similarities between Hassall's corpuscles of the human thymus and the epidermis. An investigation by electron microscopy and histochemistry. Cell Tissue Res 151: 347-368, 1974.
104. Baker JR: Cytological technique. 2nd Ed. London, Methuen, 1946, pp 1-120.
105. Baker JR: Principles of biological microtechnique. A study of fixation and dyeing. London, Methuen, 1958.
106. Jones W: On the enzyme of the thymus. Am J Physiol 10: 24-25, 1903.
107. Kouvalainen K: Some organ antibodies in congenital nephrosis. A preliminary report. Ann Paediat Fenn 7: 165-168, 1961.
108. Pearson B, Wolf P, Andrews M: The histochemical demonstration of leucine aminopeptidase by means of a new indolyl compound. Lab Invest 12: 712-720, 1963.
109. Kouvalainen K, Wallgren GR: The thymus in myasthenia gravis with special reference to the significance of Hassall's corpuscles. Ann Med Exp Biol Fenn 45: 112-116, 1967.
110. Timperley WR, Ahmed A: Histochemistry of the thymus and a thymoma. Arch Pathol 89: 405-413, 1970.
111. Timperley WR, Barson AJ, Davies S: Enzyme histochemistry of the developing human thymus. Biol Neonate 19: 42-54, 1971.
112. Vetters JM, Macadam RF: Fine structural evidence for hormone secretion by the human thymus. J Clin Pathol 26: 194-197, 1973.
113. Fennel RH Jr, Reddy CR, Vazquez JJ: Progressive systematic sclerosis and malignant hypertension. Immunohistochemical study of renal lesions. Arch Pathol 72: 209-215, 1961.
114. Arvy L: Histoenzymology of the parathyroid gland in Ovis Aries L. C R Soc Biol 155: 2094-2098, 1961.
115. Kouvalainen K: Species and organ dependence of alkaline phosphatase activity in lymphatic tissues. A histochemical, biochemical and electrophoretic study. Histochem J 3: 55-96, 1971.
116. Arvy L: Enzymology of the thymus. In: Handbuch der Histochemie. Vol. 8. Stuttgart, Gustav Fischer, 1973.
117. Navaro D: The thymus gland of birds and the arylsulfatase activity (ASA) in its parenchyma. Arch Ita Anat Embriol 84: 265-283, 1979.
118. Ruuskanen O, Toivanen A, Rackallio J: Histochemical characterization of chicken lymphoid tissue. Dev Comp Immunol 1: 231-240, 1977.
119. D'Anna F: Histochemical studies of the peroxidase activity of the bursa of Fabricius of Gallus domesticus; study of the enzymatic metabolism of that immunocompetent organ during its development. Rivista Istochim, Norm Patol 20: 48-60, 1976.
120. D'Anna F, Rossi F, Arbico R: Hassall's corpuscle: a regressive formation. Basic Appl Histochem 25: 169-181, 1981.
121. Gori P: Ferritin in the integumentary cells of Oxalis corniculata. J Ultrastruct Res 60: 95-98, 1977.
122. Gomori G: The distribution of phosphatase in normal organs and tissues. J Cell Comp Physiol 17: 71-83, 1941.
123. Gomori G: Histochemistry of enzymes. J Mt Sinai Hosp 19: 446-451, 1952.
124. Posalaky Z, Gyevai A, Bukulya B: Data on the mechanism of spermatogenesis based on studies with acid-fast dyes on lipid and succinic dehydrogenase. Kiserl Orvostud 13: 35-48, 1961.
125. Barka T, Anderson PJ: Histochemical methods for acid phosphatase using hexazonium pararosanilin as coupler. J Histochem Cytochem 10: 741-753, 1962.
126. Smith C: Studies on the thymus of the mammal. XV. Sites of aminopeptidase activity in the thymus of the mouse. Anat Rec 158: 149-161, 1967.
127. Korhonen LK, Ruponen B: Leucine aminopeptidase in the thymus. Acta Anat 50: 186-190, 1962.
128. Hess EL, Lagg SE: The chemical fractionation of rabbit and swine thymuses. J Biophys Biochem Cyto 3: 812-815, 1958.
129. Pearse AGE: Histochemistry: Theoretical and applied. London, Churchill, 1960.
130. Hirokawa K, Saitoh K, Hatakeyama S: Enzyme histochemical study on human thymus and its age change. Acta Pathol Jap 33: 275-285, 1983.
131. D'Anna F, Morano GP: Non-specific esterase activity in birds' lymphoid organs. Comparative histochemical research. Cell Molec Biol 28: 457-464, 1982.
132. Brody IA: Histochemistry of lactate dehydrogenase isozymes in intact muscle. Ann NY Acad Sci 151: 587-593, 1968.
133. Barnett RJ, Seligman AM: Histochemical demonstration of sulfhydryl and disulfide groups of proteins. J Natl Cancer Inst 14: 769-803, 1954.
134. Soderstrom KO, Rinne R, Hopsu-Havu VK, Jarvinen M, Rinne A: Identification of acid cysteine proteinase inhibitor (cystatin A) in the human thymus. Anat Rec 240: 115-119, 1994.
135. Bodey B, Calvo W: Histochemistry of the Hassall's bodies in the thymuses of dog and human fetuses during prenatal development. Abstract, 7th Internat Congress Histochem Cytochem, Helsinki, 1984.
136. Burfeind P, Hoyer-Fender S, Doenecke D, Tsaousidou S, Engel W: Expression of a histone H1 gene 9H1.1 in human testis and Hassall's corpuscles of the thymus. Thymus 19: 245-251, 1992.
137. Pertschuk LP: Immunofluorescence of Hassall's corpuscles with pemphigus serum. Confirmation of their squamous epithelial nature. Tissue Antigens 4: 446-451, 1974.
138. Pertschuk LP, Rosen Y: An immunofluorescent study of tumors with specific antisera for squamous epithelial intercellular substance and basement membrane. Am J Clin Pathol 60: 601-607, 1973.
139. Beutner EH, Lever WF, Witebsky E, Jordon RE, Chertock B: Autoantibodies in pemphigus vulgaris. Responses to an intercellular substance of epidermis. J Am Med Assoc 192: 682-688, 1965.
140. Monier JC, Dardenne M, Pléau JM, Schmitt D, Deschaux P, Bach JF: Characterization of facteur thymique sérique (FTS) in the thymus. I. Fixation of anti-FTS antibodies on thymic reticuloepithelial cells. Clin Exp Immunol 42: 470-476, 1980.
141. Schmitt D, Monier JC, Dardenne M, Pleau JM, Deschaux P, Bach JF: Cytoplasmic localization of FTS (facteur thymique sérique) in thymic epithelial cells. An immunoelectronmicroscopical study. Thymus 2: 177-186, 1980.
142. Schmitt D, Monier JC, Dardenne M, Pleau JM, Bach JF: Location of FTS (facteur thymique serique) in the thymus of normal and auto-immune mice. Thymus 4: 221-231, 1982.
143. Dalakas MC, Engel WK, McClure JE, Goldstein A, Askanas V: Immunocytochemical characterization of thymosin in thymic epithelial cells of normal and myasthenia gravis patients and in thymic cultures. J Neurol Sci 50: 239-247, 1981.
144. Savino W, Dardenne M: Thymic hormone containing cells. VI. Immunohistologic evidence for the simultaneous presence of thymulin, thymopoietin and thymosin $\alpha 1$ in normal and pathological human thymuses. Eur J Immunol 14: 987-992, 1984.

145. Drenckhahn D, von Gaudecker B, Müller-Hermelink HK, Unsicker K, Groschel-Stewart U: Myosin and actin containing cells in the human postnatal thymus. Ultrastructural and immunhistochemical findings in normal thymus and in myasthenia gravis. Virchows Arch B Cell Pathol 32: 33-45, 1979.
146. Takigawa M, Imamura S, Ofuji Sh: Demonstration of epidermis-specific heteroantigens in thymic epithelial cells. Int Arch Allergy Appl Immunology 55: 58-60, 1977.
147. Laster AJ, Itoh T, Palker TJ, Haynes BF: The human thymic microenvironment. Thymic epithelium contains specific keratins associated with early and late stages of epidermal keratinocyte maturation. Differentiation 31: 67-77, 1986.
148. Janossy G, Thomas JA, Bollum FJ, Granger S, Pizzolo G, Bradstock KF, Wong L, McMichael A, Ganeshaguru K, Hoffbrand AV: The human thymic microenvironment: an immunohistologic study. J Immunol 125: 202-212, 1980.
149. McFarland EJ, Scearce RM, Haynes BF: The human thymic microenvironment: cortical thymic epithelium is an antigenically distinct region of the thymic microenvironment. J Immunol 133: 1241-1249, 1984.
150. Haynes BF: The human thymic microenvironment. Adv Immunol 36: 87-142, 1984.
151. van Vliet E, Melis M, Ewijk W: Monoclonal antibodies to stromal cell types of the mouse thymus. Eur J Immunol 14: 524-529, 1984.
152. Kendall MD, van de Wijngaert FP, Schuurman HJ, Rademakers LH, Kater L: Heterogeneity of the human thymus epithelial microenvironment at the ultrastructural level. Adv Exp Med Biol 186: 289-297, 1985.
153. de Maagd RA, MacKenzie WA, Schuurman HJ, Ritter MA, Price KM, Broekhuizen R, Kater L: The human thymus microenvironment: heterogeneity detected by monoclonal anti-epithelial cell antibodies. Immunology 54: 745-754, 1985.
154. Ritter MA, Schuurman HJ, MacKenzie WA, de Maagd RA, Price KM, Broekhuizen R, Kater L: Heterogeneity of human thymus epithelial cells revealed by monoclonal anti-epithelial cell antibodies. Adv Exp Med Biol 186: 283-288, 1985.
155. von Gaudecker B, Steinmann GG, Hansmann ML, Harpprecht J, Milicevic NM, Müller-Hermelink HK: Immunocytochemical characterization of the thymic microenvironment. A light-microscopic and ultrastructural immunocytochemical study. Cell Tissue Res 244: 403-412, 1986.
156. Lobach DF, Haynes BF: Ontogeny of the human thymus during fetal development. J Clin Immunol 7: 81-97, 1987.
157. Colic M, Matanovic D, Hegedis Lj, Dujic A: Immunohistochemical characterization of rat thymic non-lymphoid cells. I. Epithelial and mesenchymal components defined by monoclonal antibodies. Immunology 65: 277-284, 1988.
158. Bloodworth JMB, Hiratsuka H, Hickey RC, Wu J: Ultrastructure of the human thymus, thymic tumors, and myasthenia gravis. In: Pathology Annual. Vol. 10. (Sonners SC, ed) New York, Appleton Century-Crofts, 1975, pp 329-391.
159. Nabarra B, Papiernik M: Phenotype of thymic stromal cells. An immunoelectron microscopic study with anti-IA, anti-MAC-1, and anti-MAC-2 antibodies. Lab Invest 58: 524-531, 1988.
160. Kaiserling E, Stein H, Müller-Hermelink HK: Interdigitating reticulum cells in the human thymus. Cell Tissue Res 155: 47-55, 1974.
161. Duijvestijn AM, Kohler YG, Hoefsmit ECM: Interdigitating cells and macrophages in the acute involving rat thymus. An electro-microscopic study on phagocytic activity and population development. Cell Tissue Res 224: 291-301, 1982.
162. Vicente A, Varas A, Sacedón R, Zapata AG: Histogenesis of the epithelial component of rat thymus: an ultrastructural and immunohistological analysis. Anat Rec 244: 506-519, 1996.
163. Bodey B, Bodey B Jr, Kemshead JT, Siegel SE, Kaiser HE: Identification of neural crest derived cells within the cellular microenvironment of the human thymus employing a library of monoclonal antibodies raised against neuronal tissues. IN VIVO 10: 39-47, 1996.
164. Bodey B, Bodey B Jr, Hall FL: Expression of Transforming Growth Factor-β type II receptors in the cells of the human thymic microenvironment during ontogenesis. In: Molecular Biology of Hematopoiesis. Vol. 5. (NG Abraham, S Asano, G Brittinger, R. Shadduck, eds) New York, Plenum Press, Chapter 78, 1996, pp. 645-658.
165. Bodey B, Bodey B Jr, Siegel SE, Kaiser HE: Immunocytochemical Detection of the Homeobox B3, B4, and C6 Gene Products within the Human Thymic Cellular Microenvironment. IN VIVO, 2000; 14: 419-424.
166. Bodey B, Bodey B Jr, Siegel SE, Kaiser HE: Novel Insights into the Function of the Thymic Hassall's Bodies. IN VIVO, 2000; 14: 407-418.
167. Koka T: Electron-microscopical studies on the thymus, especially on its epithelial cells. Igaku Kenkyu 30: 309-325, 1960; and Excerpta Med 15: 101, 1961.
168. Kohnen P, Weiss L: An electron microscopic study of thymic corpuscles in the guinea pig and the mouse. Anat Rec 148: 29-58, 1964.
169. Ito T, Hoshino T: Fine structure of the epithelial reticular cells of the medulla of the thymus in the golden hamster. Z Zellforsch 69: 311-318, 1966.
170. Frazier JA: Ultrastructure of the chick thymus. Z Zellforsch 136: 191-205, 1973.
171. Mandel T: The development and structure of Hassall's corpuscles in the guinea pig. A light and electron microscopic study. Z Zellforsch 89: 180-192, 1968.
172. Haar JL: Light and electron microscopy of human fetal thymus. Anat Rec 179: 463-476, 1974.
173. Hwang WS, Ho TY, Luck SC, Simon GT: Ultrastructure of the rat thymus. Lab Invest 31: 473-487, 1974.
174. Mandel T: Ultrastructure of epithelial cells in the medulla of guinea pig thymus. Austr J Exptl Biol Med Sci 46: 755-767, 1968.
175. Mandel T: Differentiation of epithelial cells in the mouse thymus. Z Zellforsch 106: 498-515, 1970.
176. Parakkal PF: An electron microscopic study of esophageal epithelium in the newborn and adult mouse. Am J Anat 121: 175-195, 1967.
177. Hirokawa K: Electron microscopic observation of the human thymus of the fetus and the newborn. Acta Pathol Jpn 19: 1-13, 1969.
178. Lavker RM, Matoltsy AG: Substructure of keratohyalin granules of the epidermis as revealed by high resolution electron microscopy. J Ultrastruct Res 35: 575-581, 1971.
179. Dearth OA: Late development of the thymus in the cat: nature and significance of the corpuscles of Hassall and cystic formations. Am J Anat 41: 321-345, 1928.
180. Parakkal PF, Matoltsy AG: An electron microscopic study of developing chick skin. J Ultrastruct Res 23: 403-416, 1968.
181. Didierjean L, Saurat JH: Epidermis and thymus. Similar antigenic properties in Hassall's corpuscle and subsets of keratinocytes. Clin Exp Dermatol 5: 395-404, 1980.
182. Frithiof L, Wersall J: A highly ordered structure in keratinizing human oral epithelium. J Ultrastruct Res 12: 371-379, 1965.

183. Kameya T, Watanabe Y: Electronmicroscopic observations on human thymus and thymoma. Acta Pathol Jap 15: 223-246, 1965.
184. Jolly J, Levin S: Sur le modifications histologiques du thymus à la suite de jeûne. C R Soc Biol (Paris) 71: 374-377, 1911.
185. Jackson CA: The effects of inanition and malnutrition upon growth and structure. McGraw-Hill, Philadelphia, 1925.
186. Boyd E: Growth of the thymus. Am J Dis Child 33: 867-879, 1927.
187. Jonson A: Studien über die Thymusinvolution. Die akzidentelle Involution bei Hunger. Arch Mikr Anat 73: 390-443, 1909.
188. Oksanen A: Thymus in acute starvation and re-feeding, an experimental histoquantitative and statistical study in mice. Ann Acad Sci Fenn (A) IV Biologica 181: 1-37, 1971.
189. Henry L: "Accidental" involution of the human thymus. J Pathol Bacteriol 96: 337-343, 1968.
190. Regaud C, Crémieu R: Sur l' involution du thymus produite par les rayons X; resultats experimentaux. Lyon Med 116: 1-18, 1911.
191. Malkina DG: Changes in Hassall's corpuscles during thymus regeneration in the condition of roentgen irradiation. Biull Eksp Biol Med 52: 104-109, 1961.
192. Jaroslow BN: Genesis of Hassall's corpuscles. Nature 215: 408-409, 1967.
193. Maor D, Alexander P: Abscopal stimulation of the thymus of rats by exposure of the head to x-rays. Nature 205: 40-42, 1965.
194. Maor D, Alexander P: Possible role of growth hormone in the stimulation of the thymus of rats following irradiation of the head. Experientia 28: 957-959, 1972.
195. Calvo W, Fliedner TM, Herbst EW, Hügl E, Bodey B: Degenerative changes and recovery of the thymus of lethally irradiated dogs, rescued by transfusion of cryopreserved autologous blood leukocytes. Exp Hematol 15: 1171-1178, 1987.
196. Ross MA, Korenchevsky V: Thymus of rat and sex hormones. J Pathol Bacteriol 52: 349-360, 1941.
197. Plagge JC: Some effects of prolonged massive estrogen treatment on rat, with special reference to thymus. Arch Pathol 42: 598-606, 1946.
198. Hume EM, Burbank R, Korenchevsky V: Some effects of administration of oestrogens on organs of castrated and non-castrated male rats partially deprived of vitamin A. J Pathol Bacteriol 49: 291-298, 1939.
199. Korenchevsky V, Hall K: Pathological changes in sex organs after prolonged administration of sex hormones to female rats. J Pathol Bacteriol 50: 295-315, 1940.
200. Gitlin D, Landing BH, Whipple A: Localization of homologous plasma proteins in tissues of young human beings as demonstrated with fluorescent antibodies. J Exp Med 97: 163-176, 1953.
201. Kouvalainen K: Significance of Hassall's corpuscles in the light of their morphological and histochemical appearance. Ann Med Exp Fenn 42: 177-184, 1964.
202. Burnet FM: The clonal selection theory of acquired immunity. Nashville, Vanderbilt Univ Press, 1959.
203. Burnet FM: Immunological recognition of self. Science 133: 307-311, 1961.
204. Burnet FM: The immunological significance of the thymus: an extension of the clonal selection theory of immunity. Aust Ann Med 11: 19-91, 1962.
205. Burnet FM: Role of the thymus and related organs in immunity. Br Med J 2: 807-811, 1962.
206. Blau JN: Antigen and antibody localization in Hassall's corpuscles. Nature 215: 1073-1075, 1967.
207. Blau JN: The dynamic behavior of Hassall's corpuscles and the transport of particulate matter in the thymus of the guinea-pig. Immunology 13: 281-292, 1967.
208. Blau JN: The localization of BCG in the guinea-pig thymus with special reference to Hassall's corpuscles. Immunology 17: 995-998, 1969.
209. Kater L, Gorp LH van: Morphological aspects of Hassall's corpuscles. Path Eur 4: 361-369, 1969.
210. Tomasi TB Jr, Yurchak AM: The synthesis of secretory component by the human thymus. J Immunol 108: 1132-1135, 1972.
211. Blau JN: A phagocytic function of Hassall's corpuscles. Nature 208: 564-566, 1965.
212. Blau JN: Histological changes and macrophage activity in the adult guinea-pig thymus. Br J Exp Pathol 52: 142-146, 1971.
213. Wassen A: Beobachtungen an Thymuskulturen in vitro. Anat Hefte 52: 279-318, 1914.
214. Tschassownikow N: Über die in vitro-Kulturen des Thymus. Arch Exp Zellforsch 3: 250-276, 1926.
215. Tschassownikow N: Einige Daten zur Frage von der Struktur der Thymusdrüse auf Grund experimenteller Forschungen (Gewebszüchtung in vitro, Röntgenbestrahlung, Vitalfärbung mit Lithiokarmin). Ztschr Zellforsch 8: 251-295, 1929.
216. Murray RG: Pure cultures of rabbit thymus epithelium. Am J Anat 81: 369-411, 1947.
217. Mandel T, Russell PJ: Differentiation of foetal mouse thymus. Ultrastructure of organ cultures and of subcapsular grafts. Immunology 21: 659-674, 1971.
218. Mandel T, Russell PJ, Byrd W: Differentiation of the thymus in vivo and in vitro. In: Cell Interactions. Proceedings of the Third Lepetit Colloquium (Silvestri LG, North Holland, eds) Amsterdam, 1972, pp 183-191.
219. Pyke KW, Gelfand EW: Morphological and functional maturation of human thymic epithelium in culture. Nature 251: 421-423, 1974.
220. Papiernik M, Nabarra B, Bach JF: In vitro culture of functional human thymic epithelium. Clin Exp Immunol 19: 281-287, 1975.
221. Kruisbeek AM: Thymic factors and T-cell maturation in vitro: a comparison of the effects of thymic epithelial cultures with thymic extracts and thymus dependent serum factors. Thymus 1: 163-185, 1979.
222. Kruisbeek AM: Effect of factors in thymic epithelial culture supernatants (TES) on in vitro maturation of thymocytes. Ann NY Acad Sci 332: 109-112, 1979.
223. Ito T, Kasahara S, Aizu S, Kato K, Takeuchi M, Mori T: Formation of Hassall's corpuscles in vitro by the thymic epithelial cell line IT-26R 21 of the rat. Cell Tissue Res 226: 469-476, 1982.
224. Jonse KH, St. Pierre RL: Analysis of cellular heterogeneity in mouse thymus cultures. In Vitro 17: 431-440, 1981.
225. Singer KH, Harden EA, Robertson AL, Lobach DF, Haynes BF: In vitro growth and phenotypic characterization of mesodermal-derived and epithelial components of normal and abnormal human thymus. Human Immunol 13: 161-176, 1985.
226. Hammar JA: Über Bakterientoxine als Anreger einer Neubildung Hassallscher Körper. Z Mikrosk Anat Forsch 9: 68-78, 1927.
227. Kouvalainen K: Immunologic features in the congenital nephrotic syndrome: a clinical and experimental study. Ann Paediat Fenn 9 (Suppl 122): 1-137, 1963.
228. Sloan HE: The thymus in myasthenia gravis with observations on the normal anatomy and histology of the thymus. Surgery 13: 154-174, 1943.
229. Wilson A, Wilson H: Thymus and myasthenia gravis. Am J Med 19: 697-700, 1955.

230. Miller JFAP: Effect of neonatal thymectomy on the immunological response of the mouse. Proc Roy Soc Biol Sci 156: 415-428, 1962.
231. White RG, Marshall AHE: The autoimmune response in myasthenia gravis. Lancet II: 120-123, 1962.
232. Goldstein G: Thymitis and myasthenia. Lancet II: 1164-1167, 1966.
233. Thurner K: Über den Einfluss des Thymus Extrakten auf die Leistungsfähigkeit und Ermüdbarkeit des Säugetiermuskels. Arch Ges Physiol 202: 444-467, 1924.
234. Torda C, Wolff HG: Effect of organ extracts and their fractions on acetylcholine synthesis. Am J Physiol 148: 417-423, 1947.
235. Parkes JD, McKinna JA: Effects of thymic extract on the neuromuscular junction. Nature 214: 1116-1117, 1967.
236. Goldstein G: The thymus and neuromuscular function. Lancet II: 119-123, 1968.
237. Trainin N, Small M: Thymic humoral factors. In: Contemporary Topics in Immunobiology. Vol. 2. (Carter RL, Davies AJS, eds) New York, Plenum Press, 1973, pp 321-337.
238. Raviola E, Raviola G: Striated muscle cells in the thymus of reptiles and birds: an electron microscopic study. Am J Anat 121: 623-646, 1967.
239. Zoltowska A: Myoid and epithelial cell differentiation in myasthenic thymuses. Thymus 17: 237-248, 1991.
240. Hsiao L, Takahashi K, Takeya M, Arao T: Differentiation and maturation of macrophages into interdigitating cells and their multicellular complex formation in fetal and postnatal rat thymus. Thymus, 17: 219-235, 1991.
241. Ardavin C, Wu L, Li Ch, Shortman K: Thymic dendritic cells and T cells develop simultaneously in the thymus from a common precursor population. Nature 362: 761-763, 1993.
242. Inaba K, Inaba M, Deguchi M, Hagi K, Yasumizu R, Ikehara S, Muramatsu S, Steinman R: Granulocytes, macrophages, and dendritic cells arise from a common major histocompatibility complex class II-negative progenitor in mouse bone marrow. Proc Natl Acad Sci USA 90: 3038-3042, 1993.
243. Arkema A: Dendritic cells: ultrastructure and function. Doctoral Thesis, Free University, Amsterdam, The Netherlands, 1992.
244. Burnet FM, Mackay IR: Lymphoepithelial structures and autoimmune disease Lancet II: 1030-1033, 1962.
245. Waldeyer W: Die Rückbildung des Thymus. Sitzungsber Akad Wissensch 25: 433-446, 1890.
246. Syk I: Ueber Altersveränderungen in der Anzahl der Hassallschen Körper, nebst einem Beitrag zum Studium der Mengenverhältnisse der Mitosen in dem Kaninchenthymus. Anat Anz 34: 560-567, 1909.
247. Oosterom R, Kater L: The thymus in the aging individual. II. Thymic epithelial function in vitro in aging and in thymus pathology. Clin Immunol Immunopathol 18: 195-202, 1981.
248. Müller-Hermelink HK, Steinmann GG: Age-related alterations of intrathymic microenvironment. In: Lymphoid cell functions in aging. Eurage Series No. III. (de Weck AL, ed) 1984, pp 75-82.
249. Steinmann GG: Changes of the human thymus during aging. In: The Human Thymus. Histophysiology and Pathology. Current Topics in Pathology. Vol. 75. (Müller-Hermelink HK, ed) Berlin-Heidelberg-New York-Tokyo, Springer, 1986, pp 43-88.
250. Good RA, Gabrielsen AE: The thymus in immunobiology. Structure, function and role in disease. New York, Harper & Row, Inc., 1964.
251. Hammar JA: Beiträge zur Konstitutionsanatomie. 1. Mikroskopische Analyse des Thymus in 25 Fällen Basedowscher Krankheit. Beitr Klin Chir 104: 469-614, 1917.
252. Hammar JA: Die Menschenthymus in Gesundheit und Krankheit. Teil II. Das Organ unter anormalen Körperverhältnissen. Zugleich Grundlagen det Theorie der Thymusfunktion. Z Mikr Anat Forsch (Leipzig) 16, 1929.
253. Marshall AH, White RG: Experimental thymic lesions resembling those of myasthenia gravis. Lancet I: 1030-1031, 1961.
254. Good RA: Aγglobulinaemia - a provocative experiment of nature. Bull Univ Minn Hosp 26: 1-19, 1954.
255. Hitzig WH, Landolt R, Muller G, Bodmer P: Heterogeneity of phenotypic expression in a family with Swiss type aγglobulinemia: observations on the acquisition of aγglobulinemia. J Pediatr 78: 968-980, 1971.
254. Joshi VV, Oleske JM: Pathological appraisal of the thymus gland in acquired immunodeficiency syndrome in children. A study of four cases and a review of the literature. Arch Pathol Lab Med 109: 142-246, 1985.
255. Suster S, Moran CA: Malignant thymic neoplasms that may mimic benign conditions. Semin Diagn Pathol 12: 98-104, 1995.
256. Rosai J, Levine GD: Tumors of the thymus. In: Atlas of Tumor Pathology, 2nd Series, Fascicle 13. Washington: Armed Forces Institute of Pathology, 1976

Chapter 6

Thymic Accessory Cells, Including Dendritic Type Antigen Presenting Cells, within the Mammalian Thymic Microenvironment

Abstract: During mammalian ontogenesis, the thymic "pure" endodermal epithelial anlage develops and differentiates into a complex cellular microenvironment. Beginning the 7-8th week of intrauterine development, thymic epithelial cells chemotactically regulate (induce) numerous waves of migration of stem cells into the thymus, including the $CD34^+$, yolk sac-derived, committed hematopoietic stem cells. *In vitro* experiments have established that $CD34^+CD38^{dim}$ human thymocytes differentiate into T lymphocytes when co-cultured with mouse fetal thymic organs. Hematopoietic stem cells for myeloid and thymic stromal dendritic cells (DCs) are present within the minute population of $CD34^+$ progenitors within the mammalian thymus. The common myeloid, DC, natural killer (NK) and T lymphocyte progenitors have been also identified within the $CD34^+$ stem cell population in the human thymus. Interactions between the endocrine and immune systems have been reported in various regions of the mammalian body including the anterior pituitary (AP), the skin, and the central (thymus) and peripheral lymphatic system. The network of bone marrow derived DCs is a part of the reticuloendothelial system (RES) and DCs represent the cellular mediators of these regulatory endocrine-immune interactions. Folliculo-stellate cells (FSC) in the AP, Langerhans cells (LCs) in the skin and lymphatic system, "veiled" cells, lympho-dendritic and interdigitating cells (IDCs) in a number of tissues comprising the lymphatic system are the cell types of the DC meshwork of "professional" antigen presenting cells (APCs). Most of these cells express the immunocytochemical markers S-100, CD1, CD45, CD54, F418, MHC class I and II antigens, Fc and complement receptors. FSCs are non-hormone-secreting cells which communicate directly with hormone producing cells, a form of neuro-endocrine-immune regulation. As a result, an attenuation of secretory responses follows stimulation of these cells. FSCs are also the cells in the AP producing IL-6, and they have also been identified as the interferon-γ responsive elements. FSCs also express lymphatic DC markers, such as DC specific aminopeptidase, leucyl-β-naphthylaminidase, non-specific esterase, MHC class I and II molecules and various other lymphatic immunological determinants [platelet derived growth factor-α chain (PDGF-α chain), CD13, CD14 and L25 antigen]. There is strong evidence that such DCs in the AP, and similar ones in the developing thymus and peripheral lymphatic tissue are the components of a powerful "professional" antigen presenting DC network. These APCs contain a specialized late endocytic compartment, MIIC (MHC class II-enriched compartment), that harbors newly synthesized MHC class II antigens *en route* to the cell membrane. The limiting membrane of MIIC can fuse directly with the cell membrane, resulting in release of newly secreted intracellular MHC class II antigen containing vesicles (exosomes). DCs possess the ability to present foreign peptides complexed with the MHC molecules expressed on their surfaces to naive and resting T cells. There are a number of "molecular couples" that influence DC and T lymphocyte interaction during antigen presentation: CD11/CD18 integrins, intercellular adhesion molecules (ICAMs), lymphocyte function associated antigen 3 (LFA-3), CD40, CD80/B7-1, CD86/B7-2, and heat-stable antigen. The "molecular couples" are involved in adhesive or costimulatory regulations, mediating an effective binding of DCs to T lymphocytes and the stimulation of specific intercellular communications. DCs also provide all of the known co-stimulatory signals required for activation of unprimed T lymphocytes. It has been defined that DCs initiate several immune responses, such as the sensitization of MHC-restricted T lymphocytes, resistance to infections and neoplasms, rejection of organ transplants, and the formation of T-dependent antibodies. In addition, DCs and specialized epithelial tissue structures (such as "the nursing" thymic epithelial cells - TNCs) may also be involved in direct, cryptocrine-type cell to cell interactions with the epithelial cells of the thymus. TNCs regulate the development of immature thymocytes into immunocompetent T lymphocytes by *emperipolesis*, a highly specialized form of cell-cell interaction in which immature thymocytes are engulfed by large thymic RE cells. TNCs *in vitro* are capable to rescue an early subset of $CD4^+CD8^+$ thymocytes from apoptosis at 32°C, the temperature at which binding and internalization were identified. This thymocyte subpopulation later matured to a characteristic IP at the double positive stage of T lymphocyte differentiation that is indicative of positive selection

Key words: Dendritic cells (DCs); Antigen Presenting Cells (APCs); Folliculo-stellate cells (FSCs); Cytokines; Interleukin-4 (IL-4) action; Interleukin-6 (IL-6) production; Anterior Pituitary (AP); Neuroendocrine immunoregulation; Major Histocompatibility Complex (MHC) class II molecules; MIIC (MHC class II-enriched compartment, exosomes);

Adhesion molecules; Interdigitating cells (IDCs); Langerhans cells (LCs); Immunophenotype (IP); Birbeck granules; anti-CD1 monoclonal antibody; GM-CSF; Langerin (CD207); Interleukin-2 (IL-2); Interleukin-12 (IL-12); Tumor Associated Antigen (TAA); Granule-associated marker (Lag); Monoclonal antibody (MoAb); Granulocyte-macrophage colony stimulating factor (GM-CSF).

1. INTRODUCTION

1.1 The Dendritic Cell Network

Immunophenotypically and functionally heterogeneous DCs are distributed throughout human lymphatic and the majority of non-lymphatic tissues and represent the most potent, "professional" antigen presenting cells (APCs), functioning as an integral part of the immune system (1-9). DCs can present complexes between peptides derived from exogenous antigens and MHC antigens expressed on their surfaces to resting T cells, thereby initiating several immune responses such as the sensitization of MHC-restricted T lymphocytes, the rejection of organ transplants, and the formation of T-dependent antibodies (10-15). DCs need to be activated in order to perform their antigen-presenting physiologic function (16). A retrovirally immortalized myeloid cell line (FSDC), generated from mouse fetal skin and bone marrow-derived DCs, was more effective in the pinocytosis of FITC-conjugated ovalbumin and dextran (FITC-OVA and FITC-DX) than B lymphocytes or macrophages, following treatment with granulocyte-macrophage colony stimulating factor (GM-CSF) and IL-4. Pretreatment with interferon-γ (IFN-γ) reduced pinocytosis, but increased the expression of MHC and co-stimulatory/cell adhesion molecules, and promoted the peptide or OVA protein presentation to naive $CD4^+$ T lymphocytes. It has been suggested that antigen uptake and antigen presentation in DCs are regulated by different cytokine signals provided by the surrounding cellular microenvironment. The cytokine IL-4, earlier named B cell growth factor (BCGF-1) or B cell stimulatory factor (BSF-1), for instance, exerts pleiotropic effects on a number of cell types (17).

IP differences among DCs have been detected employing immunocytochemistry. Bednar (18) studied the immunoreactivity of resident DCs in twenty tissues, including lymphatic, skin, bone, soft, nervous, myocardial, lung, esophagus, stomach, and intestinal, among others, employing antibodies against CD34, F XIIIa, F VIII, actin, CD68, S-100 protein, HLA-DR, CD3, and OPD4. Three characteristic immunoreactivities of resident DCs were established: a $CD34^+$ subset of cells, another $CD68^+$ subset, which were phagocytic, and an S-100^+ subset of cells. The ability of dendritic cells (DCs and LCs) in epithelial tissue structures to interact on a cell to cell level with epithelial cells was also noted (for instance in the thymic microenvironment), which may provide these cells with increased regulatory functions. DCs can also provide all of the known co-stimulatory signals required for activation of unprimed T lymphocytes and are the most effective, specialized APCs in the induction of primary T lymphocyte-mediated immune responses (19-20). Cells with a MHC class II^+/$CD1a^+$/$CD3^-$/$CD14^-$/$CD20^-$ membrane IP, typical for DCs, can be isolated from peripheral blood mononuclear cells (PBMC) *in vitro* by the addition of IL-4 and GM-CSF (21-23). IL-13 is as effective as IL-4, and combined with GM-CSF influenced the differentiation pathway of DCs in a comparable manner. Human DCs have also been generated in large numbers *in vitro* by culturing $CD34^+$ hematopoietic progenitors in GM-CSF and tumor necrosis factor-α (TNF-α) enriched medium for 12 days (24). On days 5 to 7, the DC progenitors differentiated into two subsets ($CD1a^+$ and $CD14^+$) in tissue culture and these both matured between days 12 and 14 into cells with a typical DC morphology and IP: CD80, CD83, CD86, CD58, and high HLA class II expression. $CD1a^+$ progenitors give rise to LCs, characterized by the presence of typical Birbeck granules and the expression of Lag antigen and E-cadherin. In contrast, $CD14^+$ progenitors mature into DCs lacking Birbeck granules and the other two LC markers, but expressing CD2, CD9, CD68 and the coagulation factor XIIIa, described in dermal DCs (25). Both mature DCs have been shown to be equally efficient in stimulating allogeneic $CD45RA^+$ naive T cells. $CD14^+$ progenitors are bipotent in nature, as demonstrated by their ability to differentiate into macrophage-like cells, lacking accessory function to T lymphocytes in response to M-CSF. Strobl and co-investigators (26) have recently reported that the factors necessary for DC growth *in vitro* are poorly characterized and that the cytokine combination of GM-CSF and TNF-α, and stem cell factor (SCF) in the absence of serum supplementation is inefficient in inducing DC maturation (differentiation). The authors further demonstrated that transforming growth factor-β1 (TGF-β1) supplementation is required for substantial DC development in the absence of serum. Thus, $CD34^{+}$ (best known as an endothelial cell marker [27]) stem cells in serum free conditions required the following cytokine combination for differentiation into DCs: GM-CSF + TNF-α + SCF + TGF-β. A significant number (21% ± 7%) of DCs grown in TGF-β1 supplemented, but not plasma supplemented medium showed presence of Birbeck granules and their marker, the Lag molecule. The presence of TGF-β1 in the culture medium also diminished the number of cells with monocytic features! Human DCs, with an 80-85% purity were also isolated from peripheral blood mononuclear cells (PBMCs) by negative selection for T lymphocytes, B lymphocytes, NK cells, monocytes and granulocytes (28-29).

FSCs are an essential set of DCs in the AP, which are involved in cell to cell interactions and regulations between the endocrine and immune systems (30-32). The stellate-shaped FSCs are organized in a cellular network in the AP and are positive for S-100, produce interleukin-6 (IL-6) and are in intercellular contact with hormone producing cells, their stimulation generally resulting in the increase of secretory responses (31, 33-36). The presence of a network of lymphoid DCs, expressing a lymphoid DC specific aminopeptidase and MHC class II determinants has been reported in mouse, rat and human pituitaries (37). Since S-100 immunoreactivity is also typical for lymphoid DCs (38), the subpopulation of S-100 positive pituitary FSCs may represent members of this class of DCs. This would mean that the pituitary FSCs derive from three distinct anlagen: neuroectodermal, ectodermal (oral anlage) and mesenchymal (lymphoid anlage). The multiple ontogenetic origin of the pituitary FSCs and their intermingling with lymphoid DCs allows us to distinguish a DC-FSC cell population at the level of the AP. Since these cells form a morphologically distinct cellular network and since a close functional inter-relationship between the cell groups has been documented (i.e. the synchronized increase of S-100 protein and MHC class II determinants during ontogenesis), it is probably better to consider them as a distinguishable cell clone. The increased levels of S-100 and MHC class II antigen expression in the pituitaries of autoimmune-prone BB/R rats raises further questions concerning a possible involvement of DC-FSCs in the development of autoimmune endocrine disorders (39). Several cytokines are now known to influence the release of AP hormones by acting on the hypothalamus and on the pituitary gland. IL-1, IL-2, IL-6, TNF-α and IFN-tau are the most important cytokines involved in the stimulation of the hypothalamic-pituitary-adrenal axis and the suppression of the hypothalamic-pituitary-thyroid and gonadal axes, as well as the release of growth hormone (40). The effects of acute and chronic (systemic) diseases on growth regulation, thyroid, adrenal and reproductive functions may, at least partially, be explained by the numerous important interactions between the neuroendocrine and immune systems.

1.2 Accessory Cells of the Thymic Cellular Microenvironment involved in Antigen Presentation to Developing Thymocytes

Research in molecular immunology over the last 30 years has established the significance of lymphatic, antigen presenting accessory cells of dual, mesenchymal and bone marrow origin, which migrate into the endodermal thymic epithelial anlagen as stem cell waves, during early embryonal ontogenesis (8th week of intrauterine development) in the formation of a phenotypically heterogeneous thymic cellular microenvironment (41-46). The stem cells that migrate into the epithelial thymus are initially derived from pluripotent stem cells that reside in the embryonal yolk sac and later in the liver. Stem cells from the liver begin to colonize the bone marrow during the 16th week of intrauterine ontogenesis and, by week 22; all thymic hematopoietic progenitors are derived from the bone marrow (47-49). The human thymus contains a minute, $CD34^{+}CD38^{dim}$ stem cell clone (<0.01% of the total number of thymocytes) that does not express the typical T lymphocyte lineage surface markers CD2 and CD5, nor Thy-1, but they do express high levels of CD45RA, demonstrating that this cell clone is distinct from the pluripotent $CD34^{+}CD38^{dim}$ stem cell population (49,50). *De novo* appearance of CD45RA is in direct correlation with the disappearance of Thy-1. The majority of $CD34^{high}Thy\text{-}1^{high}$ cells lack expression of CD45 and have been established as non-hematopoietic cells. *In vitro* experiments defined that $CD34^{+}CD38^{dim}$ human thymocytes differentiate into T lymphocytes when co-cultured with mouse fetal thymic organs (51-53). Hematopoietic stem cells for myeloid and dendritic cells are present within the $CD34^{+}$ thymic progenitors (54-57). When the progenitors lose CD34 expression, the presence of CD45RA is also diminished and, in fact, further differentiated (matured) thymocytes acquire CD45RO (58). It has also been determined that the human $CD34^+CD5^-$ thymocyte clone contains bi-potential cell progenitors for T- and natural killer (NK) cells (59). The earliest, low CD4 precursors are involved in the generation of T and B lymphocytes, as well as NK and dendritic cells (DCs), but not myeloid elements. The next downstream precursor cell lineage during thymic development is characterized by a $CD4^-8^-3^-44^+25^+$ IP and is still capable of forming DCs, as well as T lymphocytes, although it no longer has the ontogenetical potential to form B lymphocytes nor myeloid cells. Fink and Bevan (60, 61), Zinkernagel *et al.* (62, 63), and Beller and Unanue (64) among others held that thymic macrophages and the associated population of dendritic cells have a crucial physiological function in directing the T lymphocyte maturation pathway by having an instructional role (antigen presentation) in the generation of the T cell receptor repertoire.

On the other hand, Oliver and LeDouarin (65) with their *in vivo* experimental grafting techniques, discovered that during thymic organogenesis, accessory cells (especially macrophages and DCs) migrate into the thymus at the same time as the initial colonization of the pure epithelial thymic rudiment by lympho-hematopoietic stem cells is taking place: on days 5-6 in quail and days 6.5-8 in chick embryos. The rapid proliferation of these precursors, along with the maturation and differentiation of various cell clone derivatives, which represent relatively mature accessory cells, occurs early during ontogenetical cell differentiation, only a few days after their initial entry (66). This was originally implied by Ewert and co-workers (67, 68), who described Ia^+ cells of the monocyte/macrophage series in 12 day embryonic chick thymuses. The evaluation of the cellular constituents of quail thymuses transplanted into chick hosts immediately after the initial infiltration by stem cells established that a significant proportion (30%) of the accessory cells in the thymus after 20 days still represented the progeny of the first wave of stem cells. This indicates that a large expression of the original accessory cell precursors takes place during embryonic life. The time when these cells stop dividing is not yet known.

The presence of a substantial major histocompatibility complex (MHC) class II (Ia) positive macrophage population in the thymus follows the specific activation of T cells by foreign or self antigens. It is generally assumed that *"educational" cell to cell and humoral, in situ signals* are provided by the cells that have migrated into the thymus to cortical, immunologically immature thymocytes, allowing for their differentiation into T lymphocytes. Somewhere along the differentiation and maturation pathways T cells *learn* to recognize antigen in the context of self MHC class I and class II molecules (MHC restriction and positive selection of T lymphocytes) (60-63, 69). Furthermore, T cells are also *educated* to be tolerant to self-peptides (tolerance induction or negative selection) (70-73).

The normal physiological function of the non-lymphatic immunological accessory cells is the organization of a communication network between the thymus and other immunological parts of the mammalian organism. The recognition of an antigen by lymphocytes expressing the αβ T cell receptor (TCR) requires the presence of *1)* CD4 and CD8 co-receptors on mature helper and cytotoxic T cells, respectively and *2)* MHC class I and class II glycoproteins complexed with small peptides on the surface of antigen presenting accessory cells (74). The role of class I and II MHC molecules (also the non-classical class I HLA-E, HLA-F, and HLA-G molecules [75]) is to bind small molecular peptides of proteolytically processed intracellular or extracellular proteins so that they can be recognized as antigens by the αβ configuration of the TCR. The TCR possesses dual peptide-restricted specificity for the antigenic peptide and for the peptide bound to the MHC molecules (76). Other cell surface molecules of accessory cells involved in interactions with developing T cells have yet to be defined (69). The specific, unique and restricted role of each accessory cell type in the promotion of sequential stages of T cell maturation is poorly understood. These essential immunological mechanisms will certainly be defined in the near future with the aid of MoABs and molecular biological observations.

1.3 Thymic Interdigitating Cells (IDCs)

Interdigitating cells (IDCs) are specialized, bone marrow derived DCs, located in T cell domains of various human tissues (77-79). Von Gaudecker and Müller-Hermelink (41) identified the precursors, originally discovered in the human thymus by Kaiserling and co-workers (80), as large electronlucent cells with irregularly shaped nuclei (during the 10th week of ontogenesis). The same authors described the TEM characteristics of IDCs in an 85 mm long (third lunar month) human fetus:

> "Large electronlucent cells with irregularly shaped nuclei are found in both the mesenchymal septa and in the presumptive medullary regions of the thymus primordium. These cells appear to be precursors of the IDCs, which have been described in the medulla of the prenatal thymus."

When the already committed, lympho-hematopoietic progenitors come into contact with these IDC precursors, the IDCs develop small finger-like protrusions to assure more efficient contact between these cell types.

A histochemical osmium-zinc iodide impregnation procedure has been reported for the identification of IDCs on semithin sections (81). IDCs in such a preparation appear as large dendritic elements containing numerous cytoplasmic protrusions with established structural contacts with lymphatic cells. Differentiating thymocytes have also been observed to be in intimate contact with IDCs. Kaiserling (80) was the first to describe this significant intercellular contact:

> "The surfaces of the IDC were in close contact with those of small lymphocytes, sometimes polysomal lymphatic cells, epithelial cells, and occasionally with those of lymphatic cells containing ergastoplasm."

As stated above, IDCs are large cells with several cytoplasmic processes, which form an extensive three-dimensional network, which envelops maturating thymocytes (82). Aggregates between IDCs, maturating thymocytes and T lymphocytes represent the specialized microanatomical and functional units within the thymic microenvironment, which are crucial for T lymphocyte differentiation (82).

The most significant TEM features of IDCs are an irregularly shaped, euchromatic nucleus, with very loosely arranged chromatin except for a small rim of heterochromatin along the inner surface of the nuclear envelope, and a nucleolus (not readily visible in all cases) which is often in an eccentric position. The cytoplasm contains single or groups of flat cisternae of the rough endoplasmic reticulum. Secretory vesicles are always present near the tubules of the Golgi complex. A number of mitochondria are distributed throughout the cytoplasm. Their characteristic cytoplasmic projections regularly *"interdigitate"* with other IDCs and several other cell types in the developing thymic medulla. One of the most important characteristics of IDCs remains that these mesenchymal cells are never connected to each other or to other cells by desmosomes (83-85). Immunocytochemical observations established the expression of MHC class I and II molecules, as well as several adhesion molecules, including intercellular adhesion molecule-1 (ICAM-1), and lymphocyte function-associated molecules (LFA-3) on the surface of IDCs (86, 87).

The well determined localization of IDCs within the thymus provides significant insight into their function. IDCs have been described in all of the many observations (including our material in dogs) to be located predominantly at the level of the thymic cortico-medullary junction, as well as in the deep cortex, but never in the subcapsular region. In guinea pig thymuses, Klug and Mager (88) found IDCs located mostly in the inner cortex, with kidney-shaped nuclei (with finely dispersed chromatin) and a thin layer of marginally located heterochromatin. The nucleolus in these cells was usually eccentric and small. These IDCs also contained all of the organelles in their cytoplasm, including lysosome-like bodies, and few ergastoplasmic and tubulovesicular structures. The most interesting cytoplasmic features were large electron-dense bodies (phagosomes) containing fragments of picnotic lymphocytes. Many finger-like processes for contact with other thymic cell types were also observed. Glycoprotein synthesis by IDCs has been discussed, but contact with thymocytes has always been described as their main physiological function.

The phagocytic activity of IDCs is not of great importance when compared to macrophages, but still numerous ingested cells can be present in their cytoplasm. Klug (89) described such findings following irradiation of the rat thymus. Duijvestijn *et al.* (90-92) observed the phagocytic activity and the population development of medullary IDC and cortical macrophages following irradiation-induced acute tissue necrosis in the thymus. IDCs clearly demonstrated phagocytic activity sixteen hours after treatment, but this activity could be attributed to the fact that at this stage, the number of necrotic thymocytes was maximal and the total number of phagocytic cells was few and thus IDCs were recruited to phagocytize the necrotic material. In discussing their findings, the authors speculated on the possible similarities between IDCs and epidermal Langerhans cells. Miyazawa *et al.* (93) reported a twelve-fold increase in the thymic macrophage population in mice eight hours after 3 Gy irradiation. The mammalian cell surface is rich in acidic sugars, such as sialic, hyaluronic, and chondroitic acids. The surface charges caused by these and similar molecules play an important role in cell to cell interactions, such as phagocytosis. The negative surface charge of irradiated thymocytes was found to be enhanced within a few hours, as illustrated by the attachment of these cells to esterase positive thymic phagocytic cells through the reduction of the electrostatic repulsive forces. The irradiated thymocytes may thus be recognized as foreign proteins (antigens) and phagocytized by thymic phagocytic cells. Higley and O'Morchoe (94) published a morphometric analysis of the thymic non-lymphoid cells within the medulla of rats, between the ages of one and 65 days. They found that the largest depot of thymic non-lymphoid cells is within the medulla or the so-called central part. They described the presence of macrophages, IDCs, and Langerhans cells at this location.

Von Gaudecker and Müller-Hermelink (41) described diapedesis from the mesenchymal septa and perivascular spaces as a possible mechanism of IDC entrance into the thymus. Once inside the thymus, these IDCs construct a cellular microenvironment at the cortico-medullary junction that is necessary for the migration pattern, differentiation, and maturation of thymic T lymphocytes (41, 95, 96). MoABs reactive with thymic macrophages and IDCs have been developed using these particular cell types isolated from peripheral lymphatic organs (Mac-1, Mac-2, and ER-BMDMI for both macrophages and IDCs; F4/80, BM8, MOMA-2 exclusively for macrophages; and NLDC-145, MIDC-8, M1-8 and ER-TR6 only for IDCs). IDCs have never been identified in B lymphocyte regions

1.4 Langerhans Cells (LCs)

Another type of thymic dendritic accessory cells is the so-called LC. Paul Langerhans following gold chloride staining of skin sections (97) first described the presence of these cells in the epidermis in 1868. It was not until 1973 that the first experimental evidence for antigen presentation by LCs was reported (98). Today it is well known that human LCs are CD1a^{+} DCs that function as very potent APCs for primary and secondary immune responses (99). TEM observations of LCs were first reported by Birbeck *et al.* (100) who described these cells as DCs, which can be distinguished by a relatively clear cytoplasm, a lobulated nucleus, cored tubules and unique organelles, the so-called Birbeck or LC granules, known today as Langerin or CD207. Although the LC granule is the ultimate marker for these cells, the same granules have occasionally been detected inside phagocytic vacuoles within activated macrophages (101-104). In addition, IDCs also occasionally possess Birbeck granules and this is the basis for the similarity between LCs and IDCs. Birbeck granules originate from the cell membrane and play a role in receptor mediated endocytosis, intracellular processing and presentation of CD1a, MHC class II and other antigens (105-108). The notion that LCs may represent epidermal macrophages has also been expressed (109, 110).

Thoughts concerning the origin and function of LCs have historically been based on two differing hypotheses. One group of investigators maintained that LCs perform neural functions (111-118), while others followed Masson's (119) theory which regarded LCs as the last stage in the life cycle of an active melanocyte (the theory of the *"worn out effect"* melanocyte) (120-122). LCs have also been described as lymphatic cells which are capable of forming antibodies (123, 124). The limited degree to which LCs phagocytize foreign particles suggests that under normal conditions this does not represent their primary function. Tissue culture experiments established that murine epidermal LCs can mature into potent immunostimulatory dendritic cells (125). The necessity of GM-CSF for ensuring the *in vitro* viability and function of epidermal LCs has also been described (126). LCs have also been reported within the rat thymus (127-130). Olah *et al.* (127) described them as cells of a special type within the thymic medulla:

> "...new kind of cells, found in the medulla of the rat thymus is described. The special structure of their cytoplasm which points to intensive cellular activity, as well as the characteristic granules contained in these cells, justify their classification as a separate cell type. The cells in question should be distinguished from the cells contained in the epithelial reticulum of the thymus, from the macrophages containing phagocyted fragments and lysosomes as also from those cells whose granules include a rod-shaped dark structure. It is, on the other hand, possible that the electron-microscopically observed cells of the present study are identical with Ito's inclusion cells (1959)."

As noted previously, the presence of these same cytoplasmic granules has been a defining characteristic of LCs of the skin (131, 132).

In routine tissue sections stained with hematoxylin-eosin (HE), LCs cannot be identified. However, they can be easily demonstrated by impregnation with gold salts (130, 133, 134). Histochemically, the ATPase method has proved most useful for the detection of LCs, provided that proper fixation and cutting techniques are employed. This reaction seems to be specific for LCs.

There is strong experimental evidence that LCs are of mesenchymal origin (109, 135-137). This conclusion can be summarized based on the following major pieces of evidence:

1. the exclusion of an ontogenetic relationship of LCs to the pigmentary system,
2. the widespread occurrence of the cells in mesodermal tissues of the organism,
3. their migration from the dermis into the epidermis, and
4. the structural similarity of LCs to the cells in the pathological condition histiocytosis X. The difficulty in stating that all LCs are of mesenchymal origin is illustrated by the problem concerning the identity of the intraepidermal versus the extraepidermal LCs and their relationship to the cells in histiocytosis X.

This difficulty is centered on questions concerning the specificity and significance of the LC granules. Hashimoto (110) in his histoenzymatic study, described that the granules, not only those attached to the plasma membrane, but also those entirely enclosed within the cytoplasm, were permeated passively by lanthaum complex during a post-vital incubation procedure. These results suggest that these intracytoplasmic granules have a direct connection with the extracellular spaces. In his opinion, the granules are endocytic in nature, but he discussed that these typical organelles may also be involved in the secretory functions of LCs. The extrusion of specific molecules from the interior of the LCs into the intercellular space within the thymic microenvironment may be the mechanism by which these DCs are involved in the regulation of thymocyte differentiation and also provides a way for interaction induction between DCs and RE cells located in the medulla.

Rowden (138-140) described the expression of Ia-like antigens on the surface of LCs. The relationship between LCs and IDCs in the T cell areas of lymph nodes and spleen (141), the dendritic reticulum cells of germinal centers (142) and other areas of the spleen (143) appears to lie in their common function in antigen presentation, rather than in antigen processing since when trypsin-treated LC suspensions were observed, C3 receptors were detected on LCs in humans (138, 139). Trypsin has been reported to destroy C3 receptors on B lymphocyte surfaces, but Berman and Gigli (144) have found that, as in macrophages, this receptor is not sensitive to enzyme digestion in LCs.

It is known that in mice the H-2 system is on chromosome 17 and has two parts: an H-2K and an H-2D region, comparable to human HLA-A and HLA-B. These regions code for a set of serologically detectable cell surface alloantigens, known as the Ia antigens. These have α and β polypeptide subunits of approximately 33kD and 28kD, respectively, and they lack a common β_2-microglobulin subunit (on primate B cells, HLA-DR antigens). In mice, the distribution of Ia antigens is as follows: I-A, I-E, and I-C region products are present on B cells, macrophages, sperm, and fetal liver cells, but not on T cells (145). Certain subsets of T cells, however, appear to possess excess products of the I-J region (suppressor cells) and it has also been shown that a high percentage of the cells in the epidermis express Ia antigen (30-90%) (146). Quantitative absorption studies have found that these cells express 2-4 times less Ia antigen than B lymphocytes (146, 147). The investigations of Rowden (138, 139) demonstrated that only LCs express HLA-DR antigens on their surface in human skin. Naturally, the expression of MHC (Ia) antigens on the LCs have a physiological role, but it has yet to be studied in great detail. Perhaps in skin, LCs play a role in the recognition of viral and microbial intrusions into the body. Viruses have been found within LCs (148), but it is not known whether viral antigens are displayed in association with the Ia antigens. Recent studies identified that the LCs originate from the CD34$^+$ hematopoietic progenitor cells of the bone marrow (149). *In vitro* stimulation of these cells with GM-CSF and TNF-α led to their rapid proliferation and differentiation into a cell clone with the following IP: CD45$^+$/CD68$^+$/CD3$^-$/CD19/CD56$^-$, and also expressing CD1a, CD4 and MHC class II. The CD1a$^+$ cells included three cell populations:

1. LCs identified by the presence of Birbeck granules;
2. Birbeck granule negative DCs; and
3. CD14$^+$ monocytes. Addition of IL-4 interfered with the generation of monocytes, but also resulted in an increased percentage of CD1a$^+$ LCs (<24%), which are potent stimulators of the primary mixed leukocyte reaction and as such they are promising candidates for the generation or augmentation of host responses against different pathogens following vaccination.

The classical experiments of Katz *et al.* (150) on radiation chimeras and the investigations by Frelinger *et al.* (151) show that bone marrow grafting with appropriate stain differences in the Ir gene region, permit tracing of the arrival of Ia$^+$ cells in the epidermis. Since bone marrow transplantation is now a clinical procedure in the treatment of patients with various pathologic hematologic conditions, there must be many examples of the repopulation of LCs in the organism. A detailed study on these transplantations should prove illuminating since graft-versus-host (GvH) disease is a common complication in such procedures. Although LCs may arise from bone marrow monoblasts, nothing is known concerning the relative pool sizes of other possible precursor cells. Evidence is accumulating of the existence of cells within the lymph nodes and spleen with structural, antigenic and functional characteristics similar to LCs (95, 152-154). These localizations of LCs also represent a symbiotic union of epithelial (in origin) and mesenchymal cells, but the intimate role played in T cell maturation and in phagocytosis is not yet clear.

In conclusion the presence and interaction of IDCs and LCs with other cell types in the human and vertebrate (including mice and dogs) thymic tissues observed by us following various physiological and experimental conditions suggests that DCs play a critical role within the thymic cellular and humoral microenvironment and thus may have a profound effect on the maturation of T lymphocytes. These non-lymphatic and non-stromal cell populations and their endless number of clones are involved in the immune induction of stem cells, which is evident from the way in which the two distinct arms of the immune system developed during phylogeny. It is well established that specific immunity developed as a function necessary to complement and increase the efficacy of the uptake and elimination of antigens, while ensuring the lack of immunity to self.

In our cell culture experiments, immature cortical thymocytes, separated in bovine albumin density gradients were placed in tissue culture medium, supplemented with thymic DC culture supernatants. After 48-72 hours *in vitro*, the rapidly proliferating thymocytes were found to change their IP by acquiring MHC class I and II molecules, losing differentiation CD antigens, and thereby becoming responsive in mixed leukocyte cultures. We have already published part of our results as *in vitro* experiments on the humoral mechanisms involved in intrathymic T lymphocyte maturation (169).

During ontogenesis, the progressive maturation of T lymphocytes, following acquisition of a certain TCR repertoire, is in correlation with the number of different antigens presented to these T cells by DCs. Thus, quite immature T lymphocytes are presented with antigens necessary for the assurance that no self-reactive clones will leave the thymus, while nearly completely mature T lymphocytes may well be presented foreign antigens by these professional APCs. Certainly, the secretory products of thymic RE and dendritic cells play an additional important role in the differentiation of the various stem cell clones into mature T lymphocytes, NK cells or other dendritic elements within the thymus. One such factor is the 40kD thymic-differentiating factor (TDF), isolated by Beller and Unanue (170).

It is a fact that the intimate events of intrathymic T lymphocyte maturation regulated by DCs and other professional APCs support the action of the various putative thymic hormones, driving the immature thymocytes (prethymocytes) to a stage of maturity. Recent experiments have established extensive roles of the DC network, but one of the main and most significant ones remains the initial presentation of antigens to developing T lymphocytes as thymic accessory cells, which may occur in the periphery to an increasing degree following involution of the thymus at puberty.

1.5 Monocytes/Macrophages

One of the most important types of thymic non-lymphatic and non-stromal accessory cells are the bone marrow derived, secondary antigen presenting monocytes/macrophages. Their regular presence in the thymus, especially after any kind of thymic involution, has led to two major hypotheses concerning their origin:

1. that these macrophages are derived from the stromal cells of the RE meshwork of the thymus; and
2. that they migrate into the thymus during the initial colonization of the pure epithelial anlagen by pluripotent, as well as already committed hematopoietic stem cells (mostly lymphomyeloid) from the yolk sac, and later from the bone marrow. The cells from which the thymic monocytes/macrophages were believed to originate according to these two hypotheses were either the endodermal RE cells, in the former, or the mesodermal/mesenchymal cells, in the latter.

Kostowiecki (171), in his report on the guinea pig thymus argued:

> "...those macrophages came directly from the reticular cells of the thymic medulla.....In addition the influence of vital staining on the development of Hassall's bodies (HB) around

> the reticular macrophages is investigated. This problem is closely linked with the role of the macrophages system (the RES) in the normal thymus."

There are three hypotheses concerning the role of macrophages within the thymic microenvironment. The first is based upon the observation of foreign granular substances in the RE cells of the thymus. Some investigators have described fat within the RE cells due to fatty degeneration during physiological involution (172-179). Others also established that during acute involution in pathological conditions (*accidental involution* [180, 181]), the cytoplasm of the RE cells contains either fat particles of degenerated cells, or granules of vital stain. The interpretation of such observations was that under extreme conditions, RE cells can transform into active thymic tissue macrophages (182-191). However, during our extensive observations, we never detected this kind of cell transformation in the thymus! Furthermore, if we accept that cell transformation is possible between thymic RE cells and macrophage, then why should we not accept the RE origin of thymocytes? (192, 193)

Tschassownikow (194) reported *de novo* phagocytic activity in the thymus *in vitro*. Cultured RE cells developed phagocytic activity between 12 and 24 hours following explantation. The cells of the phagocytic system, including macrophages, therefore represent the third basic thymic compartment.

According to Aschoff (195), the RE cells of the thymus could not participate in the RES because these cells are of epithelial origin and further observations have established that these cells do not carry out the reactions characteristic of cells of the RES (196, 197).

Murray (198) concluded:

> "Cells capable of rapidly ingesting particulate matter are present throughout the thymus. It is possible that they are epithelial cells. On the other hand, pure thymus epithelium has not given rise to typical macrophages in culture."

A third hypothesis of monocyte/macrophage origin is based upon the lack of granules of vital stain within the thymic RE cells. Proponents of this view concluded that the RE cellular network should belong to the RES (199-201).

It is well known, as described previously, that systemic physical, chemical, or biological stress, such as that produced by irradiation, malnutrition, infectious diseases, chronic inflammatory diseases, neoplasms, and drug abuse results in a marked, acute/accidental involution of the thymus (181, 202-209). The most characteristic effect of acute involution is the rapid degeneration of thymocytes. After just two days, thymic disintegration is nearly completed and nuclear debris begins to disappear from the lobules. Kostowiecki (171) commented on these cases of accidental involution: *"In all of these cases the reticular cells of the medulla hypertrophy and any of them change into macrophages."* In his study, he described two types of macrophages:

1. ordinary connective tissue phagocytes, located extralobularly and
2. intralobular macrophages originating from the RE cells of the thymus.

Kostowiecki (210) classified the interlobular septa into wide, intermediate, and narrow. The wide ones are composed of connective tissue, partially or completely replaced in several regions by fat cells. Under some conditions, these fat cells may compress the cortex of the newly formed lobules, leading to regressive changes in growing animals. When the septum has a fibrous structure, the macrophages are present in single or small groups and occupy the central or the more peripheral parts of the septum. When the septum contains fat cells, the macrophages are located in the narrow spaces between these large cells. The junctional areas (where the three interlobular septa meet) have the same morphological structure and distribution of macrophages. Intermediate septa are more common and retain a fibrous structure, with few fat cells. Macrophages are located along the course of the septum and are spindle-shaped. Here they occupy the middle portion of the septum. The narrow interlobular septa are made up of mainly collagenous fibers, which are replaced by reticular fibers in very narrow septa. Macrophages are usually not present here. The blood vessels, running in the septa, contain macrophages in their adventitia (the so-called perivascular or adventitial macrophages). These cells are situated longitudinally or transversely in relation to the vessels. These vessels penetrate the cortex, on their way to the medulla.

Furthermore, in his publication concerning macrophages and formation of a second type of Hassall's bodies (HBs), Kostowiecki (210) described the presence of so-called "reticular macrophages" which arise from the medullary thymic microenvironment (probably also from the RE cells):

> "The reticular macrophages develop from the reticular cells of the thymic medulla. In previous studies, none mentioned these cells

> briefly, which appear during the development of the second type of Hassall's corpuscles. The contents of the second type of Hassall's corpuscles consists of a part of the medulla enclosed by the corpuscular wall. Hence several reticular cells and thymocytes from these contents. The relationship between these two kinds of cells will determine the kind of degeneration of the contents. When the number of reticular cells and thymocytes is approximately equal, most of the reticular cells transform into macrophages (first type of degeneration). When the thymocytes are scarce or absent in Hassall's corpuscles, the reticular cells hypertrophy and form concentric lamellae (second type of degeneration). If the thymocytes predominate in the corpuscle, all the cellular elements show rapid degeneration and the macrophages do not develop or may be abortive (third type of degeneration)...Often, free reticular macrophages may develop in quantity in Hassall's corpuscle. In one, the total number of macrophages exceeds three hundred...The ample cavity of this body contains a large number of smaller and larger free reticular macrophages and a few binucleated small giant cells. The shapes of these cells are predominantly spherical, but a few are crescent-shaped."

Macrophages were once characterized as the sole scavenger cells, with the ability to phagocytose, and thereby eliminate dead cells. Our knowledge of monocyte/macrophage physiology has increased considerably, primarily due to the significant advances in our understanding of T lymphocyte maturation (211). They play an important role in the induction and expression of both cellular and humoral immunity as accessory and regulatory cells. It is known that their secretory activities are diverse and extensive, and that they posses cell surface receptors that may allow them to carry antibodies. There have been numerous reviews and monographs written on various aspects of their function (212-225).

Within any macrophage population, there is considerable and obvious heterogeneity in size and phagocytic capacity. There is heterogeneity in other aspects, including the expression of surface MHC class II receptors. The physical, biochemical, and functional differences between macrophages under normal and pathological conditions are quite dramatic and important. The induction and expression of immunity requires the effective presentation of antigens on the surface of macrophages. This is mostly demonstrated in the cases of macrophage participation in T cell and/or B cell induction mechanisms, in the production of antibodies against thymus-dependent antigens, and in macrophage-T cell interactions in the maturation and induction of cell-mediated immunity. In addition, macrophages collaborate in cell-mediated immune responses by secreting various monokines, such as lymphocyte activating factor (LAF).

The group of Unanue (226) has conducted research into the physiological role of the macrophage system in three directions:

1. antigen presentation for antibody production and T cell stimulation;
2. secretion of lymphostimulatory molecules; and
3. regulation of T cell differentiation.

The results from over 20 years of experiments in these areas established the fundamental role of the phagocytic system in immune stimulation, in antigen recognition and in the control of lymphocyte proliferation and maturation. Detailed immunological cell to cell collaboration requires genetic compatibility between T cells and macrophages in the Ia region of the MHC. This contact during development is thymus dependent, but the intimate mechanism by which this is mediated is still not known. In macrophages, Ia antigen expression has a phenotypically unstable character: Ia^- macrophages, can become Ia^+ under the influence of lymphokines, following phagocytosis, as well as under various *in vitro* tissue culture conditions. Macrophages have also been reported to present antigen to induce IgE antibody responses (227) and thus are implicated in the direct regulation of IgE production (228).

In cancer, the nature of the association between tumor associated antigens (TAA) and MHC antigens on the tumor cell surface, may determine whether the host organism responds to the malignancy by producing highly efficient, TAA directed, cytotoxic T cells or by producing only T cells mediating delayed-type hypersensitivity, a much less efficient host defense against tumors (229-231). Macrophages consuming necrotic tumor cell masses may be able to induce the clonal expansion of specific TAA-directed cytotoxic T lymphocytes *in situ*, thereby mediating the anti-tumor immune response.

Nathan *et al.* (225) described the existence of at least 54 secretory products of macrophages (monokines), and this number may rise as new products are discovered or fall as molecules that have multiple physiological specificities are identified (i.e. lymphocyte activating factor (LAF, interleukin 1), endogenous pyrogen, and a serum amyloid A inducing factor may be identical). Macrophages function not only as professional APCs and immunoregulatory molecule secretors, but also as regulatory/suppressor cells. Indeed, activated macrophages have been shown to be quite potent inhibitors of T cell responses (229-232). They may also represent mediators between suppressor T and cytotoxic T cells (233, 234). A recent study has established that monocytes/macrophages have a more generalized MHC class I antigen processing pathway than reported previously (235).

Macrophages also regulate the number of natural killer (NK) cells, which are involved in the first line of resistance to cancer (236-239), and control their own numbers via autocrine mechanisms. They produce both colony stimulating factors, which promote monocytopoiesis, and, when appropriately stimulated, macrophages also release prostaglandin (PGE 2) which inhibits it (240-244). It is also possible that macrophages can regulate erythropoiesis (245), but results have been published which also implicate fibroblast proliferation in the control of erythropoiesis (246, 247).

In our observations, we found extralobular macrophages in the capsule, and the intralobular and interlobular septa. The capsule contained macrophages, which were quite numerous in the depressions between lobules. These were continuous with the macrophages of the interlobular and intermediate septa, but terminated at the narrow septa. From these depressions they spread in the capsule, but were rare or absent at the periphery of the lobules. These macrophages were also present in the connective tissue and contained small, round granules, detected following incubation with trypan blue as the vital stain. The macrophages varied in size, and had an oval, spindle-shaped or pyriform outline depending on the direction of the cell in the cut section. The cytoplasm of the capsular macrophages was clear and translucent, and following vital staining was observed to consist of small and rather oval granules of trypan blue. In larger ones of these extralobular macrophages, the dispersion of light passing through the blue granules led to the observation of numerous *"homogeneously"* stained cells.

In our bone marrow transfusion experiments, following total body irradiation (TBI) (3x6 Gy on three consecutive days) and transfusion of stem cells of various origins, the thymus looked as if it had undergone acute involution within 72 hours after TBI, and the presence of numerous macrophages was always detected (248). We also observed DNA granules and particles in the RE cells, as well as the presence of several free macrophages within the thymic medulla, in *de novo* formed thymic cysts, and in HBs, but the origin of these cells was not clear. Intrathymic macrophages are less active functionally, compared to those found in other sites within the body. Despite this fact, they do participate significantly in the complex intrathymic T lymphocyte differentiation cascade, which ultimately results in the production of functionally mature T lymphocyte subpopulations.

2. ANTIGEN PRESENTATION BY DCS

According to the research group of Shortman (249-257), experimental results suggest a "dual" DC differentiation model, demonstrating the existence of both myeloid-derived (with characteristic IF: $CD11b^+$, $CD11c^+$, $CD8\alpha^-$ and $DEC205^+$) and lymphoid-derived DCs (showing $CD11b^-$, $CD11c^-$, $CD8\alpha^+$ and $DEC205^+$ IF). DCs, including Interdigitating cells (IDCs) and Langerhans cells (LCs), are characterized by dendritic morphology, high migratory mobility and are the most effective, "professional" cells for antigen presentation in primary immune responses. Most of the DCs express immunocytochemically detectable antigens like: S-100, CD1a, CD40 receptor, adhesion molecules (ICAM-1 or CD54, LFA-1 and LFA-3), integrins (CD11a, CD11c and CD18), CD45, CD54, co-stimulatory molecules (B7-1 or CD80, B7-2 or CD86), F418, MHC class I and II, and DEC-205, multilectin receptor, immunostimulatory cytokine (IL-12) and, of course, Fc and complement receptors. Following recognition and uptake of antigens, mature dendritic cells (DCs) migrate to the T lymphocyte rich area of draining lymph nodes, display an array of antigen-derived peptides on the surface of major histocompatibility complex (MHC) molecules and acquire the cellular specialization to select and activate naive antigen-specific T lymphocytes. Dendritic cells (DCs) patrol throughout the peripheral blood, lymph and peripheral tissues, including secondary lymphatic organs (258). Processing exogenous and endogenous proteins for presentation by MHC molecules to T lymphocyte's receptor repertoire is the defining function of "professional" antigen-presenting cells (APC) as major regulatory cells in antigen specific immune responses (259, 260). DCs responds to two types of "signals":

1. direct recognition of foreign antigen, pathogens (called "danger signal") through specific receptors (named pattern recognition receptors) and
2. indirectly sensed inflammatory mediators such as TNF-α, IL-1β, PGE-2 of infection.

Both of these two patways induce a well integrated program called "maturation process" which transforms peripheral DCs into the most efficient APCs and T lymphocyte activators (261-263). In a detailed rewiev the research group of Guermonprez (258) functionally characterized five types of surface receptors triggering the maturation process in DCs:

1. toll-like receptors - TLR2 to TLR4 (264, 265);
2. cytokine receptors;
3. TNF receptor family molecules;
4. receptors for immunoglobulins (most of FcR) by immune complexes or specific antibodies (266-268); and
5. sensors for cell death. T cells through CD40 dependent and independent signaling and endothelial cells *via* cell to cell contacts and secretion of cytokines are helping in the maturation of DCs (269, 270).

The complex process of antigen uptake, internalization by receptor-mediated endocytosis, degradation and specific loading on MHC class I and II molecules was named "antigen presentation". DCs present complexes between peptides derived from exogenous antigens and MHC antigens expressed on their surfaces to resting T cells, thereby initiating several immune responses such as the sensitization of MHC-restricted T lymphocytes, the rejection of organ transplants, and the formation of T-dependent antibodies (14, 15, 271-276).

The T lymphocytes are capable of recognizing a number of self and non-self antigens, but they also need an immunostimulatory microenvironment to become activated and initiate an immune response (277-280). $CD4^+$ and $CD8^+$ T lymphocytes can initiate a systemic immune response when the TAAs are presented to them in the form of short peptides connected to the surface of MHC class I and II molecules together with costimulatory B7 or other molecules (277, 281, 282). MHC class II-restricted antigen presentation to $CD4^+$ T lymphocytes is achieved by an essentially common, multimolecular pathway, the so-called "immunological synapse" (283-286). This multimolecular interaction is subject to variation with regard to the location and extent of degradation of protein antigens and the site of peptide binding to MHC class II molecules (260). To provoke a $CD4^+$ lymphocyte activation, antigens connected to MHC class II molecules (cytoskeletally accumulated) should be presented together with costimulatory molecules (B7.1, B7.2) in case of APC interaction with CD28 in lymphocytes and adhesion molecules (usually ICAM-1) if the APC lymphocyte interaction is fulfilled *via* lymphocyte function associated antigens 1 and 3 (LFA-1 and LFA-3) (287). At the same time, activated $CD4^+$ lymphocytes are able "to help" or activate maturing (licensing) APCs, a functionally ready stage, directly to initiate $CD8^+$ lymphocytes. This activation step requires the presence of CD40 receptor on the APCs and the CD40-ligand (CD40-L or CD154) expressed on the surface of $CD4^+$ lymphocytes (283, 284). The same results can be achieved employing a similar interaction between TRANCE, present on the surface of activated $CD4^+$ T lymphocytes and the TNF superfamily receptor RANK, expressed on APCs (288, 289). There are a number of "molecular couples" that influence DC and T lymphocyte interaction during antigen presentation: CD11/CD18 integrins, CD80/B7-1, CD86/B7-2, and heat-stable antigen (289). Antigens presented by MHC class I molecules on previously activated APCs to $CD8^+$ lymphocytes is recognized by the T lymphocyte receptor (TCR). After the TCR-antigen interaction, cytotoxic lymphocytes (CTLs) recirculate through lymphatic organs and peripheral tissues to initiate the death of cells expressing the same antigen (289-291).

DEC-205 (gp200MR6) is a type I protein, the multilectin receptor for adsorptive endocytosis mostly expressed on DCs (including IDCs and LCs) in T lymphocyte areas of lymphatic tissues, but also, at low levels, on macrophages and T lymphocytes and on the cortical reticulo-epithelial (RE) cells of the thymic cellular microenvironment (292-295). The results indicate that thymic RE cells participate in clearance of apoptotic thymocytes through the DEC-205 protein. DEC-205 belongs to a family of C-type multilectins that also include the macrophage mannose receptor (MMR) and the phospholipase A_2 receptor (PLA_2R) (296). The antigen-presentation function is associated with the high-level expression of DEC-205, an integral membrane protein homologous to the MMR and related receptors that are able to bind carbohydrates and mediate endocytosis. DEC-205 and MMR can serve as potential antigen-uptake receptors. After binding to DEC-205, antigens (proteins) are internalized, processed, and presented in a complex with MHC class II molecules to $CD4^+$ lymphocytes (297). A number of receptors for endocytosis recycle enter into and out of cells through early endosomes. New research defined that the DEC-205 receptor targets late endosomes or lysosomes rich in MHC II products, whereas the homologous macrophage mannose receptor (MMR), as expected, is found in more peripheral endosomes.

> It is well known that not all C-type (type II) lectins on DCs serve as antigen receptors recognizing pathogens through carbohydrate structures. The ICAM-3-grabbing, non-integrin, type II lectin DC-SIGN is unique in that it regulates adhesion processes of interstitial DCs, such as DC trafficking and T-cell synapse formation, as well as antigen capture (298-300). DC-SIGN does not only capture HIV-1 but instead protects it in early endosomes, allowing HIV-1 transport by DCs to lymphoid tissues, where it enhances transinfection of T lymphocytes. This recent article discusses the carbohydrate/protein recognition profile and other features of DC-SIGN that contribute to the potency of DCs to control immunity. Another type II lectin, the langerin induces the formation of a unique endocytic compartment of LCs, the so-called Birbeck granules (298).

Proteasomes are multisubunit enzyme complexes that reside in the cytoplasm (mostly in the endoplasmic reticulum) and nucleus of eukaryotic cells. Employing selective protein degradation, proteasomes regulate a number of cytoplasmic biochemical functions including MHC class I antigen processing (301). Three constitutively expressed catalytic subunits are responsible for proteasome regulated proteolysis. These subunits are exchanged for three homologous subunits, the immunosubunits, in IFNγ-exposed cells and in cells with "professional" and specialized antigen presenting function. Both constitutive and immunoproteasomes degrade the endogenous proteins into small peptide fragments that are capable of binding *via* specialized transporters TAP1 and TAP2 to MHC class I molecules for presentation on the cell surface to cytotoxic T lymphocytes. However, immunoproteasomes seem to fulfill this function more efficiently. IFNγ further induces the expression of a proteasome activator, PA28, which can also enhance antigenic peptide production by proteasomes. The ubiquitin-proteasome system also plays an important role in antigen presentation. However, unfortunately, neoplasms employ different ways to target the proteasome system to avoid MHC class I presentation of their CAAs.

2.1 Neuroendocrine-immune interactions in antigen presentation

During mammalian ontogenesis, the thymic "pure" endodermal epithelial anlage develops and differentiates into a complex cellular and humoral microenvironment. Beginning the 7-8th week of intrauterine development, thymic reticulo-epithelial (RE) cells chemotactically regulate (induce) numerous waves of migration of stem cells into the thymus, including the $CD34^+$, yolk sac-derived, pluripotent hematopoietic stem cells (302). *In vitro* experiments have established that $CD34^+CD38^{dim}$ human thymocytes differentiate into T lymphocytes when co-cultured with mouse fetal thymic organs. Hematopoietic stem cells for myeloid and thymic DCs, IDCs and LCs are present within the minute population of $CD34^+$ progenitors in the mammalian thymus. The RE cells express the DEC-205 receptor that is so important for antigen presentation. In addition, DCs and specialized epithelial tissue structures (such as "the nursing" thymic epithelial cells - TNCs) may also be involved in direct, cryptocrine-type cell to cell interactions with the epithelial cells of the thymus. TNCs regulate the development of immature thymocytes into immunocompetent T lymphocytes by *emperipolesis*, a highly specialized form of cell-cell interaction in which immature thymocytes are engulfed by large thymic reticulo-epithelial (RE) cells. TNCs *in vitro* are capable of rescuing an early subset of $CD4^+CD8^+$ thymocytes from programmed cell death at 32°C, the temperature at which binding and internalization were identified. This thymocyte subpopulation later matured into a characteristic IP at the double positive stage of T lymphocyte differentiation that is indicative of positive selection. Antigen presentation by thymic RE cells to thymocytes that undergo differentiation is one of the most important events in the selection of the T lymphocyte repertoire (303).

Interactions between the neuroendocrine and immune systems have been reported in various regions of the mammalian body including the anterior pituitary (AP), the skin, and the central (thymus) and peripheral lymphatic tissue (304). The network of bone marrow derived DCs is part of the reticuloendothelial system (RES), with DCs representing the cellular mediators of these regulatory endocrine-immune interactions. Folliculo-stellate cells (FSCs) are non-hormone secreting cells that communicate directly with hormone producing cells, which is a form of neuro-endocrine-immune regulation. As a result, an attenuation of secretory responses follows stimulation of these cells. FSCs are also the cells in the AP producing interleukin-6 (IL-6) and have been identified as the IFN-γ responsive elements. FSCs express lymphatic DC markers, such as DC specific aminopeptidase, leucyl-β-naphthylaminidase, non-specific esterase, MHC class I and II molecules and various other lymphatic immunological determinants [platelet derived growth factor-α chain (PDGF-α chain), CD13, CD14 and L25 antigen]. There is strong evidence that such DCs in the AP, and similar ones in the developing thymus and peripheral lymphatic tissue, are the components of a powerful "professional" antigen presenting DC network. These APCs contain a specialized late endocytic compartment, MIIC (MHC class II-enriched compartment), that harbors newly synthesized MHC class II antigens *en route* to the cell membrane. The limiting membrane of MIIC can fuse directly with the cell membrane, resulting in the release of newly secreted intracellular MHC class II antigen containing vesicles (exosomes). DCs possess the ability to present foreign peptides complexed with the MHC molecules expressed on their surfaces to naive and resting T cells. There are a number of "molecular couples" that influence DC and T lymphocyte interaction during antigen presentation: CD11/CD18 integrins, intercellular adhesion molecules (ICAMs), lymphocyte function associated antigen 3 (LFA-3), CD40, CD80/B7-1, CD86/B7-2, and heat-stable antigen. The "molecular couples" are involved in adhesive or costimulatory regulations, mediating an effective binding of DCs to T lymphocytes and the stimulation of specific intercellular communications. DCs also provide all of the known co-stimulatory signals required for activation of unprimed T lymphocytes. It has been defined that DCs initiate several immune responses, such as the sensitization of MHC-restricted T lymphocytes, resistance to infections and neoplasms, rejection of organ transplants, and the formation of T-dependent antibodies.

Neuroimmunologic aspects of skin inflammation are also regulated by several interacting systems (305). The modulating influence of autonomic and sensory nerves has been known for a long time. Neurokinines derived from these nerves have recently been shown to interact with antigen presentation in dermal LCs and other key functions of allergic skin disease. While some connections between the afferent function and local reflex are known, the nature of efferent regulatory effects (from brain to periphery) remains to be discovered. New research topics include the involvement of the brain-derived neurotrophic factor, in addition to, the autonomic nervous system in mental stress response and insight in the immunomodulation by proopiomelanocortins.

3. THE SIGNIFICANCE OF DCS IN ANTI-NEOPLASTIC IMMUNOTHERAPY

The growth and metastatic spread of neoplasms, to a large extent, depends on their capacity to evade host immune surveillance and overcome host defenses. All tumors express antigens that are recognized, to a variable extent, by the immune system, but in many cases, an inadequate immune response is elicited because of partial antigen masking or ineffective activation of effector cells (306, 307). Tumor Associated Antigens (TAAs) presented in the context of MHC class I complexes on either the neoplastic cell itself or on antigen-presenting cells are only inducing immunological tolerance but not the production of CAA specific cytotoxic T lymphocytes (277). The presence of co-stimulatory molecules, such as B7-1 and B7-2, on antigen-presenting cells and the secretion of IL-2 promote the differentiation of recruited $CD8^+$ lymphocytes into cytotoxic T lymphocytes (308).

Dynamic and permanent changes in the IP of cells following their neoplastic transformation have been explored in numerous immunocytochemical studies (309-315). Expression of novel antigens not usually expressed in the surrounding normal cells and re-expression of developmental antigens has also been well established. Many further alterations in the physiology of cells following neoplastic transformation are contained within the cytoplasm and nucleus of the cell itself, thus making molecules associated with such changes less than ideal targets for immunotherapy. Furthermore, these regulatory molecules are of quite low immunogenicity due to the fact that their presence usually represents more of a quantitative dysregulation than a qualitative one as such antigens are components of normal cells. The downregulation of major histocompatibility complex (MHC) molecule expression and antigen processing in such cells has also been described. The establishment and maintenance of a humoral milieu unfavorable for *in situ* immune activation and neoplastic cell lysis poses yet another difficulty. Neoplastic cell escape from the host's immune effectors is most often caused by weak immunogenicity of TAAs, antigen masking, or overall immunosuppression, a characteristic of advanced stage neoplastic disease. Failure of antigen processing or binding to MHC molecules, inadequate or low-affinity binding of MHC complexes to T lymphocyte receptors, or inadequate expression of co-stimulatory adhesion molecules in conjunction with the antigen-presenting MHC complex may all lead to poor immunogenicity of neoplasm associated peptides and impaired antitumor response (259, 316, 317). Neoplasm induced defects are known to occur in all major branches of the immune system (318). Antigen presentation to T lymphocytes appears to be one of the steps that is highly deficient in a number of human neoplasms, thereby making it that much more difficult to generate specific cellular mediators of the host anti-neoplastic immune response (319).

DCs are crucial orchestrators of the adaptive immune response (320). Antigen presentation is a very important regulatory element for the induction of cellular immune responses, which is why one of the main goals of tumor immunotherapy is to control and enhance tumor antigen presentation (321). Thus, the basic immunobiology of DCs (322) is still being widely investigated to allow the development of effective DC based immunotherapy protocols for the treatment of human malignancies (323, 324).

Efficient MHC class I and II presentation is developed when antigens are synthesized endogenously by the DCs. This is fully achieved in patients with acute or chronic myelogenous leukemia, which upon cytokine stimulation, undergo cellular differentiation into DCs that contain the full repertoire of neoplastic antigens (325-327). In other neoplasms, desired immunotherapy antigens must be delivered to the DCs *ex vivo*. The identification of a large number of CAAs has suggested new possibilities for more effective, individualized antineoplastic immunotherapy. However, multiple mechanisms may contribute to the ability of tumors to escape antitumor immune responses. Tumor antigen heterogeneity, modulation of HLA expression, and immune suppressive mechanisms may occur at any time during tumor cell progression, and can affect the outcome of therapeutic immune intervention. In particular, the appearance of altered HLA class I phenotypes during the progression of neoplasms may have important biological and medical implications due to the role of these molecules in T and NK cell functions. Exhaustive tumor tissue studies are necessary before deciding whether a particular patient is suitable for inclusion in T cell-based immunotherapy protocols. However, immune suppressive mechanisms may occur at any time during neoplastic cell progression, and can affect the outcome of immunotherapeutic intervention (328). To overcome this deficit, it is possible to employ "professional" antigen presenting DCs. DCs can be purified from the spleen, bone marrow and peripheral or cord blood. This *in vitro* propagation of tumor specific CTLs is hampered by the necessity for large amounts of professional APCs used for periodical cycles of restimulation (329). The principle of the treatment is to prime the purified DCs with CAAs, and to reinject them into the neoplasm-bearing patient. The sensitization to the CAAs could be obtained employing the crude extract of neoplastic cells or purified antigen, which will lead to MHC class II restricted antigen presentation to $CD4^{+}$ T lymphocytes (306). Unfortunately these peptides have rapid turnover on the surface of DCs and their efficacy is subsequently limited (330, 331). Recently, several computer assisted approaches have been developed to predict CTL epitopes within larger protein sequences based on proteasome cleavage specificity. The availability of such programs, as well as, a general insight into the proteasome mediated steps in MHC class I antigen processing, provides us with a rational basis for the design of new anti-neoplastic T lymphocyte vaccines (301). Almost all types of neoplasms contain survivin, a TAA and inhibitor of the apoptosis protein family, making this protein a useful target for the development of broadly applicable tumor vaccines (332). Interestingly, the authors revealed that upregulation of survivin by DCs was carried out upon stimulation with TNF-α. Employment of neopastic cell derived exosomes (cell secretory compartment), defined source of tumor rejection antigens released in membrane vesicles for vaccination could be the most efficient approach (333). Exosomes transfer a number of neoplastic cell specific antigens to DCs and induce peptide specific, MHC class I restricted presentation to T lymphocytes clones and induce neoplastic antigens specific CTL responses in cancer patients.

Passive immunotherapeutic interventions involve efforts to augment the neoplastic cells' antigenicity through vaccination with immunogenic peptides, administration of tumor infiltrating immune effector cells *in vitro* expanded and activated, *in vivo* effector cell expansion with cytokine therapies, or genetic modification of either immune effectors or neoplastic cells with cytokine genes or genes encoding costimulatory molecules to effectively activate the immune response (334-338). A variety of adoptive cellular strategies, aimed at boosting the patient's immune system, have been tested in the management of human neoplastic diseases. Despite the drawbacks associated with *ex vivo* cell manipulation and upscaling, several such approaches have been assessed in the clinic (339, 340).

The disclosure of the human genome sequence and rapid advances in genomic expression profiling have revolutionized our knowledge about molecular changes in neoplastic diseases (341). Rapidly growing gene expression databases and improvements in bioinformatics tools set the stage for new approaches using large-scale molecular information to develop specific therapeutics in cancer (342). On the one hand, the ability to detect clusters of genes differentially expressed in normal and malignant tissue may lead to widely applicable targeting of defined molecular structures. On the other hand, analyzing the "molecular fingerprint" of an individual neoplasm raises the possibility of developing customized immunotherapeutical regiments. One approach to using the emerging new datasets for the development of novel therapeutics is to identify genes that are specifically expressed in neoplasms as targets for immune intervention employing the method of "reverse immunology" for the screening of potential candidate genes by bioinformatics in order to identify the immunogenicity of candidate CAAs (341). Gene transfer can also be employed in the case of a cloned antigen (like the co-stimulatory molecule B7.1) and would lead to the MHC class I restricted priming of $CD8^+$ CTLs (329). Unfortunately, following gene transfer, the efficacy of transduction is still very low, but with an increase in our understanding, this difficulty has the possibility of being overcome. Reinjection of DCs primed with neoplastic cell lysate leads to the protection of mice against a tumor challenge. The mode and place of administration, the nature of employed DCs, and the technology of sensitization may all depend on the malignancy and metastatic potential.

What kind of whole tumor cells, apoptotic or necrotic, are more efficient for tumor vaccines? A group of scientists in Hannover, Germany compared the effect of apoptosis versus necrosis on the effective antigen presentation by DCs, employing various tumor models (343). Their experimental data determined that only apoptotic whole tumor cell vaccines were capable of inducing a DC mediated, potent antineoplastic immune response. In contrast, necrotic cell vaccines produced a strong, localized macrophage response.

Gene-engineered DCs are currently being tested in antineoplastic immunotherapy throughout the world. Genetic immunotherapy with DCs can also be engineered to express tumor antigens, having the potential advantages of endogenous epitope presentation by both major histocompatibility complex (MHC) class I and II molecules. Another advantage of using DCs is that DCs can be gene-modified to express immunostimulatory molecules that further enhance their antigen-presenting function (282).

Numerous routes to employing DCs for antineoplastic immunotherapy are being tested, ranging from direct *in situ* expansion and activation of DCs to adoptive transfer of *ex vivo* generated DCs. Numerous techniques have also been designed to optimize DC maturation and their migratory abilities, for effective tumor antigen delivery to DCs, and induction of tumor-specific, as well as, helper immune responses, *in vivo*. However, the results of recent preclinical studies and the diversity of the clinical phase I trials that are currently underway indicate that little is still known about the exact mechanisms by which DCs modulate tumor immunity. This lack of knowledge brings with it the concern that premature clinical trials might not yield the desired results and might even be harmful to, rather than promote, the concept of DC based tumor immunotherapy.

DCs have also been used to help improve the efficacy of tumor vaccines (344). Certainly we are still far from the "ideal vaccination" which should be cell free, employing immunorelevant, neoplasm specific antigens, regardless of patient's HLA haplotype and the production allow large volume manufacturing. Recently clinical investigators have developed genetically modified tumor cell vaccines (345, 346). Irradiated tumor cells transduced with and expressing various cytokines, such as IL-2, IL-6, IL-12, lymphotactin, IFN-γ, or GM-CSF, or co-stimulatory molecules, such as B7-1, were capable of eliciting their trafficking into lymph nodes and their interaction with T lymphocytes (347-349). These gene-modified DC vaccines also produced a direct regression of pre-existing neoplasms, thereby curing the experimental animals. Induction of strong immune responses in neoplasm-bearing animals against non-immunogenic or weakly immunogenic malignancies supports the research view that active immunization of cancer patients deserves further consideration.

Another possibility to ensure endogenous antigen synthesis has been to fuse DCs to neoplastically transformed cells. Gong and co-workers (350) demonstrated that murine DCs fused to colon cancer cells enhanced the formation of neoplasm specific CTL both in *in vitro* and *in vivo*. In a recent experimental study, a transgenic murine model expressing polyomavirus middle T oncogene and mucin 1 TAA was used to determine the preventive effect of a vaccine containing DCs fused to breast carcinoma cells (351). The animals developed mammary carcinoma between the ages of 65 to 108 days with 100 percent penetrance. No spontaneous CTL were identified. The prophylactic vaccination of these genetically modified mice induced polyclonal CTL activity and rendered 60 percent of the animals free of neoplastic disease by the end of the 180 days experiment. These results indicate that prophylactic vaccination with dendritic/tumor fusion cells is capable of inducing sufficient anti-neoplastic immunity to counter the process of oncogenesis of powerful oncogenic products. A similar type of tumor vaccine was successful in the prevention of human cervical cancer (352). The fusion cell vaccine experimental approach has been applied in several other antineoplastic immunotherapy protocols, including the immunoprevention of colorectal cancer, confirming the immunogenicity of such tumor vaccines (353, 354).

In vivo DCs are found in a number of tissues and reside in direct proximity to extracellular matrix proteins (355-357). Since extracellular matrix proteins affect differentiation and location of cells in tissues, a number of research observations investigate potential effects of extracellular matrix proteins on differentiation and maturation of dendritic cells. DCs were isolated and enriched from $CD34^+$ human cord blood stem cells in the presence of GM-CSF and tumor necrosis factor (TNF)-α for 6-d and subsequently cultured for an additional 6-d period on tissue culture plates coated with various extracellular matrix proteins. Among the extracellular matrix proteins tested, exposure to fibronectin stimulated DC/LC cell differentiation as indicated by the 50% increase of the number of cells expressing the Birbeck granule-associated marker Lag and displaying numerous Birbeck granules. Adhesion on fibronectin was shown to be specifically mediated by the integrin $\alpha 5\beta$. Because laminin and collagen were unable to cause similar changes in LC development, these results suggest that fibronectin may cause changes affecting cellular differentiation of progenitors. Hematopoietic progenitors may exhibit maturational regulated differences in response to both matrix molecules and cytokines.

Human cord blood, $CD34^+$ progenitors cultured in the presence of GM-CSF and TNF-α generate a heterogeneous population of DCs including Langerhans-like DCs (LLDCs) and monocytes (358). The authors noticed that IL-4 exerts different effecs in cultures according to the cells considered. Thus, IL-4 favors DC components at the expense of monocytic development, and permits long-time persistence of DCs that can be maintained up to one month in culture. These results show an IL-4-dependent inhibition of proliferation and emergence of $CD14^+$ cells. Notably, however, IL-4 also acts on the DC precursors. Thus, IL-4 enhances survival and delays maturation of LLDCs from $CD1a^+$ $CD14^-$ precursors. In addition, IL-4 also favors orientation of $CD14^+CD1a^-$ DC/monocyte precursors towards dermal-type $CD1a^+$ DC. DCs recovered from IL-4 treated cultures displayed reduced allostimulatory capacity, but this function is restored upon IL-4 weaning. A significant discovery suggested that a short (48h) IL-4 pulse is sufficient to favor DC development. These *in vitro* experiments demonstrate that IL-4 positively regulates DC development at several levels on distinct precursor cells.

3.1 DC-based tumor vaccines in clinical trials

DC-based immunotherapy was employed in clinical trials in melanoma, lymphoma, myeloma, renal and prostate cancer patients (359-363). Numerous clinical trials testing DC-based vaccines against neoplasms are in progress and partial clinical efficacy has been already proved. In some cases, bone marrow-derived DCs have been employed to treat established experimental neoplasms by unleashing a cellular immune response against TAAs (364). An *ex vivo* gene transfer technique with viral and non-viral vectors provided such antigens into DCs' antigen-presenting molecules. This gene transfer technique is often used to obtain expression of TAAs and hence, used thereby to formulate the antineoplastic immunotherapeutic vaccines. Efficacy of the approaches is greatly enhanced if DCs are transfected with more than one gene encoding immunostimulating factors. In some cases, such as with IL-12, IL-7 and CD40L genes, injection inside experimental malignancies of thus transfected DCs induces complete tumor regression in experimental animal models. In this case, TAAs are captured by DCs by still unclear mechanisms and transported to lymphatic organs where productive antigen presentation to T lymphocytes takes place. Transfection of genes will further strengthen the immunogenicity of CAAs and the antigen presentation efficacy of DCs. New immunotherapeutical strategies will soon join the clinical research being conducted.

Because of cases mostly resistant to chemotherapy, renal cell carcinoma (RCC) has been a testing ground for immunotherapy in decades (365, 366). The approval of IL-2 for immuno- treatment of RCC was a landmark "proof of principle" showing that agents working solely *via* the immune system can cause long lasting neoplasm remission. *In vitro* strategies to expand and load DCs with antigens have now led to human vaccine trials in RCC (at the University of California at Los Angeles, a phase 1 clinical trial) and a number of other malignancies.

Malignant lymphomas (MLs) are clonal neoplasms of lymphatic origin. By definition, all cells of the malignant clone have undergone the same rearrangement of antigen receptor genes and express identical antigen receptor molecules (immunoglobulin for B lymphocytes, T cell receptor for T lymphocyte MLs). The hypervariable that stretches within the variable regions of these receptors are considered true tumor-specific antigens ('idiotypes') (367). In several animal models, protective humoral or cellular immunity was induced against the ML by vaccination with the neoplasm-derived idiotype. Successful experimental immunization strategies in animals include idiotype protein vaccines combined with various adjuvants, genetically or immunologically modified ML cells, idiotype-presenting DCs, idiotype-encoding viral vectors, and DNA immunization. Firm evidence for the induction of lymphoma-specific immunity has also been obtained from human idiotype vaccination trials. Furthermore, some trials have provided strong but hitherto formally unproven evidence for clinical benefit of idiotype-vaccinated patients. Alternative vaccination approaches are based on immunologically modified neoplastic cells. Future research efforts should involve efforts for identifying the most efficacious vaccination route, on definitive proof of clinical efficacy, and on the development of new protocols to produce individually designed idiotype vaccines.

Over the last decade, the incidence of malignant melanomas (MMs) has been continuously increasing worldwide (368). Surgery is a treatment of choice in the early stages of primary lesions. Advanced MM, however, is resistant to chemotherapy and radiotherapy. Therefore, there is an essential need for new, possibly more effective treatments. In the last few years, biotherapy such as immunotherapy, has been receiving quite a bit of attention. Unfortunately, systemic administration of immunostimulatory factors is very often associated with severe side effects. Thus, concepts of specific immunotherapies, such as immunogene therapy, have been developed. Currently, various gene therapy strategies of MM are being evaluated in multiple clinical trials carried out all over the world (359, 369). As was mentioned before the new tendencies include gene modified tumor vaccines (GMTV) modified with genes encoding cytokines or co-stimulatory molecules and DCs modified with genes encoding CAA or immunostimulatory factors. Since January 1996, in the Department of Cancer Immunology USOMS, (at Great Poland Cancer Center in Poznan, Poland), a GMTV has been tested in MM patients. More than 220 patients were enrolled in this study of GMTV consisting of melanoma cells modified with genes encoding IL-6 and its agonistic soluble receptor (sIL-6R). More than 25% of the patients were observed to have objective clinical responses and significant life extensions. The encouraging results formed a basis for design of a phase III prospective, randomized clinical study.

4. CONCLUSIONS

During the last twenty years, several essential advances have been reported in the area of early T lymphocyte maturation and in the regulation signaling pathway of intrathymic positive and negative selection. The roles of both thymic cortical and medullar RE cells in regulation and induction, as well as the involvement of matrix molecules in the differentiation of T lymphocytes and a further elucidation of the diverse signaling pathways involved in this process have recently become clearer (370). The ultimate challenge remains the generation of immunocompetent T lymphocytes from hematopoietic stem cells under experimental conditions *in vitro*, whereby all of the steps involved in intrathymic lymphopoiesis may be defined. Antigen presentation by thymic RE cells to thymocytes that undergo differentiation is one of the most important events in the selection of the T lymphocyte repertoire (371). Mizuochi and co-investigators (372) reported that medullary RE cell lines, established from newborn C57BL/6 (H-2b) mice possess the ability to present a soluble antigen (ovalbumin - OVA) to OVA-specific, I-Ab restricted helper T lymphocyte lines *in vitro*, but cortical RE cell lines are not. Recently, the distribution of both MHC class II molecules and invariant chain (Ii) in both RE cell subpopulations, as well as the expression of mRNA encoding H-2Ma or H-2Mb was observed by laser confocal scanning microscopic analysis and immunoprecipitation. The results established that both RE cell subpopulations are able to present peptide antigens to peptide-specific, I-Ab restricted helper T cell hybridoma and are also able to present class II MHC alloantigens to an I-Ab-specific T lymphocyte line, and that SDS-stable MHC class II molecules, onto which peptides were loaded, are formed in both RE cells. These molecules were more rapidly degraded in the medullary RE cell subpopulation. In addition, two Ii-derived polypeptides of 21kD and 10kD were precipitated by MoAB Y3P, produced against MHC class II. Both RE cells contain the 10 kD polypeptide, but the 21 kD molecule was only detected in cortical RE cells. Finally, β chains of MHC class II molecules were successfully precipitated from the cell surface of cortical RE cells with a lower concentration of sialyted oligosaccharides. It seems that there are substantial differences in the antigen-presenting pathways employed by the cortical or medullary subpopulations of thymic RE cells, and that these differences may underlie the specificity of intrathymic T lymphocyte selection events. Recently, an experimental model in which the immune system of nude (athymic) mice was reconstituted at birth with subcutaneous grafts of embryonic thymic RE cells, removed from 10 days allogeneic embryos was observed (373). The RE graft was colonized by the hematopoietic cells of the nude mouse (host), which eventually matured into immunocompetent T lymphocytes. Such T lymphocyte reconstituted nude mice were able to reject third party skin grafts and were tolerant to skin of their own haplotype, but also the RE cell H-2 type. This tolerance induction can also be transferred by $CD4^+$ peripheral T lymphocytes selected on allogeneic RE cells! These selected T lymphocytes were also able to control the effector activity of peripheral T cells derived from normal mice. This represents a negative regulatory activity induced by RE cells. To make the process of intrathymic differentiation even more confusing, B lymphocytes have also been detected within the thymic microenvironment! Thymic RE cells are also involved in the regulation of intrathymic negative selection (374). By using two transgenic mice, the presentation of nuclear β-galactosidase (β-gal) to CD4 T lymphocytes by medullary RE cells was observed, resulting in tolerance induction. *In vitro* experiments have further substantiated this antigen presenting ability of thymic medullary RE cells, while bone marrow derived thymic DCs were found incapable of inducing tolerance. Another recent study established that these same thymic medullary RE cells also induce a complete tolerance against MHC class I restricted self peptides presented on their own cell surface (375).

Publications concerning the constructive role of B lymphocytes in the development of the T lymphocyte repertoire have once again raised questions concerning the possibility of intrathymic differentiation of B lymphocytes and the potential regulatory role of the thymic RE cells in this process (376). Although the double negative ($CD4^-CD8^-$) thymocytes produce a wide spectrum of cytokines in response to a T lymphocyte receptor independent stimulus as they approach the double positive $CD4^+CD8^+$ maturation stage, they lose their cytokine secreting ability (377). After thymocytes become single positive (either $CD4^+$ or $CD8^+$), they reacquire the ability to produce cytokines. AP-1-luciferase and the newly developed NFAT-luciferase transgenic mice were employed to analyze the transcriptional and DNA binding activities of these transcription factors involved in the regulation of cytokine gene expression. It was determined that both AP-1 and NFAT activity are not inducible in the great majority of double positive thymocytes. The cytokine producing ability was reacquired at the end of double positive stage of maturation, before the complete down-regulation of one of the receptors. Indeed AP-1 inducibility is reacquired during a specific maturation stage in double positive thymocytes: in the transition from a TCR^{low}, $CD69^{low}$, heat stable antigen $(HSA)^{high}$ IP to a TCR^{high}, $CD69^{high}$, HSA^{high} IP, which is considered to represent the initial signal of intrathymic positive selection. This could be a preventive mechanism in the T lymphocyte differentiation pathway to assure that the activation of double positive thymocytes will only occur after positive selection. Employing a long term, thymic RE cell dependent cell culture system, certain stages of B lymphocyte differentiation from triple negative ($CD4^-CD8^-CD3/TCR^-$) thymocytes was revealed. Some thymic RE cell lines maintained cells with B lymphocyte characteristics, but could not fully support their differentiation, while others efficiently supported the development of surface IgM expressing B lymphocytes. The results also revealed the functional heterogeneity of thymic RE (stromal) cells in their ability to induce different stages of B lymphocyte generation.

The problem of elucidating novel extra-thymic pathways of T cell maturation to explain the lack of change in lymphopoiesis following involution of the thymus at puberty has been one of the most fundamental questions of cellular immunology. The establishment of a bodywide network of DCs, which are capable of antigen presentation and specific cytokine secretion, is, in our opinion, the answer to this dilemma. It is quite logical that following sufficient time for the development of the full DC arsenal, the roles of the thymus in providing a primary site of maturation are better served by the secondary lymphatic organs (i.e. lymph nodes), distributed throughout the body and which are continually infiltrated by DCs. These cells possess the ability to interact with other cell types, to secrete regulatory cytokines and to present "self" and "non-self" antigens, to aid in the proper, and efficient development of T lymphocytes at many sites throughout the body. This antigen presenting capability of classical DCs has also been expanded. The expression of MHC class I molecules on the surface of DCs has been observed to be important in inducing transplant related immune responses (378). Furthermore, the ability of the dendritic cell line D2SC/1 to present exogenous viral peptides and thereby stimulate MHC class I restricted, $CD8^+$ cytotoxic T lymphocytes has also been established (379).

The notion that the presentation of antigens is a unique functional capability of "professional" APCs has recently been modified. Fibroblasts transfected with viral proteins have been shown to directly induce antiviral cytotoxic T lymphocyte (CTL) responses *in vivo* (380). Fibroblasts are not professional APCs in the sense that they do not express all of the necessary co-stimulatory molecules to stimulate the clonal expansion of antigen-specific CTL clones. Although only low-level T cell activation via MHC class I interaction was observed *in situ*, fibroblasts were found to be very effective APCs in the cytokine-rich milieu of lymphatic organs such as the spleen (6). Thus, the network of accessory cells involved in mediating and regulating the immune response should be expanded to include the possible roles that connective tissue elements may play. The DC network, composed of both resident and migratory cellular elements, may in fact secrete cytokines also necessary for aiding this functional role of fibroblasts in the various tissues of the body.

Antigen presentation to T lymphocytes appears to be one of the steps which is highly deficient in a number of malignancies, thereby making it that much more difficult to generate specific cellular mediators of the anti-neoplastic immune response (381). To overcome this deficit, it is possible to employ "professional" antigen presenting DCs. DCs can be purified from the spleen, bone marrow and peripheral or cord blood. This *in vitro* propagation of tumor specific CTLs is hampered by the necessity for large amounts of professional APCs used for periodical cycles of restimulation (382). The principle of the treatment is to prime the purified DCs with cancer specific or associated antigens, and to reinject them into the tumor-bearing patient. The sensitization to the cancer related antigen could be obtained using crude neoplastic extract or purified antigen, which will lead to MHC class II restricted antigen presentation to $CD4^+$ T cells. Gene transfer can also be employed in the case of a cloned antigen (like the costimulatory molecule B7.1) and would lead to the MHC class I restricted priming of $CD8^+$ CTLs (382). Unfortunately, following gene transfer, the efficacy of transduction is still very low, but with an increase in our understanding, this difficulty should be overcome. Reinjection of DCs primed with neoplastic cell lysate leads to the protection of mice against a tumor challenge. The mode and place of administration, the nature of employed DCs, and the technology of sensitization may all depend on the malignancy and metastatic potential. Recent investigators have employed genetically modified tumor cell vaccines (383, 384). Irradiated tumor cells transduced with and expressing various cytokines, such as IL-2, IL-6, IFN-γ, or GM-CSF, or co-stimulatory molecules, such as B7-I, were capable of eliciting direct regression of pre-existing tumors, thereby curing the experimental animals (385). Induction of strong immune responses in neoplasm-bearing animals against non-immunogenic or weakly immunogenic malignancies supports the research view that active immunization of cancer patients deserves further consideration (386).

And finally:

1. Dendritic cells (DCs) are the most effective, "professional" antigen presenting cells that capture antigens in the periphery, migrate centrally, and present the processed antigens in the context of MHC and appropriate co-stimulatory molecules to T lymphocytes for the initiation of an immune response.
2. Dendritic cells (DCs) and thymic reticulo-epithelial (RE) cells are capable to perform important immunoregulatory functions by presenting antigens in the form of peptides bound to cell-surface MHC molecules to T lymphocytes. It is a fact that the intimate events of intrathymic T lymphocyte maturation regulated by DCs and other professional APCs support the action of the various putative thymic hormones, driving the immature thymocytes (prethymocytes) to a stage of maturity.
3. Antigen presentation is a critical regulatory element for the induction of cellular immune responses. Therefore, one of the principal current goals of anti-neoplastic immunotherapy is to control and enhance tumor antigen presentation. In this respect, dendritic cells (DC) are now being widely investigated as

immunotherapeutic agents for the treatment of disseminated malignancies.

4. During the progression of neoplastic disease and the constant IP changes of tumor cells, the appearance of altered HLA class I phenotypes may have important immunobiological and immunotherapeutical implications due to the role of these molecules in T and NK cell functions.
5. The fundamental immunobiology of DCs is still being widely investigated to allow the development of effective DC based immunotherapy protocols for the treatment of human malignancies. In the last decade, experimental protocols for *in vitro* growth and maturation of large quantities of DCs from their $CD34^+$ bone marrow derived stem cells have been carried out employing several cytokine cocktails. This strategy enables the generation of functionally mature DCs even from advanced neoplasm patients whose antigen presentation is suppressed by neoplasm derived molecules. The ability to culture autologous DCs *ex vivo* has influenced the development of protocols of genetically engineering them. The era of active, specific immunotherapy is born.
6. There is scientific evidence showing that DCs and thymic RE cells share surface receptors that play a crucial role in antigen presentation. There are also tight regulation connections between the neuroendocrine and the immune system that should be investigated in the process of antigen presentation. Without neuroendocrine regulation, there is no proper immune function. Neuroendocrine regulation is currently not encompassed in the tumor vaccine design, which is a suggestion that we have that future research into the neuroendocrine question could improve the quality of DC based tumor vaccines.
7. In vitro strategies to expand and load DCs with CAA antigens have now led to human vaccine trials in renal cell carcinoma and a number of other malignancies such as malignant melanomas, breast carcinomas and lymphomas. We are sure that new immunotherapeutical strategies and protocols will soon join the clinical research being conducted worldwide.

REFERENCES

1. Hoefsmit EC, Duijvestijn AM, Kamperdijk EW: Relation between Langerhans cells, veiled cells, and interdigitating cells. Immunobiology *161:* 255-265, 1982.
2. Becker Y: Anticancer role of dendritic cells (DC) in human and experimental cancers - a review. Anticancer Res *12:* 511-520, 1992.
3. Neefjes JJ, Ploegh HL: Intracellular transport of MHC class II molecules. Immunol Today *13:* 179-184, 1992.
4. Roake JA, Austyn JM: The role of dendritic cells and T cell activation in allograft rejection. Exp Nephrol 1*:* 90-101, 1993.
5. Austyn JM, Hankins DF, Larsen CP, Morris PJ, Rao AS, Roake JA: Isolation and characterization of dendritic cells from mouse heart and kidney. J Immunol 152*:* 2401-2410, 1994.
6. Sprent J: Antigen-presenting cells. Professionals and amateurs. Current Biology *5:* 1095-1097, 1995.
7. Tambur AR, Gebel HM: Alloantigen processing and presentation. J Heart & Lung Transpl *14:* 1031-1037, 1995.
8. Finkelman FD, Lees A, Birnbaum R, Gause WC, Morris SC: Dendritic cells can present antigen in vivo in a tolerogenic or immunogenic fashion. J Immunol *157:* 1406-1414, 1996.
9. Raposo G, Nijman HW, Stoorvogel W, Liejendekker R, Harding CV, Melief CJ, Geuze HJ: B lymphocytes secrete antigen-presenting vesicles. J Exp Med *183:* 1161-1172, 1996.
10. Steinman RM, Nussenzweig MC: Dendritic cells: features and functions. Immunol Rev *53:* 127-148, 1980.
11. Steinman RM: The dendritic cell system and its role in immunogenicity. Annual Rev Immunol *9:* 271-296, 1991.
12. Monaco JJ: Structure and function of genes in the MHC class II region. Curr Opin Immunol *5:* 17-20, 1993.
13. Neefjes JJ, Momburg F: Cell biology of antigen presentation. Curr Opin Immunol *5:* 27-34, 1993.
14. Germain RN: MHC-dependent antigen processing and peptide presentation: providing ligands for T lymphocyte activation. Cell *76:* 287-299, 1994.
15. Ossevoort MA, Kleijmeer MJ, Nijman HW, Geuze HJ, Kast WM, Melief CJM: Functional and ultrastructural aspects of antigen processing by dendritic cells. Adv Exp Med Biol *378:* 227-231, 1995.
16. Lutz MB, Assmann CU, Girolomoni G, Ricciardi-Castagnoli P: Different cytokines regulate antigen uptake and presentation of a precursor dendritic cell line. Eur J Immunol *26:* 586-594, 1996.
17. Salmon-Ehr V, Gillery P, Kalis B, Banchhereau J, Maquart FX: L'interleukine-4: du lymphocyte B au fibroblaste. Pathol Biol *42:* 262-268, 1994.
18. Bednar B: Dendritic resident cells and their immunohistologic determination. Ceskoslov Patologie *31:* 9-16, 1995.
19. Paglia P, Girolomoni G, Robbiati F, Granucci F, Ricciardi-Castagnoli P: Immortalized dendritic cell line fully competent in antigen presentation initiates primary T cell responses *in vivo*. J Exp Med *178:* 1893-1901, 1993.
20. Lenz A, Heine M, Schuler G, Romani N: Human and murine dermis contain dendritic cells, Isolation by means of a novel method and phenotypical and functional characterization. J Clin Invest *92:* 2587-2596, 1993.
21. Piemonti L, Bernasconi S, Luini W, Trobonjaca Z, Minty A, Allavena P, Mantovani A: IL-13 supports differentiation of dendritic cells from circulating precursors in concert with GM-CSF. Eur Cytokine Network *6:* 245-252, 1995.
22. Barratt-Boyes SM, Henderson RA, Finn OJ: Chimpanzee dendritic cells with potent immunostimulatory function can be propagated from peripheral blood. Immunology *87:* 528-534, 1996.
23. Hanada K, Tsunoda R, Hamada H: GM-CSF-induced *in vivo* expansion of splenic dendritic cells and their strong costimulation activity. J Leukocyte Biol *60:* 181-190, 1996.
24. Caux C, Vanbervliet B, Massacrier C, Dezutter-Dambuyant C, de Saint-Vis B, Jacquet C, Yoneda K, Imamura S, Schmitt D, Banchereau J: $CD34^+$ hematopoietic progenitors from human cord blood differentiate along two independent dendritic cell pathways in response to GM-CSF and TNF α. J Exp Med *184:* 695-706, 1996.
25. Hochrein H, Jahrling F, Kreyschh HG, Sutter A: Immunophenotypical and functional characterization of bone marrow derived dendritic cells. Adv Exp Med Biol *378:* 61-63, 1995.
26. Strobl H, Riedl E, Scheinecker C, Bello-Fernandez C, Pickl WF, Rappersberger K, Majdic O, Knapp W: TGF-β 1 promotes in vitro development of dendritic cells from $CD34^+$ hemopoietic progenitors. J Immunol *157:* 1499-1507, 1996.
27. Yamazaki K, Eyden BP: Ultrastructural and immunohistochemical observations on intralobular fibroblasts of human breast, with observations on the CD34 antigen. J Submicr Cytol Pathhol *27:* 309-323, 1995.
28. Nijman HW, Kleijmeer MJ, Ossevoort MA, Oorschot VM, Vierboom MP, van de Keur M, Kenemans P, Kast WM, Geuze HJ, Melief CJ: Antigen capture and major histocompatibility class II compartments of freshly isolated and cultured human blood dendritic cells. J Exp Med *182:* 163-174, 1995.
29. Davoust J, Banchereau J: Naked antigen-presenting molecules on dendritic cells. Nat Cell Biol *2:* E46-E48, 2000.
30. Cocchia D, Miani N: Immunocytochemical localization of the brain-specific S-100 protein in the pituitary gland of adult rat. J Neurocytol *9:* 771-782, 1980.
31. Baes M, Allaerts W, Denef C: Evidence for functional communication between folliculo-stellate cells and hormone-secreting cells in perfused anterior pituitary cell aggregates. Endocrinology *120:* 685-691, 1987.
32. Vankelecom H, Carmeliet P, van Damme J, Billiau A, Denef C: Production of interleukin-6 by folliculo-stellate cells of the anterior pituitary gland in a histiotypic cell aggregate culture system. Neuroendocrinology *49:* 102-106, 1989.
33. Nakajima T, Yamaguchi H, Takahashi K: S-100 protein in folliculo-stellate cells of the rat of the pituitary anterior lobe. Brain Res *191:* 523-531, 1980.
34. Allaerts W, Denef C: Regulatory activity and topological distribution of folliculo-stellate cells in rat anterior pituitary cell aggregates. Neuroendocrinology *49:* 409-418, 1989.
35. Allaerts W, Jeucken PHM, Hofland LJ, Drexhage HA: Morphological, immunohistochemical and functional homologies between pituitary folliculo-stellate cells and lymphoid dendritic cells. Acta Endocrinol *125:* 92-97, 1991.
36. Carmeliet P, Vankelecom H, van Damme J, Billiau A, Denef C: Release of interleukin-6 from anterior pituitary cell aggregates: developmental pattern and modulation by glucocorticoids and forskolin. Neuroendocrinology *53:* 29-34, 1991.
37. Allaerts W, Jeucken PHM, Bosman FT, Drexhage HA: Relationship between dendritic cells and folliculo-stellate cells in the pituitary: immunohistochemical comparison between mouse, rat and human pituitaries. *In:* Dendritic Cells in Fundamental and Clinical Immunology

(Kamperdijk et al, eds), Plenum Press, New York, 1993, pp 637-642.

38. Takahashi K, Yamaguchi H, Ishizeki J, Nakajima T, Nakazato Y: Immunohistochemical and immunoelectron microscopic localization of S-100 protein in the interdigitating reticulum cells of the human lymph node. Virchows Arch [Cell Pathol] *37:* 125-135, 1981.
39. Allaerts W, Jeucken PH, Bosman FT, Drexhage HA: Relationship between dendritic cells and folliculo-stellate cells in the pituitary: immunohistochemical comparison between mouse, rat and human pituitaries. Adv Exp Med Biol *329:* 637-642, 1993.
40. Jones TH, Kennedy RL: Cytokines and hypothalamic-pituitary function. Cytokine *5:* 531-538, 1993.
41. Gaudecker B von, Müller-Hermelink HK: Ontogeny and organization of the stationary non-lymphoid cells in the human thymus. Cell Tissue Res *207:* 287-306, 1980.
42. Ardavin C, Martinez del Hoyo G, Martin P, Anjuère, F, Arias CF, Marin AR, Ruiz S, Parrillas V, Hernandez H: Origin and differentiation of dendritic cells. Trends in Immunology *22:*691-700, 2001.
43. Ardavin C, Wu L, Li C-L, Shortman K: Thymic dendritic cells and T cells develop simultaneously in the thymus from a common precursor population. Nature *362:* 761-763, 1993.
44. Haynes BF, Heinly CS: Early human T cell development: analysis of the human thymus at the time of initial entry of hematopoietic stem cells into the fetal thymic microenvironment. J Exp Med *181:* 1445-1458, 1995.
45. Wu L, Vremec D, Ardavin C, Winkel K, Suss G, Georgiou H, Maraskovsky E, Cook W, Shortman K: Mouse thymus dendritic cells: kinetics of development and changes in surface markers during maturation. Eur J Immunol *25:* 418-425, 1995.
46. Wu L, Li C-L, Shortman K: Thymic dendritic cell precursors: relationshhip to the T lymphocyte lineage and phenotype of the dendritic cell progeny. J Exp Med *184:* 903-911, 1996.
47. Gale RP: Development of the immune system in human fetal liver. Thymus *10:* 45-56, 1987.
48. Lobach DF, Haynes BF: Ontogeny of the human thymus during fetal development. J Clin Immunol 7: 81-97, 1987.
49. Res P, Martinez-Caceres E, Jaleco AC, Staal F, Noteboom E, Weijer K, Spits H: $CD34^{+}$CD38dim cells in the human thymus can differentiate into T, natural killer, and dendritic cells but are distinct from pluripotent stem cells. Blood 87: 5196-5206, 1996.
50. Craig W, Kay R, Cutler RL, Lansdorp P: Expression of Thy-1 on human hematopoietic progenitor cells. J Exp Med 177: 1331-1342, 1993.
51. Merkenschlager M, Fisher AG: CD45 isoform switching precedes the activation-driven death of human thymocytes by apoptosis. Int Immunol *3:* 1-7, 1991.
52. Yeoman H, Gress RE, Bare CE, Leary AG, Boyse EA, Bard J, Schultz LD, Harris DT, DeLuca D: Human bone marrow and umbilical cord blood cells generate $CD4^{+}$ and $CD8^{+}$ single-positive T cells in murine fetal thymus organ culture. Proc Natl Acad Sci USA *90:* 10778-10782, 1993.
53. Plum J, De Smedt M, Defresne M-P, Leclercq G, Vanderkerckhove B: Human $CD34^{+}$ fetal liver stem cells differentiate to T cells in a mouse thymic microenvironment. Blood *84:* 1587-1593, 1994.
54. Kurtzberg J, Denning SM, Nycum LM, Singer KH, Haynes BF: Immature human thymocytes can be driven to differentiate into nonlymphoid lineages by cytokines from thymic epithelial cells. Proc Natl Acad Sci USA 86*:* 7575-7579, 1989.
55. Fairchild PF, Austyn JM: Thymic dendritic cells: Phenotype and function. Int Rev Immunol *6:* 187-196, 1990.
56. Barcena A, Galy AHM, Punnonen JJ, Muench MO, Schols D, Roncarola MG, de Vries JE, Spits H: Lymphoid and myeloid differentiation of fetal liver $CD34^{+}$ lineage-cells in human thymic organ culture. J Exp Med *180:* 123-132, 1994.
57. Marquez C, Trigueros C, Fernandez E, Toribio ML: The development of T and non-T cell lineages from $CD34^{+}$ human thymic precursors can be traced by the differential expression of CD44. J Exp Med *181:* 475-483, 1995.
58. Deans JP, Wilkins JA, Caixia S, Pruski E, Pilarski LM: Prolonged expression of high molecular mass CD45RA isoform during the differentiation of human progenitor thymocytes to $CD3^{+}$ cells in vitro. J Immunol *147:* 4060-4068, 1991.
59. Sanchez M-JJ, Muench MO, Roncarolo MG, Lanier L, Phillips JH: Identification of a common T/natural killer cell progenitor in human fetal thymus. J Exp Med *180:* 569-576, 1994.
60. Fink PJ, Bevan MJ: H-2 antigens of the thymus determine lymphocyte specificity. J Exp Med *148:* 766-775, 1978.
61. Bevan MJ, Fink PJ: The influence of thymus H-2 antigens on the specificity of maturing killer and helper cells. Immunol Rev *42:* 3-19, 1978.
62. Zinkernagel RM, Callahan GN, Klein J, Dennert G: Cytotoxic T cells learn specificity for self H-2 during differentiation in the thymus. Nature *271:* 251-253, 1978.
63. Zinkernagel RM, Callahan GN, Althage A, Cooper S, Klein PA, Klein J: On the thymus in the differentiation of "H-2 self-recognition" by T cells: evidence for dual recognition? J Exp Med *147:* 882-896, 1978.
64. Beller DI, Unanue ER: Ia antigens and antigen-presenting function of thymic macrophages. J Immunol *124:* 1433-1440, 1980.
65. Oliver PD, LeDouarin NM: Avian thymic accessory cells. J Immunol *132:* 1748-1755, 1984.
66. Beller DI, Kiely JM, Unanue ER: Regulation of macrophage populations. I. Preferential induction of Ia-rich peritoneal exudates by immunologic stimuli. J Immunol *124:* 1426-1432, 1980.
67. Ewert DL, Gilmour DG, Briles WE, Cooper MD: Genetics of Ia-like alloantigens in chickens and linkage with B major histocompatibility complex. Immunogenetics *10:* 169-174, 1980.
68. Ewert DL, Munchus MS, Chen CL, Cooper MD: Analysis of structural properties and cellular distribution of avian Ia antigen by using monoclonal antibody to monomorphic determinants. J Immunol *132:* 2524-2530, 1984.
69. Brekelmans P, van Ewijk W: Phenotypic characterization of murine thymic microenvironments. Sem Immunol *2:* 13-24, 1990.
70. Kappler JW, Roehm N, Marrack P: T cell tolerance by clonal elimination in the thymus. Cell *49:* 273-280, 1987.
71. MacDonald HR, Lees RK, Schneider R, Zinkernagel RM, Hengartner H: Positive selection of $CD4^{+}$ thymocytes controlled by MHC class II gene products. Nature *336:* 471-473, 1988.
72. Kisielow P, Teh HS, Bluthmann H, von Boehmer H: Positive selection of antigen-specific T cells in thymus by restricting MHC molecules. Nature *335:* 730-733, 1988.
73. Kisielow P, Bluthmann H, Staerz UD, Steinmetz M, von Boehmer H: Tolerance in T-cell-receptor transgenic mice involves deletion of nonmature $CD4^{+}8^{+}$ thymocytes. Nature *333:* 742-746, 1988.
74. Kisielow P, von Boehmer H: Negative and positive selection of immature thymocytes: timing and the role of the ligand for $\alpha\beta$ T cell receptor. Sem Immunol *2:* 35-44, 1990.

75. Le Bouteiller P, Lenfant F: Antigen-presenting function(s) of the non-classical HLA-E,-F and -G class I molecules: the beginning of a story. Res Immunol *147:* 301-313, 1996.
76. Grey HM, Buus S, Colon S, Miles C, Sette A: Structural requirements and biological significance of interactions between peptides and the major histocompatibility complex. Phil Trans Royal Soc London. Series B: Biol Sci *323:* 545-552, 1989.
77. Steinman RM, Cohn ZA: Identification of a novel cell type in peripheral lymphoid organs of mice. I. Morphology, quantitation, tissue distribution. J Exp Med *137:* 1142-1162, 1973.
78. Markgraf R, von Gaudecker B, Müller-Hermelink HK: The development of the human lymph node. Cell Tissue Res *225:* 387-413, 1982.
79. Witmer MD, Steinman RM: The anatomy of peripheral lymphoid organs with emphasis on accessory cells: light-microscopic immunocytochemical studies of mouse spleen, lymph node and Peyer's patch. Am J Anat *170:* 4655-4681, 1984.
80. Kaiserling E, Stein H, Müller-Hermelink HK: Interdigitating reticulum cells in the human thymus. Cell Tiss Res *155:* 47-55, 1974.
81. Crivellato E, Mallardi F, Basa M, Zweyer M: Osmium-zinc iodide reacts with interdigitating cells in the mouse lymph nodes and spleen. Z mikroskop-anat Forsch *104:* 476-484, 1990.
82. Crivellato E, Baldini G, Basa M, Fusaroli P: The three-dimensional structure of interdigitating cells. Italian J Anat Embryol *98:* 243-258, 1993.
83. Kelly RH, Balfour BM, Armstrong JA, Griffiths S: Functional anatomy of lymph nodes. II. Peripheral lymph-borne mononuclear cells. Anat Rec *190:* 5-22, 1978.
84. Kamperdijk EWA, de Leeuw JHS, Hoefsmit ECM: Lymph node macrophages and reticulum cells in the immune response; the secondary response to paratyphoid vaccine. Cell Tissue Res *227:* 277-290, 1982.
85. Fossum S, Vaalard JL: The architecture of rat lymph nodes. I. Combined light and electronmicroscopy of lymph node cell types. Anat Embryol *167:* 229-246, 1983.
86. Hart DNJ, Fabre JW: Demonstration and characterization of Ia-positive dendritic cells in the interstitial connective tissue of rat heart and other tissues, but not brain. J Exp Med *154:* 347-361, 1981.
87. Hart DNJ, McKenzie JL: Interstitial dendritic cells. Int Rev Immunol *6:* 128-149, 1990.
88. Klug H, Mager B: Ultrastructure and function of interdigitating cells in the guinea pig thymus. Acta Morph Acad Sci Hung *27:* 11-9, 1979.
89. Klug H: Elektronenmikroskopische Untersuchungen zur Phagocytose strahlengeschädigter Lymphozyten im Thymus von Ratten. Z Zellforsch *68:* 43-56, 1965.
90. Duijvestijn AM, Kamperdijk EW: Birbeck granules in interdigitating cells of thymus and lymph node. Cell Biol Int Rep *6:* 655, 1982.
91. Duijvestijn AM, Sminia T, Kohler YG, Janse EM, Hoefsmit EC: Rat thymus micro-environment: an ultrastructural and functional characterization. Adv Exp Med Biol *149:* 441-446, 1982.
92. Duijvestijn AM, Kohler YG, Hoefsmit EC: Interdigitating cells and macrophages in the acute involuting rat thymus. An electron-microscopic study on phagocytic activity and population development. Cell Tiss Res *224:* 291-301, 1982.
93. Miyazawa T, Sato C, Kojima K: Thymic phagocytosis and reduction in the negative surface charge of thymocytes after X irradiation. Radiat Res *79:* 622-629, 1979.
94. Higley HR, O'Morchoe CC: Morphometric analysis of thymic medullary non-lymphoid cell changes during postnatal development. Dev Comp Immunol *8:* 711-719, 1984.
95. Ewijk W van, Verzijden JH, Kwast TH van der, Luijcx-Meijer SW: Reconstitution of the thymus dependent area in the spleen of lethally irradiated mice. A light and electron microscopical study of the T-cell microenvironment. Cell Tiss Res *149:* 43-60, 1974.
96. Heusermann U, Stutte HJ, Müller-Hermelink HK: Interdigitating cells in the white pulp of the human spleen. Cell Tiss Res *153:* 415-7, 1974.
97. Langerhans P: Uber die Nerven der menschlichen Haut. Virchow's Arch A (Pathol Anat) *44:* 325-338, 1868.
98. Silberberg I: Apposition of mononuclear cells to Langerhans cells in contact allergic reactions: an ultrastructural study. Acta Dermatol Venereol *53:* 1-12, 1973.
99. Strunk D, Rappersberger K, Egger C, Strobl H, Kromer E, Elbe A, Maurer D, Stingl G: Generation of human dendritic cells/Langerhans cells from circulating $CD34^+$ hematopoietic progenitor cells. Blood *87:* 1292-1302, 1996.
100. Birbeck MS, Breathnach AS, Everall JD: An electron microscope study of basal melanocytes and high-level clear cells (Langerhans cells) in vitiligo. J Invest Dermatol *37:* 51-64, 1961.
101. Silberberg I, Baer RL, Rosenthal SA; Circulating Langerhans cells in a dermal vessel. Acta Dermato-Venereol *54:* 81-85, 1974.
102. Silberberg-Sinakin I, Fedorko ME, Baer RL, Rosenthal SA, Berezowsky V, Thorbecke GJ: Langerhans cells: target cells in immune complex reactions. Cell Immunol *32:* 400-416, 1977.
103. Silberberg-Sinakin I, Gigli I, Baer RL, Thorbecke GJ: Langerhans cells: role in contact hypersensitivity and relationship to lymphoid dendritic cells and to macrophages. Immunol Rev *53:* 203-232, 1980.
104. Bucana CD, Munn CG, Song MJ, Dunner K Jr, Kripke ML: Internalization of Ia molecules into Birbeck granule-like structures in murine dendritic cells. J Invest Dermatol *99:* 365-373, 1992.
105. Henkes W, Syha J, Reske K: Nucleotide sequence of rat invariant γ chain cDNA clone pLRγ34.3. Nucleic Acids Res *16:* 11822, 1988.
106. Bakke O, Dobberstein B: MHC class II associated invariant chain contains a sorting signal for endosomal compartments. Cell *63:* 707-716, 1990.
107. Lotteau V, Teyton L, Peleraux A, Nilsson T, Karlsson L, Schmid SL, Quaranta V, Peterson PA: Intracellular transport of class II MHC molecules directed by invariant chain. Nature *348:* 600-605, 1990.
108. Naujokas MF, Morin M, Anderson MS, Peterson M, Miller J: The chondroitin sulfate form of invariant chain can enhance stimulation of T cell responses through interaction with CD44. Cell *74:* 257-268, 1993.
109. Hashimoto K, Tarnowski WM: Some new aspects of the Langerhans cell. Arch Dermatol *97:* 450-464, 1968.
110. Hashimoto K: Langerhans' cell granule. An endocytotic organelle. Arch Dermatol *104:* 148-160, 1971.
111. Ferreira-Marques J: Systema sensitivum intra-epidermicum. Die Langerhansschen Zellen als Rezeptoren des hellen Schmerzes: Doloriceptores. Arch Dermatol Syph *193:* 191-250, 1951.
112. Niebauer G: Über die interstitiellen Zellen der Haut. Hautarzt *7:* 123-126, 1956.
113. Richter R: Studien zur Neurohistologie der nervösen vegetativen Peripherie der Haut bei verschiedenen chronischen infektiösen Granulomen mit besonderer Berücksichtigung der Langerhansschen Zellen; Tuberkolosen der Haut. Arch Klin Exp Dermatol *202:* 466-495, 1956.

114. Richter R: Studien zur Neurohistologie der nervösen vegetativen Peripherie der Haut bei verschiedenen chronischen infektiösen Granulomen mit besonderer Berücksichtigung der Langerhansschen Zellen; tertiäre Syphilide der Haut. Arch Klin Exp Dermatol *202:* 496-508, 1956.
115. Richter R: Studien zur Neurohistologie der nervösen vegetativen Peripherie der Haut bei verschiedenen chronischen infektiösen Granulomen mit besonderer Berücksichtigung der Langerhansschen Zellen; Leishmaniosis cutis. Arch Klin Exp Dermatol *202:* 509-517, 1956.
116. Richter R: Studien zur Neurohistologie der nervösen vegetativen Peripherie der Haut bei verschiedenen chronischen infektiösen Granulomen mit besonderer Berücksichtigung der Langerhansschen Zellen; Lepra. Arch Klin Exp Dermatol *202:* 518-555, 1956.
117. Niebauer G: Über die Dendritenzellen bei Vitiligo. Dermatologica *130:* 317-324, 1965.
118. Niebauer G, Sekido N: Über die Dendritenzellen der Epidermis. Eine Studie über die Langerhans-Zellen in der normalen und ekzematösen Haut des Meerschweinchens. Arch Klin Exp Dermatol *222:* 23-42, 1965.
119. Masson P: My conception of cellular nevi. Cancer *4:* 9-38, 1951.
120. Billingham RE, Medawar PB: "Desensitization" to skin homografts by injections of donor skin extracts. Ann Surg *137:* 444-449, 1953.
121. Fan J, Hunter R: Langerhans cells and the modified technic of gold impregnation by Ferreira-Marques. J Invest Dermatol *31:* 115-121, 1958.
122. Fan J, Schoenfeld RJ, Hunter R: A study of the epidermal clear cells with special references to their relationship to the cells of Langerhans. J Invest Dermatol *32:* 445-450, 1959.
123. Billingham RE, Silvers WK: Re-investigation of the possible occurrence of maternally induced tolerance in guinea pigs. J Exp Zool *160:* 221-224, 1965.
124. Billingham RE, Silvers WK: Some biological differences between thymocytes and lymphoid cells. Wistar Inst Sympos Monogr *2:* 41-51, 1964.
125. Schuler G, Steinman RM: Murine epidermal Langerhans cells mature into potent immunostimulatory dendritic cells *in vitro*. J Exp Med *161:* 526-546, 1985.
126. Witmer-Pack MD, Olivier W, Valinsky J, Schuler G, Steinman RM: Granulocyte/macrophage colony-stimulating factor is essential for the viability and function of cultured murine epidermal Langerhans cells. J Exp Med *166:* 1484-1498, 1987.
127. Olah I, Dunay C, Rohlich P, Toro I: A special type of cells in the medulla of the rat thymus. Acta Biol Acad Sci Hung *19:* 97-113, 1968.
128. Haelst U van: Light and electron microscopic study of the normal and pathological thymus of the rat. I. The normal thymus. Zeitschr Zellforsch Mikroskop Anat *77:* 534-553, 1967.
129. Haelst U van: Light and electron microscopic study of the normal and pathological thymus of the rat. II. The acute thymic involution. Zeitschr Zellforsch Mikroskop Anat *80:* 153-182, 1967.
130. Warchol JB, Brelinska R, Jaroszewski J: Granules of Langerhans cells in the thymus contain gold. Experientia *40:* 75-76, 1984.
131. Zelickson AS: The Langerhans cell. J Invest Dermatol *44:* 201-212, 1965.
132. Breathnach AS, Wyllie LM: Electron microscopy of melanocytes and Langerhans cells in human fetal epidermis at fourteen weeks. J Invest Dermatol *44:* 51-60, 1965.
133. Wolff K: The fine structure of the Langerhans cell granule. J Cell Biol *35:* 468-473, 1967.
134. Wolff K: The Langerhans cell. Curr Prob Dermatol *4:* 79-145, 1971.
135. Kiistala U, Mustakallio KK: Electronmicroscopic evidence of synthetic activity in Langerhans cells of human epidermis. Zeitschr Zellforsch Mikroskop Anat *78:* 427-440, 1967.
136. Kiistala U, Mustakallio KK: The presence of Langerhans cells in human dermis with special reference to their potential mesenchymal origin. Acta Dermato-Venereol *48:* 115-122, 1968.
137. Prunieras M: Interactions between keratinocytes and dendritic cells. J Invest Dermatol *52:* 1-17, 1969.
138. Rowden G: Immuno-electron microscopic studies of surface receptors and antigens of human Langerhans cells. Br J Dermatol *97:* 593-608, 1977.
139. Rowden G, Lewis MG, Sullivan AK: Ia antigen expression on human epidermal Langerhans cells. Nature *268:* 247-248, 1977.
140. Rowden G: Expression of Ia antigens on Langerhans cells in mice, guinea pigs, and man. J Invest Dermatol *75:* 22-31, 1980.
141. Veerman AJ: On the interdigitating cells in the thymus-dependent area of the rat spleen: a relation between the mononuclear phagocyte system and T-lymphocytes. Cell Tiss Res *148:* 247-257, 1974.
142. Nossal GJ, Abbot A, Mitchell J, Lummus Z: Antigens in immunity. XV. Ultrastructural features of antigen capture in primary and secondary lymphoid follicles. J Exp Med *127:* 277-290, 1968.
143. Steinman RM, Witmer MD: Lymphoid dendritic cells are potent stimulators of the primary mixed leukocyte reaction in mice. Proc Natl Acad Sci USA *75:* 5132-5136, 1978.
144. Berman B, Gigli I: Complement receptors on guinea pig epidermal Langerhans cells. J Immunol *124:* 685-690, 1980.
145. Hammerling GJ, McDevitt HO: Antigen-binding structures on the surface of T lymphocytes. Israel J Med Sci *11:* 1331-1341, 1975.
146. Klein J, Hauptfeld V: Ia antigens: their serology, molecular relationships, and their role in allograft reactions. Transplant Rev *30:* 83-100, 1976.
147. Shreffler DC, David CS: The H-2 major histocompatibility complex and the I immune response region: genetic variation, function, and organization. Adv Immunol *20:* 125-195, 1975.
148. Nagao S, Inaba S, Iijima S: Langerhans cells at the sites of vaccinia virus inoculation. Arch Dermatol Res/Archiv fur Dermatol Forsch *256:* 23-31, 1976.
149. Strunk D, Rappersberger K, Egger C, Strobl H, Kromer E, Elbe A, Maurer D, Stingl G: Generation of human dendritic cells/Langerhans cells from circulating $CD34^+$ hematopoietic progenitor cells. Blood *87:* 1292-1302, 1996.
150. Katz SI, Tamaki K, Sachs DH: Epidermal Langerhans cells are derived from cells originating in bone marrow. Nature *282:* 324-326, 1979.
151. Frelinger JG, Hood L, Hill S, Frelinger JA: Mouse epidermal Ia molecules have a bone marrow origin. Nature *282:* 321-323, 1979.
152. Rausch E, Kaiserling E, Goos M: Langerhans cells and interdigitating reticulum cells in the thymus-dependent region in human dermatopathic lymphadenitis. Virchows Archiv - B Cell Pathol *25:* 327-343, 1977.
153. Hoffman-Fezer G, Rodt H, Thierfelder S: Immunohistochemical identification of T- and B-lymphocytes delineated by the unlabeled antibody enzyme method. II. Anatomical distribution of T- and B-cells in

lymphoid organs of nude mice. Beitr Pathol *161:* 17-26, 1977.
154. Hoffman-Fezer G, Rodt H, Götze D, Thierfelder S: Anatomical distribution of T and B lymphocytes identified by immunohistochemistry in the chicken spleen. Int Arch Allergy Appl Immunol *55:* 86-95, 1977.
155. Wekerle H, Ketelsen U-P, Ernst M: Thymic nurse cells. Lymphoepithelial cell complexes in murine thymuses: morphological and serological characterization. J Exp Med *151:* 925-944, 1980.
156. Ritter MA, Sauvage CA, Cotmore SF: The human thymus microenvironment: in vivo identification of thymic nurse cells and other antigenically-distinct subpopulations of epithelial cells. Immunology *44:* 439-446, 1981.
157. Kyewski BA, Kaplan H.S: Lymphoepithelial interactions in the mouse thymus: phenotypic and kinetic studies on thymic nurse cells. J Immunol *128:* 2287-2294, 1982.
158. Janckila AJ, Yam LT, Li C-Y: Immunoalkaline phosphatase cytochemistry. Am J Clin Pathol *84:* 476-480, 1985.
159. Cattoretti G, Berti E, Schiro R, D' Amato L, Valeggio C, Rilke F: Improved avidin-biotin-peroxidase complex (ABC) staining. Histol J *20:* 75-80, 1988.
160. Bodey B, Zeltzer PM, Saldivar V, Kemshead J: Immunophenotyping of childhood astrocytomas with a library of monoclonal antibodies. Int J Cancer *45:* 1079-1087, 1990.
161. Yam LT, Janckila AJ, Epremian BE, Li C-Y: Diagnostic significance of levamisole-resistant alkaline phosphatase in cytochemistry and immunocytochemistry. Am J Clin Pathol *91:* 31-36, 1989.
162. Strasburger CJ, Amir-Zaltsman Y, Kohen F: The avidin-biotin reaction as an universal amplification system in immunoassays. Prog Clin Biol Res *285:* 79-100, 1988.
163. Wilchek M, Bayer EA: Introduction to avidin-biotin technology. Methods Enzymol *184:* 5-13, 1990.
164. Duhamel RC, Whitehead JS: Prevention of nonspecific binding of avidin. Methods Enzymol *184:* 201-207, 1990.
165. Diamandis EP, Christopoulos TK: The biotin-(strept) avidin system: principles and applications in biotechnology. Clin Chem *37:* 625-636, 1991.
166. Hsu SM, Raine L, Fanger H: Use of avidin-biotin-peroxidase complex (ABC) in immunoperoxidase techniques: a comparison of ABC and unlabeled antibody (PAP) procedure. J Histochem Cytochem *29:* 577-581, 1981.
167. Battifora H: The multitumor (sausage) tissue block: novel method for immunohistochemical antibody testing. Lab Invest *55:* 244-248, 1986.
168. Battifora H, Mehta P: The checkerboard tissue block. An improved multitissue control block. Lab Invest *63:* 722-724, 1990.
169. Bodey B, Bodey B Jr, Kaiser HE: Cell culture observations of Postnatal Thymic Epithelium: An In Vitro Model for Growth and Humoral Influence on Intrathymic T Lymphocyte Maturation. IN VIVO *10:* 515-526, 1996.
170. Beller DI, Unanue ER: Thymic maturation in vitro by a secretory product from macrophages. J Immunol *118:* 1780-1787, 1977.
171. Kostowiecki M: The thymic macrophages. Z Mikr-Anat Forsch *69:* 585-614, 1963.
172. Herxheimer G: Fettinfiltration des Thymus. Verh Dtsch Path Ges *6 Tagg*, 1903-1904.
173. Holmström R: Über das Vorkommen von Fett und fettähnlichen Substanzen im Thymusparenchym. Arch Mikrosk Anat *77:* 323-345, 1911.
174. Barbano C: Die normale Involution der Thymus. Virchows Arch *207:* 1-27, 1912.
175. Hart C: Thymusstudien. I. Über das Auftreten von Fett in der Thymus. Virchows Arch *207:* 27-55, 1912.
176. Fulci F: Die Natur der Thymusdrüse nach Untersuchungen über ihre Regenerationsfähigkeit bei Säugetieren. Deutsch Med Wschenschr *39:* 1776-1780, 1913.
177. Hammar JA, Lagergren KA: Beiträge zur Konstitutionsanatomie 5. Verhalten der Thymus bei akuten Infektionen: mikroskopische Analyse der Thymus in Fällen von Diptheria. Ztschr Angew Anat Konstitutionsl *4:* 314-398, 1918.
178. Hammar JA: Beiträge zur Konstitutionsanatomie 7. Mikroskopische Analyse der Thymus in einigen Fällen von Lues congenita. Beitr Path Anat Allg Path *66:* 37-91, 195-258, 1919.
179. Day AJ: Lipid metabolism by macrophages and its relationship to atherosclerosis. Adv Lipid Res *5:* 185-207, 1967.
180. Hammar JA: Zur Frage der Histogenese der Thymusdrüse. Zbl Path *33:* 505-513, 1923.
181. Hammar JA: Die Menschenthymus in Gesundheit und Krankheit: Ergebnisse der numerischen Analyse von mehr als tausend menschlichen Thymusdrüsen. Teil I. Das normale Organ - zugleich eine kritische Beleuchtung der Lehre des "Status thymicus". Z Mikr Anat Forsch (Leipzig) *6:* 1-570, 1926.
182. Rudberg H: Studien über die Thymusinvolution. I. Die Involution nach Röntgenbestrahlung. Arch Anat *Suppl:* 123-174, 1907.
183. Lubarsch O: Zur pathologischen Anatomie der Erschöpfungs- und Ernährungskrankheiten. Zieglers Beitr *69:* 242-251, 1921.
184. Saito H: Beiträge zur pathologischen Anatomie und Histologie der Ernährungsstörungen der Säuglinge. Virchows Arch *250:* 69, 1924.
185. Lewin JE: Involution und Regeneration des Thymus unter dem Einfluss von Benzol. Virchows Arch Pathol Anat *268:* 1-16, 1928.
186. Babès A: Sur la nature du réticulum du thymus. Compt Rend Soc Biol *101:* 196, 1929.
187. Marine D: Some effects of suprarenal injury on natural and acquired resistance. Libmann Anniv Vols *2:* 773-780, 1932.
188. Crotti A: Research on etiology of endemic goiter. Tr Int Coll Surg *1:* 141-154, 1938.
189. Downey H: Cytology of rabbit thymus and regeneration of its thymocytes after irradiation, with some notes on the human thymus. Blood *3:* 1315-1341, 1948.
190. Baillif RN: Thymic involution and regeneration in albino rat, following injection of acid colloidal substances. Am J Anat *84:* 457-510, 1949.
191. Loewenthal LA, Smith C: Studies in the thymus of the mammal. IV. Lipid-laden foamy cells in the involuting thymus of the mouse. Anat Rec *112:* 1-15, 1952.
192. Bodey B, Calvo W, Prummer O, Fliedner TM, Borysenko M: Development and histogenesis of the thymus in dog. A light and electronmicroscopical study. Dev Comp Immunol *11:* 227-238, 1987.
193. Bodey B: Development of the lymphopoiesis as a function of the thymic microenvironment. Use of $CD8^+$ cytotoxic T lymphocytes for cellular immunotherapy of human cancer. IN VIVO *8:* 915-943, 1994.
194. Tschassownikow N: Über die *in vitro*-Kulturen des Thymus. Arch Exp Zellforsch *3:* 250-276, 1926.
195. Aschoff L: Das reticulo-endotheliale System. Erg Inn Med *26:* 1-118, 1924.
196. Teploff I: Zur Methodik der Blutmikroanalyse. Biochem Ztschr *202:* 14-17, 1928.
197. Seki M: Zur Kenntnis der intra- und supravitalen Färbung; färberischer Beweis für die Reichlichkeit von basischen

Substanzen in den Histiocyten und Retikuloendothelien. Ztschr Zellforsch Mikr Anat *19:* 238-265, 1933.
198. Murray RG: Pure cultures of rabbit thymus epithelium. Am J Anat *81:* 369-411, 1947.
199. Kiyono K: Zur Frage der histiozytären Blutzellen. Folia Hematol *18:* 149-170, 1914.
200. Popoff NW: Histogenesis of the thymus as shown by tissue cultures. Arch Exp Zellforsch *4:* 395-418, 1927.
201. Baginski S, Borsuk J: Le thymus et le système réticulo-endothélial. Bull d'Histol Appliq Physiol *16:* 105-109, 1939.
202. Hammar JA: Die normal-morphologische Thymusforschung im letzten Vierteljahrhundert. Analyse und Synthese. Leipzig, Barth, 1936.
203. Deanesly R: Experimental studies on histology of mammalian thymus. Quart J Micr Sc *72:* 247-275, 1928.
204. Dontigny P, Beland E, Hall E, Selye H: Influence de l'adrénalectomie sur l'action néphrosclérotique des préparations hypophysaires. Rev Canad Biol *5:* 356-358, 1946.
205. Deane HW, Greep RD: Cytochemical study of adrenal cortex in hypo- and hyperthyroidism. Endocrinology *41:* 243-257, 1947.
206. Savard K, Homburger F: Thymic atrophy and lymphoid hyperplasia in mice bearing sarcoma 180. Proc Soc Exp Biol Med *70:* 68-70, 1949.
207. Selye H: Thymus and adrenals in the response of the organism to injuries and intoxications. Br J Exp Pathol *17:* 234-248, 1936.
208. Smith C, Holst EA: Studies of the thymus of the mammal. VII. Lipids in the thymuses of irradiated mice. Anat Rec *116:* 123-137, 1953.
209. Smith C: The thymus in immunobiology. *In:* The Thymus in Immunobiology. Structure, function and role in disease. (Good RA, Gabrielsen AE, eds) New York, Harper & Row, Inc., 1964.
210. Kostowiecki M: Development and degeneration of the second type of Hassall's corpuscles in the thymus of guinea pig. Anat Rec *142:* 195-203, 1962.
211. Nelson DS: Macrophages: progress and problems. Clin Exp Immunol *45:* 225-233, 1981.
212. Carr I: The fine structure of microfibrils and microtubules in macrophages and other lymphoreticular cells in relation to cytoplasmic movement. J Anat *112:* 383-389, 1972.
213. Carr I: The reticulum cell and the reticular cell in the mouse popliteal lymph node. An electron microscopic autoradiographic study. Virchows Arch B Cell Pathol *15:* 1-10, 1973.
214. van Furth R, van der Meer JW: Reticulum cell not a haematopoietic stem cell. Br Med J *3:* 371, 1975.
215. Shevach EM: The function of macrophages in antigen recognition by guinea pig T lymphocytes. III. Genetic analysis of the antigens mediating macrophage-T lymphocyte interaction. J Immunol *116:* 1482-1489, 1976.
216. Unanue ER: Secretory function of mononuclear phagocytes: a review. Am J Pathol *83:* 396-417, 1976.
217. Unanue ER: The macrophage as a regulator of lymphocyte function. Hosp Pract *14:* 61-64, 69-74, 1979.
218. Unanue ER: Cooperation between mononuclear phagocytes and lymphocytes in immunity. N Engl J Med *303:* 977-985, 1980.
219. Cline MJ: Monocytes, macrophages, and their diseases in man. J Invest Dermatol *71:* 56-58, 1978.
220. Cline MJ, Lehrer RI, Territo MC, Golde DW: UCLA Conference. Monocytes and macrophages: functions and diseases. Ann Int Med *88:* 78-88, 1978.
221. Cohn ZA: Activation of mononuclear phagocytes: fact, fancy, and future. J Immunol *121:* 813-816, 1978.
222. Persson U, Hammarstrom L, Moller E, Moller G, Smith CI: The role of adherent cells in B and T lymphocyte activation. Immunol Rev *40:* 78-101, 1978.
223. Hopper KE, Wood PR, Nelson DS: Macrophage heterogeneity. Vox Sanguinis *36:* 257-274, 1979.
224. Hopper KE, Harrison J, Nelson DS: Partial characterization of anti-tumor effector macrophages in the peritoneal cavities of concomitantly immune mice and mice injected with macrophage-stimulating agents. J Reticuloendo Soc *26:* 259-271, 1979.
225. Nathan CF, Murray HW, Cohn ZA: The macrophage as an effector cell. N Engl J Med *303:* 622-626, 1980.
226. Unanue ER: The regulation of lymphocyte functions by the macrophage. Immunol Rev *40:* 227-255, 1978.
227. Holt PG, Batty JE: Macrophage regulation of the IgE response. I. Studies on the immunogenicity and antigenicity of macrophage-associated antigen. Int Arch Allergy App Immunol *63:* 73-82, 1980.
228. Katz DH: Recent studies on the regulation of IgE antibody synthesis in experimental animals and man. Immunology *41:* 1-24, 1980.
229. Nelson M, Nelson DS: Macrophages and resistance to tumours: influence of agents affecting macrophages and delayed-type hypersensitivity on resistance to tumours inducing concomitant immunity. Australian J Exp Biol Med Sci *56:* 211-223, 1978.
230. Nelson M, Nelson DS: Macrophages and resistance to tumours. I. Inhibition of delayed-type hypersensitivity reactions by tumour cells and by soluble products affecting macrophages. Immunology *34:* 277-290, 1978.
231. Nelson DS, Hopper KE, Blanden RV, Gardner ID, Kearney R: Failure of immunogenic tumors to elicit cytolytic T cells in syngeneic hosts. Cancer Lett *5:* 61-67, 1978.
232. Kirchner H, Glaser M, Holden HT, Fernbach BR, Herberman RB: Suppressor cells in tumor bearing mice and rats. Biomedicine *24:* 371-374, 1976.
233. Ptak W, Zembala M, Gershon RK: Intermediary role of macrophages in the passage of suppressor signals between T-cell subsets. J Exp Med *148:* 424-434, 1978.
234. Kallenberg CG, de Gast GC, The TH: Suppression of DNA synthesis by con A-activated human lymphocytes: role of monocytes in con A-induced suppression. Clin Exp Immunol *41:* 583-590, 1980.
235. Harris PE, Colovai AI, Maffei A, Liu Z, Foca NS: Major histocompatibility complex class I presentation of exogenous and endogenous protein-derived peptides by a transfected human monocyte cell line. Immunology *86:* 606-611, 1995.
236. Herberman RB, Holden HT: Natural cell-mediated immunity. Adv Cancer Res *27:* 305-377, 1978.
237. Herberman RB, Holden HT, Djeu JY, Jerrells TR, Varesio L, Tagliabue A, White SL, Oehler JR, Dean JH: Macrophages as regulators of immune responses against tumors. Adv Exp Med Biol *121B:* 361-379, 1979.
238. Herberman RB, Djeu J, Kay HD, Ortaldo JR, Riccardi C, Bonnard GD, Holden HT, Fagnani R, Santoni A, Puccetti P: Natural killer cells: characteristics and regulation of activity. Immunol Rev *44:* 43-70, 1979.
239. Herberman RB, Holden HT: Natural killer cells as antitumor effector cells. J Natl Cancer Inst *62:* 441-445, 1979.
240. Kurland J, Moore MA: Modulation of hemopoiesis by prostaglandins. Exp Hematol *5:* 357-373, 1977.
241. Kurland JI, Bockman R: Prostaglandin E production by human blood monocytes and mouse peritoneal macrophages. J Exp Med *147:* 952-957, 1978.
242. Kurland JI, Broxmeyer HE, Pelus LM, Bockman RS, Moore MA: Role for monocyte-macrophage-derived colony-stimulating factor and prostaglandin E in the positive and

negative feedback control of myeloid stem cell proliferation. Blood *52:* 388-407, 1978.
243. Kurland JI, Bockman RS, Broxmeyer HE, Moore MA: Limitation of excessive myelopoiesis by the intrinsic modulation of macrophage-derived prostaglandin E. Science *199:* 552-555, 1978.
244. Kurland JI, Pelus LM, Ralph P, Bockman RS, Moore MA: Induction of prostaglandin E synthesis in normal and neoplastic macrophages: role for colony-stimulating factor(s) distinct from effects on myeloid progenitor cell proliferation. Proc Natl Acad Sci USA *76:* 2326-2330, 1979.
245. Kurland JI, Meyers PA, Moore MA: Synthesis and release of erythroid colony- and burst-potentiating activities by purified populations of murine peritoneal macrophages. J Exp Med *151:* 839-852, 1980.
246. Leibovich SJ, Ross R: A macrophage-dependent factor that stimulates the proliferation of fibroblasts in vitro. Am J Pathol *84:* 501-14, 1976.
247. Leibovich SJ: Production of macrophage-dependent fibroblast-stimulating activity (M-FSA) by murine macrophages. Effects on BALBc 3T3 fibroblasts. Exp Cell Res *113:* 47-56, 1978.
248. Calvo W, Fliedner TM, Herbst EW, Hugl E, Bodey B: Degenerative changes and recovery of the thymus of lethally irradiated dogs, rescued by transfusion of cryopreserved autologous blood leukocytes. Exp Hematol 15: 1171-1178, 1987.
249. Wu L, Scollay R, Egerton M, Pearse M, Spangrude GJ, Shortman K: CD4 expressed on earliest T-lineage precursor cells in the adult murine thymus. Nature 349: 71-74, 1991.
250. 250. Wu L, Li CL, Shortman K: Thymic dendritic cell precursors: relationship to the T lymphocyte lineage and phenotype of the dendritic cell progeny. J Exp Med *184:* 903-911, 1996.
251. 251. Shortman K, Caux C: Dendritic cells development: multiple pathways to nature's adjuvants. Stem Cells 15: 409-419, 1997.
252. Shortman K, Wu L: Parentage and heritage of dendritic cells. Blood *97:* 3325, 2001.
253. Wu L, D'Amico A, Hochrein H, O'Keeffe M, Shortman K, Lucas K: Development of thymic and splenic dendritic cell populations from different hemopoietic precursors. Blood 98: 3376-3382, 2001.
254. Shortman K, Liu YJ: Mouse and human dendritic cell subtypes. Nat Rev Immunol *2:* 151-161, 2002.
255. O'Keeffe M, Hochrein H, Vremac D, Scott B, Hertzog P, Tatarczuch L, Shortman K: Dendritic cell precursor populations of mouse blood: identification of the murine homologues of human blood plasmocytoid pre-DC2 and CD11+ DC1 precursors. Blood *101:* 1453-1459, 2003.
256. Tan PS, Gavin AL, Barnes N, Sears DW, Vremec D, Shortman K, Amigorena S, Mottram PL, Hogarth PM: Unique monoclonal antibodies define expression of FcγRI on macrophages and mast cell lines and demonstrate heterogeneity among subcutaneous and other dendritic cells. J Immunol *170:* 2549-2556, 2003.
257. Kim YJ, Broxmeyer HE: 4-IBB ligand stimulation enhances myeloid dendritic cell maturation from human umbilical cord blood CD34 (+) progenitor cells. J Hematother Stem Cell Res *11:* 895-903, 2002.
258. Guermonprez P, Valladeau J, Zitvogel L, Thery C, Amigorena S: Antigen presentation and T cell stimulation by dendritic cells. Annu Rev Immunol *20:* 621-667, 2002.
259. Steinman RM: Dendritic cells and immune-based therapies. Exp Hematol *24:* 859-862, 1996.
260. Robinson JH, Delvig AA: Diversity in MHC class II antigen presentation. Immunology 105: 252-262, 2002.
261. Kalinski P, Hilkens CM, Wierenga EA, Kapsenberg ML: T-cell priming by type-1 and type-2 polarized dendritic cells: the concept of a third signal. Immunol Today *20:* 561-567, 1999.
262. Banchereau J, Briere F, Caux C, Davoust J, Lebecque S, Liu YJ, Pulendran B, Palucka K: Immunobiology of dendritic cells. Annu Rev Immunol *18:* 767-811, 2000.
263. Lutz MB, Assmann CU, Girolomoni G, Ricciardi-Castagnoli P: Different cytokines regulate antigen uptake and presentation of a precursor dendritic cell line. Eur J Immunol *26:* 586-594, 1996.
264. Aderem A, Ulevitch RJ: Toll-like receptors in the induction of the innate immune response. Nature *406:* 782-787, 2000.
265. Visintin A, Mazzoni A, Spitzer JH, Wyllie DH, Dower SK, Segal DM: Regulation of Toll-like receptors in human monocytes and dendritic cells. J Immunol *166:* 249-255, 2001.
266. Jurgens M, Wollenberg A, Hanau D, de la Salle H, Bieber T: Activation of human epidermal Langerhans cells by engagement of the high affinity receptor for IgE, Fc epsilon RI. J Immunol 155: 5184-5189, 1995.
267. Regnault A, Lankar D, Lacabanne V, Rodriguez A, Thery C, Rescigno M, Saito T, Verbeek S, Bonnerot C, Ricciardi-Castagnoli P, Amigorena S: Fcγ receptor-mediated induction of dendritic cell maturation and major histocompatibility complex class I-restricted antigen presentation after immune complex internalization. J Exp Med 189: 371-380, 1999.
268. Geissmann F, Launay P, Pasquier B, Lepelletier Y, Leborgne M, Lehuen A, Brousse N, Monteiro RC: A subset of human dendritic cells expresses IgA Fc receptor (CD89), which mediates internalization and activation upon cross-linking by IgA complexes. J Immunol *166:* 346-352, 2001.
269. Thery C, Amigorena S: The cell biology of antigen presentation in dendritic cells. Curr Opin Immunol *13:* 45-51, 2001.
270. Galluci S, Matzinger P: Danger signals: SOS to the immune system. Curr Opin Immunol *13:* 114-119, 2001.
271. Steinman RM, Nussenzweig MC: Dendritic cells: features and functions. Immunol Rev 53: 127-148, 1980.
272. Steinman RM: The dendritic cell system and its role in immunogenicity. Annual Rev Immunol 9: 271-296, 1991.
273. Monaco JJ: Structure and function of genes in the MHC class II region. Curr Opin Immunol 5: 17-20, 1993.
274. Neefjes JJ, Momburg F: Cell biology of antigen presentation. Curr Opin Immunol 5: 27-34, 1993.
275. Braciale TJ, Braciale VL: Antigen presentation: structural themes and functional variations. Immunol Today *12:* 124-129, 1991.
276. Engering AJ, Cella M, Fluitsma DM, Hoefsmit EC, Lanzavecchia A, Pieters J: Mannose receptor mediated antigen uptake and presentation in human dendritic cells. Adv Exp Med Biol *417:* 183-187, 1997.
277. Matzinger P: Tolerance, danger, and the extended family. Annu Rev Immunol *12:* 991-1045, 1994.
278. Nanda NK, Sercarz E: A truncated T cell receptor repertoire reveals underlying immunogenicity of an antigenic determinant. J Exp Med *184:* 1037-1043, 1996.
279. Lee PP, Yee C, Savage PA, Fong L, Brockstedt D, Weber JS, Johnson D, Swetter S, Thompson J, Greenberg PD, Roederer M, Davis MM: Characterization of circulating T cells specific for tumor-associated antigens in melanoma patients. Nat Med *5:* 677-685, 1999.
280. Pardoll DM: Inducing autoimmune disease to treat cancer. Proc Natl Acad Sci USA *96:* 5340-5342, 1999.
281. Bodey B, Bodey B Jr., Siegel SE, Kaiser HE: Failure of cancer vaccines: the significant limitations of this approach

to immunotherapy. Anticancer Research *20:* 2665-2676, 2000.
282. Ribas A, Butterfield LH, Glaspy JA, Economou JS: Cancer immunotherapy using gene-modified dendritic cells. Curr Gene Ther *2:* 57-78, 2002.
283. Bennett SR, Carbone FR, Karamalis F, Flavell RA, Miller JF, Heath WR: Help for cytotoxic-T-cell responses is mediated by CD40 signalling. Nature *393:* 478-480, 1998.
284. Ridge JP, Di Rosa F, Matzinger P: A conditioned dendritic cell can be a temporal bridge between a CD4+ T-helper and a T-killer cell. Nature *393:* 474-478, 1998.
285. Grakoui A, Bromley SK, Sumen C, Davis MM, Shaw AS, Allen PM, Dustin ML: The immunological synapse: a molecular machine controlling T cell activation. Science *285:* 221-227, 1999.
286. Malissen B: Dancing the immunological two-step. Science *285:* 207-208, 1999.
287. Wulfing C, Davis MM: A receptor/cytoskeletal movement triggered by costimulation during T cell activation. Science *282:* 2266-2269, 1998.
288. Bachmann MF, Wong BR, Josien R, Steinman RM, Oxenius A, Choi Y: TRANCE, a tumor necrosis factor family member critical for CD40 ligand-independent T helper cell activation. J Exp Med *189:* 1025-1031, 1999.
289. Lu Z, Yuan L, Zhou X, Sotomayor E, Levitsky HI, Pardoll DM: CD40-independent pathways of T cell help for priming of CD8(+) cytotoxic T lymphocytes. J Exp Med *191:* 541-550, 2000.
290. Chinnaiyan AM, Hanna WL, Orth K, Duan H, Poirier GG, Froelich CJ, Dixit VM: Cytotoxic T-cell-derived granzyme B activates the apoptotic protease ICE-LAP3. Curr Biol *6:* 897-899, 1996.
291. Froelich CJ, Dixit VM, Yang X: Lymphocyte granule-mediated apoptosis: matters of viral mimicry and deadly proteases. Immunol Today *19:* 30-36, 1998.
292. Inaba K, Swiggard WJ, Inaba M, Meltzer J, Mirza A, Sasagawa T, Nussenzweig MC, Steinman RM: Tissue distribution of the DEC-205 protein that is detected by the monoclonal antibody NLDC-145. I. Expression on dendritic cells and other subsets of mouse leukocytes. Cell Immunol *163:* 148-156, 1995.
293. Guo M, Gong S, Maric S, Misulovin Z, Pack M, Mahnke K, Nussenzweig MC, Steinman RM: A monoclonal antibody to the DEC-205 endocytosis receptor on human dendritic cells. Hum Immunol *61:* 729-738, 2000.
294. Kato M, Neil TK, Fearnley DB, McLellan AD, Vuckovic S, Hart DN: Expression of multilectin receptors and comparative FITC-dextran uptake by human dendritic cells. Int Immunol *12:* 1511-1519, 2000.
295. Small M, Kraal G: In vitro evidence for participation of DEC-205 expressed by thymic cortical epithelial cells in clearance of apoptotic thymocytes. Int Immunol *15:* 197-203, 2003.
296. Jiang W, Swiggard WJ, Heufler C, Peng M, Mirza A, Steinman RM, Nussenzweig MC: The receptor DEC-205 expressed by dendritic cells and thymic epithelial cells is involved in antigen processing. Nature *375:* 151-155, 1995.
297. Mahnke K, Guo M, Lee S, Sepulveda H, Swain SL, Nussenzweig M, Steinman RM: The dendritic cell receptor for endocytosis, DEC-205, can recycle and enhance antigen presentation via major histocompatibility complex class II-positive lysosomal compartments. J Cell Biol *151:* 673-684, 2000.
298. Valladeau J, Ravel O, Dezutter-Dambuyant C, Moore K, Kleijmeer M, Liu Y, Duvert-Frances V, Vincent C, Schmitt D, Davoust J, Caux C, Lebecque S, Saeland S: Langerin, a novel C-type lectin specific to Langerhans cells, is an endocytic receptor that induces the formation of Birbeck granules. Immunity *12:* 71-81, 2000.
299. Geijtenbeek TB, Krooshoop DJ, Bleijs DA, van Vliet SJ, van Duijnhoven GC, Grabovsky V, Alon R, Figdor CG, van Kooyk Y: DC-SIGN-ICAM-2 interaction mediates dendritic cell trafficking. Nat Immunol *1:* 353-357, 2000.
300. Geijtenbeek TB, Engering A, Van Kooyk Y: DC-SIGN, a C-type lectin on dendritic cells that unveils many aspects of dendritic cell biology. J Leukoc Biol *71:* 921-931, 2002.
301. Sijts A, Zaiss D, Kloetzel PM: The role of the ubiquitin-proteasome pathway in MHC class I antigen processing: implications for vaccine design. Curr Mol Med *1:* 665-676, 2001.
302. Bodey B, Neuroendocrine influence on thymic haematopoiesis via the reticulo-epithelial cellular network. Expert Opinion Therapeutical Targets *6:* 57-72, 2002.
303. Kasai M, Hirokawa K, Kajino K, Ogasawara K, Tatsumi M, Hermel E, Monaco JJ, Mizuochi T: Difference in antigen presentation pathways between cortical and medullary thymic epithelial cells. Eur J Immunol *26:* 2101-2107, 1996.
304. Bodey B, Bodey B Jr., Kaiser HE: Dendritic type, accessory cells within the mammalian thymic microenvironment. Antigen presentation in the dendritic neuro-endocrine-immune cellular network. IN VIVO *11:* 351-370, 1997.
305. Darsow U, Ring J: Neuroimmune interactions in the skin. Curr Opinion Allergy Clin Immunol *1:* 435-439, 2001.
306. Timmerman JM, Levy R: Dendritic cell vaccines for cancer immunotherapy. Annu Rev Med *50:* 507-529, 1999.
307. Foss FM: Immunologic mechanisms of antitumor activity. Semin Oncol *29:* 5-11, 2002.
308. Porgador A, Gilboa E: Bone marrow-generated dendritic cells pulsed with a class I-restricted peptide are potent inducers of cytotoxic T lymphocytes. J Exp Med 182: 255-260, 1995.
309. Bodey B, Bodey B Jr., Kaiser HE: Apoptosis in the mammalian thymus during its normal histogenesis and under various in vitro and in vivo experimental conditions. IN VIVO *12:* 123-134, 1998.
310. Bodey B, Bodey B Jr., Siegel SE, Kaiser HE: Over-expression of endoglin (CD105): A marker of breast carcinoma-induced neo-vascularization. Anticancer Research *18:* 3621-3628, 1998.
311. Bodey B, Bodey B Jr., Siegel SE, Kaiser HE: Immunophenotypical (IP) analysis and immunobiology of childhood primary brain tumors. Anticancer Research *19:* 2973-2992, 1999.
312. Bodey B, Bodey B Jr., Siegel SE, Kaiser HE: Immunocytochemical detection of MMP-3 and -10 expression in hepatocellular carcinomas. Anticancer Research *20:* 4585-4590, 2000.
313. Bodey B, Bodey B Jr., Gröger AM, Siegel SE, Kaiser HE: Immunocytochemical detection of Homeobox B3, B4, and C6 gene product expression in lung carcinomas. Anticancer Research *20:* 2711-2716, 2000.
314. Bodey B, Bodey B Jr., Siegel SE, Kaiser HE: Matrix metalloproteinase expression in malignant melanomas: tumor-extracellular matrix interactions in invasion and metastasis. IN VIVO *15:* 57-64, 2001.
315. Bodey B, Siegel SE, Kaiser HE: MAGE-1, a Cancer-Testis Antigen, Expression in Childhood Astrocytomas as an Indicator of Tumor Progression. IN VIVO *16:* 583-588, 2002.
316. Nair SK, Snyder D, Rouse BT, Gilboa E: Regression of tumors in mice vaccinated with professional antigen-presenting cells pulsed with tumor extracts. Int J Cancer *70:* 706-715, 1997.
317. Paquette RL, Hsu NC, Kiertscher SM, Park AN, Tran L, Roth MD, Glaspy JA: Interferon-α and granulocyte-

macrophage colony-stimulating factor differentiate peripheral blood monocytes into potent antigen-presenting cells. J Leukocyte Biol *64:* 358-367, 1998.
318. Carbone JE, Ohm DP: Immune dysfunction in cancer patients. Oncology (Huntingt) *16:* 11-18, 2002.
319. Pioche C, Salomon B, Klatzmann D: Cellules dendritiques et therapie cellulaire antitumorale. Pathologie Biologie 43: 904-909, 1995.
320. Turnbull E, Macpherson G. Immunobiology of dendritic cells in the rat. Immunol Rev *184:* 58-68, 2001.
321. Gunzer M, Grabbe S: Dendritic cells in cancer immunotherapy. Crit Rev Immunol 21: 133-145, 2001.
322. Gallucci S, Lolkema M, Matzinger P: Natural adjuvants: endogenous activators of dendritic cells. Nat Med *5:* 1249-1255, 1999.
323. Gabrilovich DI, Ciernik IF, Carbone DP: Dendritic cells in antitumor immune responses. I. Defective antigen presentation in tumor-bearing hosts. Cell Immunol *170:* 101-110, 1996.
324. Gabrilovich DI, Corak J, Ciernik IF, Kavanaugh D, Carbone DP: Decreased antigen presentation by dendritic cells in patients with breast cancer. Clin Cancer Res *3:* 483-490, 1997.
325. Eibl B, Ebner S, Duba C, Bock G, Romani N, Erdel M, Gachter A, Niederwieser D, Schuler G: Dendritic cells generated from blood precursors of chronic myelogenous leukemia patients carry the Philadelphia translocation and can induce a CML-specific primary cytotoxic T-cell response. Genes Chromosomes Cancer *20:* 215-223, 1997.
326. Choudhury BA, Liang JC, Thomas EK, Flores-Romo L, Xie QS, Agusala K, Sutaria S, Sinha I, Champlin RE, Claxton DF: Dendritic cells derived in vitro from acute myelogenous leukemia cells stimulate autologous, antileukemic T-cell responses. Blood *93:* 780-786, 1999.
327. Charbonnier A, Gaugler B, Sainty D, Lafage-Pochitaloff M, Olive D: Human acute myeloblastic leukemia cells differentiate in vitro into mature dendritic cells and induce the differentiation of cytotoxic T cells against autologous leukemias. Eur J Immunol *29:* 2567-2578, 1999.
328. Ruiz-Cabello F, Cabrera T, Lopez-Nevot MA, Garrido F: Impaired surface antigen presentation in tumors: implications for T cell-based immunotherapy. Semin Cancer Biol *12:* 15-24, 2002.
329. Iezzi G, Protti MP, Rugarli C, Bellone M: B7.1 expression on tumor cells circumvents the need of professional antigen presentation for in vitro propagation of cytotoxic T cell lines. Cancer Res *56:* 11-15, 1996.
330. Amoscato AA, Prenovitz DA, Lotze MT: Rapid extracellular degradation of synthetic class I peptides by human dendritic cells. J Immunol *161:* 4023-4032, 1998.
331. Ludewig B, McCoy K, Pericin M, Ochsenbein AF, Dumrese T, Odermatt B, Toes RE, Melief CJ, Hengartner H, Zinkernagel RM: Rapid peptide turnover and inefficient presentation of exogenous antigen crutically limit the activation of self-reactive CTL by dendritic cells. J Immunol *166:* 3678-3687, 2001.
332. Schmidt SM, Schag K, Mueller MR, Weck MM, Appel S, Kanz L, Gruenebach F, Brossart P: Survivin is a shared tumor associated antigen expressed in a broad variety of malignancies and recognized by specific cytotoxic T-cells. Blood Feb 6, 2003.
333. Wolfers J, Lozier A, Raposo G, Regnault A, Thery C, Masurier C, Flament C, Pouzieux S, Faure F, Tursz T, Angevin E, Amigorena S, Zitvogel L: Tumor-derived exosomes are a source of shared tumor rejection antigens for CTL cross-priming. Nat Med *7:* 297-303, 2001.
334. Kaplan JM, Yu Q, Piraino ST, Pennington SE, Shankara S, Woodworth LA, Roberts BL.: Induction of antitumor immunity with dendritic cells transduced with adenovirus vectorencoding endogenous tumor-associated antigens. J Immunol *163:* 699-707, 1999.
335. Kirk CJ, Mule JJ: Gene-modified dendritic cells for use in tumor vaccines. Hum Gene Ther 11: 797-806, 2000.
336. Furumoto K, Arii S, Yamasaki S, Mizumoto M, Mori A, Inoue N, Isobe N, Imamura M: Spleen-derived dendritic cells engineered to enhance interleukin-12 production elicit therapeutic antitumor immune responses. Int J Cancer *87:* 665-672, 2000.
337. Hirschowitz EA, Weaver JD, Hidalgo GE, Doherty DE: Murine dendritic cells infected with adenovirus vectors show signs of activation. Gene Ther *7:* 1112-1120, 2000.
338. Jenne L, Schuler G, Steinkasserer A: Viral vectors for dendritic cell-based immunotherapy. Trends Immunol 22: 102-107, 2001.
339. Paul S, Calmels B, Acres RB: Improvement of adoptive cellular immunotherapy of human cancer using ex-vivo gene transfer. Curr Gene Ther *2:* 91-100, 2002.
340. Bodey B: Spontaneous regression of neoplasms: new possibilities for immunotherapy. Expert Opinion Biological Therapy *2:* 459-476, 2002.
341. Maecker B, Von Bergwelt-Baidon, Anderson KS, Vonderheide RH, Schultze JL: Linking genomics to immunotherapy by reverse immunology--'immunomics' in the new millennium. Curr Mol Med *1:* 609-619, 2001.
342. Onaitis M, Kalady MF, Pruitt S, Tyler DS: Dendritic cell gene therapy. Surg Oncol Clin N Am *11:* 645-660, 2002.
343. Scheffer SR, Nave H, Korangy F, Schlote K, Pabst R, Jaffee EM, Manns MP, Greten TF: Apoptotic, but not necrotic, tumor cell vaccines induce a potent immune response in vivo. Int J Cancer *103:* 205-211, 2003.
344. Steinman RM, POPE M: Exploiting dendritic cells to improve vaccine efficacy. J Clin Invest *109:* 1519-1526, 2002.
345. Gilboa E: Immunotherapy of cancer with genetically modified tumor vaccines. Semin Oncol *23:* 101-107, 1996.
346. Topf N, Schmiegel WH: Immuntherapie mit genetisch modifizierten Tumorzellen. Internist *37:* 374-381, 1996.
347. Zhang W, He L, Yuan Z, Xie Z, Wang J, Hamada H, Cao X: Enhanced therapeutic efficacy of tumor RNA-pulsed dendritic cells after genetic modification with lymphotactin. Hum Gene Ther *10:* 1151-1161, 1999.
348. Klein C, Bueler H, Mulligan RC: Comparative analysis of genetically modified dendritic cells and tumor cells as therapeutic cancer vaccines. J Exp Med *191:* 1699-1708, 2000.
349. Armstrong TD, Jaffee EM: Cytokine modified tumor vaccines. Surg Oncol Clin N Am *11:* 681-696, 2002.
350. Gong J, Avigan D, Chen D, Wu Z, Koido S, Kashiwaba M, Kufe D: Activation of antitumor cytotoxic T lymphocytes by fusions of human dendritic cells and breast carcinoma cells. Proc Natl Acad Sci USA *97:* 2715-2718, 2000.
351. Xia J, Tanaka Y, Koido S, Liu C, Mukherjee P, Gendler SJ, Gong J.: Prevention of spontaneous breast carcinoma by prophylactic vaccination with dendritic/tumor fusion cells. J Immunol 170: 1980-1986, 2003.
352. Schultz J: Success of vaccine offers promise of cervical cancer prevention. J Natl Cancer Inst *95:* 102-104, 2003.
353. Shu S, Cohen P: Tumor-dendritic cell fusion technology and immunotherapy strategies. J Immunother *24:* 99-100, 2001.
354. Lollini PL, De Giovanni C, Nicoletti G, Di Carlo E, Musiani P, Nanni P, Forni G: Immunoprevention of colorectal cancer: a future possibility? Gastroenterol Clin N Am *31:* 1001-1014, 2002.
355. Bell D, Young JW, Banchereau J: Dendritic cells. Adv Immunol *72:* 255-324, 1999.

356. Banchereau J, Briere F, Caux C, Davoust J, Lebecque S, Liu YJ, Pulendran B, Palucka K: Immunobiology of dendritic cells. Annu Rev Immunol *18:* 767-811, 2000.
357. Staquet MJ, Jacquet C, Dezutter-Dambuyant C, Schmitt D: Fibronectin upregulates in vitro generation of dendritic Langerhans cells from human cord blood CD34+ progenitors. J Invest Dermatol *109:* 738-743, 1997.
358. Rougier N, Schmitt D, Vincent C: IL-4 addition during differentiation of CD34 progenitors delays maturation of dendritic cells while promoting their survival. Eur J Cell Biol *75:* 287-293, 1998.
359. Nestle FO, Alijagic S, Gilliet M, Sun Y, Grabbe S, Dummer R, Burg G, Schadendorf D: Vaccination of melanoma patients with peptide- or tumor lysate-pulsed dendritic cells. Nat Med *4:* 328-332, 1998.
360. Zitvogel L, Angevin E, Tursz T: Dendritic cell-based immunotherapy of cancer. Ann Oncol *11:* 199-205, 2000.
361. Panelli MC, Wunderlich J, Jeffries J, Wang E, Mixon A, Rosenberg SA, Marincola FM: Phase I study in patients with metastatic melanoma of immunization with dendritic cells presenting epitopes derived from the melanoma-associated antigens MART-1 and gp100. J Immunother *23:* 487-498, 2000.
362. Small EJ, Fratesi P, Reese DM, Strang G, Laus R, Peshwa MV, Valone FH.: Immunotherapy of hormone-refractory prostate cancer with antigen-loaded dendritic cells. J Clin Oncol *18:* 3894-3903, 2000.
363. Cohen L, De Moor C, Parker PA, Amato RJ: Quality of life in patients with metastatic renal cell carcinoma participating in a phase I trial of an autologous tumor-derived vaccine. Urol Oncol *7:* 119-124, 2002.
364. Tirapu I, Rodriguez-Calvillo M, Qian C, Duarte M, Smerdou C, Palencia B, Mazzolini G, Prieto J, Melero I.: Cytokine gene transfer into dendritic cells for cancer treatment. Curr Gene Ther *2:* 79-89, 2002.
365. Kugler A, Stuhler G, Walden P, Zoller G, Zobywalski A, Brossart P, Trefzer U, Ullrich S, Muller CA, Becker V, Gross AJ, Hemmerlein B, Kanz L, Muller GA, Ringert RH.: Regression of human metastatic renal cell carcinoma after vaccination with tumor cell-dendritic cell hybrids. Nat Med *6:* 332-336, 2000.
366. Gitlitz BJ, Figlin RA, Pantuck AJ, Belldegrun AS: Dendritic cell-based immunotherapy of renal cell carcinoma. Curr Urol Rep *2:* 46-52, 2001.
367. Veelken H, Osterroth F: Vaccination strategies in the treatment of lymphomas. Oncology *62:* 187-200, 2002.
368. Perales MA, Wolchok JD: Melanoma vaccines. Cancer Invest *20:* 1012-1026, 2002.
369. Wysocki PJ, Karczewska A, Mackiewicz A: Gene modified tumor vaccines in therapy of malignant melanoma. Otolaryngol Pol *56:* 147-153, 2002.
370. wen JJ, Moore NC: Thymocyte-stromal-cell interactions and T-cell selection. Immunol Today *16:* 336-338, 1995.
371. Kasai M, Hirokawa K, Kajino K, Ogasawara K, Tatsumi M, Hermel E, Monaco JJ, Mizuochi T: Difference in antigen presentation pathways between cortical and medullary thymic epithelial cells. Eur J Immunol *26:* 2101-2107, 1996.
372. Mizuochi T, Kasai M, Kokuho T, Kakiuchi T, Hirokawa K: Medullary but not cortical thymic epithelial cells present soluble antigens to helper T cells. J Exp Med *175:* 1601-1605, 1992.
373. Thomas-Vaslin V, Coltey M, Salaun J: On the mechanisms of thymic epithelium induced tolerance. Comptes Rendus Acad Sciences - Serie Iii, Sciences Vie *319:* 401-404, 1996.
374. Oukka M, Colucci-Guyon E, Tran PL, Cohen-Tannoudji M, Babinet C, Lotteau V, Kosmatopoulos K: CD4 T cell tolerance to nuclear proteins induced by medullary thymic epithelium. Immunity *4:* 545-553, 1996.
375. Oukka M, Cohen-Tannoudji M, Tanaka Y, Babinet C, Kosmatopoulos K: Medullary thymic epithelial cells induce tolerance to intracellular proteins. J Immunol *156:* 968-975, 1996.
376. Montecino-Rodriguez E, Johnson A, Dorshkind K: Thymic stromal cells can support B cell differentiation from intrathymic precursors. J Immunol *156:* 963-967, 1996.
377. Rincon M, Flavell RA: Regulation of AP-1 and NFAT transcription factors during thymic selection of T cells. Molec Cellular Biol *16:* 1074-1084, 1996.
378. Lenz P, Elbe A, Stingl G, Bergstresser PR: MHC class I expression on dendritic cells is sufficient to sensitize for transplantation immunity. J Invest Dermatol *107:* 844-848, 1996.
379. Bachmann MF, Lutz MB, Layton GT, Harris SJ, Fehr T, Rescigno M, Ricciardi-Castagnoli P: Dendritic cells process exogenous viral proteins and virus-like particles for class I presentation to $CD8^+$ cytotoxic T lymphocytes. Eur J Immunol *26:* 2595-2600, 1996.
380. Kündig TM, Bachmann MF, DiPaolo C, Simard JJL, Battegay M, Lother H, Gessner A, Kühlcke K, Ohashi PS, Hengartner H, Zinkernagel RM: Fibroblasts as efficient antigen-presenting cells in lymphoid organs. Science *268:* 1343-1347, 1995.
381. Pioche C, Salomon B, Klatzmann D: Cellules dendritiques et therapie cellulaire antitumorale. Pathologie Biologie *43:* 904-909, 1995.
382. Iezzi G, Protti MP, Rugarli C, Bellone M: B7.1 expression on tumor cells circumvents the need of professional antigen presentation for in vitro propagation of cytotoxic T cell lines. Cancer Res *56:* 11-15, 1996.
383. Gilboa E: Immunotherapy of cancer with genetically modified tumor vaccines. Semin Oncol *23:* 101-107, 1996.
384. Topf N, Schmiegel WH: Immuntherapie mit genetisch modifizierten Tumorzellen. Internist *37:* 374-381, 1996.
385. Drexhage HA, Mooy P, Jansen A, Kerrebijn J, Allaerts W, Tas MP: Dendritic cells in tumor growth and endocrine diseases. Adv Exp Med Biol *329:* 643-650, 1993.
386. Brodsky FM, Lem L, Bresnahan PA: Antigen processing and presentation. Tissue Antigens *47:* 464-471, 1996.

Chapter 7

Involution of the Mammalian Thymus and Its Role in the Overall Aging Process

Abstract: During the last century of research concerning the thymus, the fact that every mammalian thymus undergoes marked morphological changes during the complex process of aging has been defined as a basic histogenetical rule. In characterizing the physiological (i.e. chronic) involution of the mammalian thymus, the term "Altersinvolution" referring to age-related involution is used. All other types of thymic involution are associated with an initial trigger and a relatively "acute" mechanism. In all of these factor-dependent cases of thymic involution, we use the term "akzidentelle involution" (i.e. acute accidental thymic involution). Temporary thymic involution occurs during pregnancy, with a full restoration of the cellular microenvironment at the end of lactation. It is now clear that pregnancy alters the well-established adaptational homeostasis between the neuroendocrine and immune axes. Such non-progressive involution has also been observed during various seasons in various animals (i.e. seasonal involution). Changes characteristic of thymic involution begin during or soon after the first year after birth, and continue progressively throughout the entire life span. The 3% to 5% annual reduction rate of the cells of the human thymic microenvironment continues until middle age, when it slows down to less than 1% per year. According to the extrapolation of these results total loss of thymic reticulo-epithelial tissue and the associated thymocytes should occur only at age of 120 years in humans. This serious reduction of the thymic cellular microenvironment is a well controlled physiological process and is presumably under both local and global regulation by the cells of the RE meshwork and by the neuroendocrine, respectively. In humans, the age related decline in serum facteur thymique serique (FTS) levels begins after 20 years of age and FTS completely disappears from the blood between the 5th and 6th decade of life. In contrast, the serum levels of thymosin-$\alpha 1$ and thymopoietin seem to decline earlier, starting as early as 10 years of age. The influences of a variety of other hormones on the involution of the thymus have also been characterized: testosterone, estrogen and hydrocortisone treatment results in marked involution, cortisone and progesterone administration causes slight to moderate, while use of desoxycorticosterone has no effect. The experimental administration of thyroxine yielded dose dependent results: low doses resulted in thymic hypertrophy, higher doses produced a slight hypertrophy and the highest employed doses caused thymic atrophy. The atrophy was of apicnotic type, very different from that detected after treatment with corticoid hormones. Thymus transplantation experiments indicate that age-related, physiological thymic involution has been genetically preprogrammed. Grafting of the thymus from one week old C3H leukemic strain mice into 6 month old hosts resulted in changes in thymic weight and an involution pattern that were synchronous in all recipients, in direct correlation with the glands in the donor, but not in the host. These data strongly suggest that the stimulus for thymus cell proliferation and differentiation is genetically determined within the organ implant. Since the thymus is the primary T-lymphopoietic organ during ontogenesis in the mammalian organism, its age-related involution with the already mentioned morphological alterations can be held responsible only for a decline in antigen-specific T lymphocyte immune functions. Thymic involution and diminished T lymphocyte proliferation can be partially restored by thymic tissue transplantation or use of thymic hormones. The leading physiological role of the thymic cellular microenvironment as a "clock" of the mammalian aging process is also discussed.

> "If present cells have come from pre-existing cells, then all cells can trace their ancestry back to the first formed cell in an unbroken line of descent." –
>
> Rudolf Virchow, 1858 (1)

> "I have never noticed the absence of the thymus gland except in cases of true acephalism."
>
> Simon, 1845 (2)

Key words: Genetically regulated thymic involution; Types of thymic involution; Acute and chronic thymic involution; Immunological consequences of thymic involution; Seasonal morphological thymic alterations; Pregnancy-related temporary thymic involution; Acute, stress related thymic involution; Thymic involution and breeding; Aging process; Hypotheses of aging; Immunology of senescence; Age related changes in T lymphocyte dependent immunity.

1. INTRODUCTION

Detailed morphological and histological observations of the mammalian and especially the human thymus had been conducted even before its main physiological function as the primary and central organ of the cellular immune system was revealed by Miller (3-28). The ontogenetical and phylogenetical development of the two arms of the immune system were well characterized:

> "A distinct dichotomy of the immune system into thymus-dependent and thymus-independent components has been demonstrated in amphibians, birds and mammals, suggesting that the B cell / T cell dichotomy of the vertebrate immune system dates back in phylogeny at least some 250 million years to the early tetrapods." (29)

During the last century, the plethora of reports on the structure and function of the thymus gland have led to the conclusion that every mammalian thymus undergoes serious morphological changes during the complex process of aging. Changes characteristic of a process termed involution commence soon after the first year after birth and continue throughout the rest of life (30-38). Contradicting the popular belief that maturation is related to puberty, the thymus reaches its maximal size (about 25 cm^3) at the end of the first year of postnatal development. Thereafter, the volume of *real* thymic tissue decreases progressively, excluding the oversized perivascular spaces, and the invading adipose and fibrous types of connective tissue which serve as substitutes for the lost tissue (39-43). Similar events of adipose tissue invasion also occur in other hematopoietic organs (the bone marrow and lymph nodes) and this suggests that lipid may be a basic component in these body systems. The 3% to 5% annual reduction in the number cells comprising the thymic microenvironment continues until middle age, at which time the process slows down to less than 1% per year. Total loss of thymic reticulo-epithelial tissue and associated thymocytes should occur at the age of 120 years in humans. Kendall and co-workers (33) conducted 574 autopsies on subjects aged 3 months to 98 years and concluded the following:

1. thymus weight decreases with aging;
2. between the age of 10 and 60 years, total weight of the thymus varied little (at 10 years the range was 21g to 30g, while at 58 years it was 19g to 29g);
3. thymic water content drops from 72% at 10 years to 30% at age 50; and + lipid and collagen content increases during the aging process.

Steinmann and co-investigators (36) examined 136 postnatal thymic tissues from accident victims and another 68 removed during open-heart surgery, from subjects aged one month to 107 years. Their two most important conclusions are still valid today: *"...age dependent involution is not related to puberty"* and *"...even in 100+ year old humans thymic epithelial tissue remains with cortical lymphocytes present"*. Authors of classical thymology had already reported that the maximum size of the thymus is attained at one year of age, followed by a slow steady drop until the age of 15, at which time adipose tissue begins replacing thymic cells (44). There is no doubt that thymic involution is influenced by a variety of exo- and endogenous factors.

The reduction of the thymic cellular microenvironment is a well controlled physiological process and is presumably regulated locally by peptides (i.e. thymic hormones) secreted by the cells of RE meshwork, as well as more globally by a number of gonadal peptides, in addition to the numerous cellular and humoral interactions within the neuro-endocrine-immune axis (45-63). Sensitivity of the thymus to sex hormones and the establishment of its connection to the hypothalamic-gonadal-immune regulation pathway develop during postnatal thymic histogenesis, and this has been well documented at the cellular level in the partially involuted thymus (64-68). The immune system, by virtue of its physiological association with reproductive functions, can also contribute to evolution (69-76). Immune regulation has long been suspected of having a direct influence on neurological functions and the observation of cytokines within the central nervous system (CNS) has led to a new field of research (77, 78). Cytokines such as Interleukin 1 (IL-1), IL-2, IL-6 and tumor necrosis factor [TNF] have been shown to play a role in the growth of nerve cells and astrocytes, while others like IL-1, Interferons (IFNs) and PAF are involved in aiding the secretion and action of neurohormones, mainly of CRF and ACTH (63,79). There are a number of cytokines (IL-1α, IL-1β, IL-2, IL-6 and TNFα - for review see [80]) released during the activation of immunological effector cells which are also capable of interfering with the hypothalamo-gonadal axis (81-84). Significant alterations in sexual functions have been observed in cases of congenital absence of the thymus in humans and in *nude* animals, born without a thymic cellular microenvironment and therefore lacking the development of the T lymphocyte dependent immune system (85, 86). Such mice, as well as neonatally thymectomized ones, display a delay in the onset of puberty and an altered sexual cycle. These sexual defects were effectively corrected by transplantation of thymic tissue upon birth, but neither immunological reconstitution with T lymphocytes nor germ-free living conditions were able to induce normal sexual development. Obviously, the lack of thymic tissue rather than immunodeficiency is the cause of this sexual deficiency (81, 87). The forces of natural selection forces seem to thus impede the reproduction of immunodeficient animals, while favoring the reproduction of healthy individuals capable of developing an efficient immune system. The castration of aged males resulted in substantial growth of the thymus, but thymic growth could be prevented by administration of testosterone (88, 89).

In describing physiological (i.e. chronic) involution of the thymus, we usually use the term *"Altersinvolution"* or age related involution (the phenomenon first described and the term coined by Charleton in 1659 [90]). All other types of thymic involution have an initial trigger and a relatively "acute" mechanism (e.g. after irradiation, starvation, hormonal influence, chronic systematic diseases such as AIDS, tuberculosis, congenital syphilis, etc.). Hormones influencing thymic involution are numerous and correspond to various intensities of organ involution: testosterone, estrogen, and hydrocortisone treatment result in marked involution, cortisone and progesterone administration caused slight to moderate, while use of desoxycorticosterone had no effect on thymic tissue (91-95). The experimental administration of thyroxine was associated with alterations in the structure of the thymus in a dose dependent manner: low doses resulted in thymic hypertrophy, higher doses produced slight hypertrophy, while the highest doses observed caused thymic atrophy. The atrophy was of apicnotic type, quite different from that detected after treatment with corticoid hormones (91, 96). Thyroxine injections have also been shown to augment antibody production by B lymphocytes and thymulin secretion by thymic RE cells in aged mice (97).

Temporary involution occurs during pregnancy, with restoration of the entire cellular microenvironment at the end of lactation, as well as during different seasons (seasonal involution) (98, 99). Pregnancy results in a change in the adaptational homeostasis between the neuroendocrine and immune systems. Essential changes in the function of endocrine glands take place in the mother (61, 100). The most significant is the presence of the placenta, a gland of internal secretion, which develops *de novo* and produces choriogonadotropin, as well as great amounts of other steroid and protein hormones. Advanced pregnancy produces one-hundred fold increase in the quantity of estrogen and a ten fold increase in the levels of progesterone and prolactin. In addition, the anterior lobe of the hypophysis produces more adrenocorticotropic hormone (ACTH), while the adrenal cortex increases its secretion of cortisol (101-103).

In lizards, marked seasonal morphological alterations have been reported in the formation of intercellular cystic structures (104-106). Observations of the thymus of fawn, yearling and adult female mule deer (*Odocoileus hemionus hemionus*) established that the volume of the organ was different in each group depending on the particular season (107). Thymic volume was lowest during the winter (December to February), independent of whether the deer were pregnant or not, and was highest during spring and early summer (May to July). The fawn thymus reached its highest volume during the summer, although pregnant females had lower values than their male counterparts, but even in these females, the weight of the thymus gland was still significantly higher than in the winter. It has been proposed that the length of exposure to daylight may be the determining environmental factor in these cases.

In all factor-dependent (secondary) thymic involution cases we employ the term "acute, accidental" thymic involution originally termed *"akzidentelle Involution"* (29, 108-112). The unfortunate Chernobyl nuclear power plant accident was the greatest radiation correlated "experiment" on humans. Changes in the cellular immune response were detectable in subjects who had been exposed to a one Gy or higher dose of radiation. A significant decrease in the number of $CD3^+CD4^+$ and $CD5^+CD8^+$ lymphocytes was detected in persons who worked within a 30 km radius of the nuclear power plant following the accident (113). A much lower level of thymosin-α1 was determined in the blood of persons working within this zone for 4.5 to 5 years. Autoantibodies against the cells of thymic RE network had developed after working within this 30 km radius for 3 to 3.5 years. A significant decrease in the numbers of helper and suppressor T lymphocytes in individuals in regular contact with small doses of radiation has also been reported.

Thymus transplantation experiments indicate that age-related, physiological thymic involution has been genetically preprogrammed (114, 115). The thymus glands from one week old mice of the C3H leukemic strain were grafted into 6 month old hosts, and it was found that the changes in thymic weight and the involution patterns in the hosts were synchronous and in direct correlation with the glands of the donor, and not the host. Therefore, Metcalf suggested that the stimulus for thymus cell growth and proliferation is genetically determined within the implant. When multiple thymic grafts were placed in the same host, each graft was found to grow to a normal, full size (116). The transplantation of neonatal thymic tissue into aged mice had no effect on their age-related decline in graft vs. host and other cellular immunological responses (117). Cells transplanted from old to young mice retain their original characteristics and are never rejuvenated (118). The best results were achieved in the transplantation of GH_3 pituitary adenoma cells, which secrete growth hormone and prolactin, into 16 to 22 month old female rats, resulting in high level *in situ* regeneration of already involuted thymic tissue. The rats were sacrificed after two months following the subcutaneous pituitary adenoma cell implantation. The observed thymuses contained a great number of cortical thymocytes and the original thymic architecture was restored (119). The reappearance of a histologically normal thymus in aged rats was accompanied by a restoration of the proliferative abilities of peripheral T lymphocytes when exposed to lectins and the synthesis of IL-2.

> "In adults the thymus is transformed into mass of adipose tissue containing scattered islands of lymphatic tissue composed of thymic epithelial regions where endodermal epithelial cells form a network, and of reticular connective tissue with argyrophilic fibers. The epithelial regions of the thymus do not disappear completely, even in old age." - Hammar, 1926 (16) (cited by von Gaudecker [32])

2. MORPHOLOGICAL CHANGES IN THYMIC ARCHITECTURE DURING INVOLUTION

Acute thymic involution, which is predominantly characterized by massive loss of cortical thymocytes, was detected in our total body irradiation (TBI) and stem cell transplantation experiments on dogs (Fig. 1 and 2) and mice. The formation of other morphological structures, identified as thymic cysts, was also noted between days 10 and 20 (112). The thymic cysts demonstrated the presence of secretory products (probably concentrated thymic hormones), and thus may represent the hormonal microenvironment for thymocyte maturation within the involuting thymus (Fig. 3). A number of degenerative Hassall's bodies (HBs) also demonstrated some microcystic changes (Fig. 4). In our experiments, the regeneration of thymic lymphopoiesis was impaired even after transplantation of hematopoietic stem cells. The same results were reported in lethally irradiated mice rescued by transplanted bone marrow from aged donors (120-124).

Age-related, physiological thymic involution is morphologically characterized by a progressive and continuous loss of predominantly immature, cortical lymphatic cells, and changes (a kind of redevelopment) in the organ architecture. As a dominant pattern of its involution, the thymus first loses the great majority of its cortical, immunologically immature thymocytes (up to 95%) and then its typical lobular structure, resulting in the re-mixing of the cortical and medullar areas (inverted thymus).

The extremely high intrathymic mitotic rate (highest in mice at birth [117]) is significantly lowered with advancing age in mammals (125). In our experimental studies in aged humans, thymocyte differentiation and maturation is slow and an increase in the amount of space between reticulo-epithelial (RE) cells is always present. During the process of involution, the immunocytochemical expression of differentiation antigens did not change on single thymocytes. We confirm the earlier reported increase of CD1 and significant decrease of CD6 expression in immature thymocytes. In the thymic medulla, an increased number of $CD8^+$, cytotoxic/suppressor T lymphocyte subset were detected (126-128). A constant velocity of lymphopoiesis emerges during the first decade of life, and decrease in rate thereafter (121, 129). The immunocytochemical observation of terminal transferase (TdT) resulted in the confinement of its expression to the 6th and 7th decades of life in humans (127). In the postnatal thymus, CD40-ligand positive ($CD40\text{-}L^+$) T lymphocytes were present in the medullary region. Overexpression of CD40-L on thymocytes altered the architecture of the thymic microenvironment, as reflected by a dramatic loss of cortical RE cells, expansion of the medullary region, and extensive infiltration of the thymic capsule with $CD3^+$ cells, B lymphocytes, macrophages and dendritic cells (130). The mentioned events also predict a serious shortage in the intrathymic maturation of double positive (DP) T lymphocytes with the IP TCRαβ^{hi}CD4$^+$CD8$^+$CD69$^+$ (131, 132). Immature CD4$^+$CD8$^+$ DP thymocytes expressing self MHC-restricted TCR are positively selected by the thymic cellular microenvironment in response to TCR signals to survive and mature into immunologically competent CD4$^+$ or CD8$^+$ single positive (SP) T lymphocytes. The DP precursors expressing autoreactive TCR are clonally deleted (133, 134). The hallmarks of positive selection include increased expression of CD5 and Bcl-2, termination of recombination activation gene (RAG-1) and pre-Tα gene expression, and a switch in *lck* promoter employment. Signals mediated *via* CD4 or CD28 also synergize with TCR signals to induce these outcomes. Other well-known events of maturation such as changes in the expression of Thy-1, HSA, MHC class I and CD45-RB are not included in this early selection process. Conditioned thymic RE cell culture medium decreased the density of Thy-1 antigen on thymocytes, especially the peanut agglutinin (PNA) reactive immature thymocyte subpopulation (120, 135).

The volumes of lymphocytic perivascular spaces, connective tissue, and the cellular organization of Hassall's bodies (HBs) reach a maximum not at puberty, but, as was already mentioned, by the end of the first postnatal year, and thereafter begin to decrease, most likely due to the growing neuro-endocrine influence.

Our immunocytochemical results in the human thymus, as well as our TEM and SEM, and electron-immunocytochemical observations of other thymologists on mammalian thymic RE microenvironments identified a heterogeneous IP for RE cells on the basis of their MHC-coded surface antigen expression (136-145). Steinmann (127) described two distinct architectural patterns for the aged thymic RE cell meshwork: *a)* as a delicate reticular network, with an axis perpendicular to the thymic capsule organization of RE cells; and *b)* depleted, with complete loss of reticular HLA-DR$^+$ RE structures, organized in island-like structures of varying sizes (tissue remnants), mostly within the thymic cortex (146-148).

CD60$^+$A$_2$B$_5^+$ neuroectodermally derived RE cells, such as thymic nurse cells (TNCs) located in the subcapsular cortex, also remain within the involuting thymic cellular microenvironment (149-156). The ability of TNCs to provide an enclosed intracellular and humoral microenvironment for maturation and differentiation of cortical thymocytes has already been established. The distribution of CD40 antigen and CD40-L was also determined in the murine thymus during prenatal organogenesis, as well as in aged animals (130). Before birth, CD40 was almost exclusively localized to foci of medullary RE cells. After birth, a dramatic upregulation in the cortical RE cellular meshwork, accompanied by a consolidation of medullary CD40$^+$ foci was reported.

A variety of thymic hormones and protein products (such as TsIF, a 75 kD protein, which is able to induce differentiation and maturation of T suppressor lymphocytes, with the Thy-1$^+$CD3$^+$CD8$^+$ IP, from bone marrow progenitors - patented by Ventrex Laboratories Inc., Portland, ME [157]) are produced by the heterogeneous subpopulations of thymic RE cells (145, 158-160), as was demonstrated by immunofluorescent and immunocytochemical techniques (thymosin-α1 [161]; facteur thymique serique [FTS]/thymulin [162-166]). In humans, the age related decline in serum FTS levels was observed to occur after 20 years of age, and FTS completely disappears from circulation between the 5th and 6th decades of life (167-174). In contrast the serum levels of thymosin-α1 and thymopoietin seems to decline earlier than that of thymulin, starting as early as 10 years of age (175, 176).

During childhood, the human thymus is separated into lobules by thin septa of connective tissue. During life, especially in elderly people, the interlobular septa have greatly expanded and mostly contain fat cells, while adipose connective tissue develops under the thymic capsule, further separating the true thymic tissue from the capsule (20, 31, 129, 177-179). During and after the 2nd decade of life, only small cell-tissue remnants of true thymic microenvironment remain, as medullary RE cells surrounding a centrally located blood vessel (112, 127, 180). In mice, the loss of real thymic tissue is not accompanied by the overgrowth of connective tissue; the gross thymic size reduction is extremely significant. The smaller thymus is also less efficient in the production of T lymphocytes; only an estimated 0.7% of the newborn T lymphocyte number was detected in a 24 month old mouse (181).

In conclusion:

1. The well determined, dramatic loss of ontogenetically developed cells (stromal) of the thymic microenvironment and their basic regulatory-inductive functions during involution is unique to the mammalian organism;
2. In humans, the total amount of lymphopoietic thymic tissue decline is in direct correlation

with age during the first 1 to 10 years of life (phase I, 5% per annum [127]);

3. When age-related groups of both sexes are compared, no statistically significant differences could be determined; and
4. The functional neuro-endocrine-immune morphological units of the thymus represented by the cells of the RE meshwork, the associated non-lymphatic stromal elements (dendritic cells, macrophages, Langerhans cells, interdigitating cells), and a limited amount of cortical thymocytes, are never completely lost during the various types of thymic involution.

3. HYPOTHESES ABOUT THE MAMMALIAN AGING PROCESS

Aging is an extremely complex physiological phenomenon of great significance. Despite hundreds of observations and diverse hypotheses, no unifying or proven hypotheses have emerged. However, aging is a multifactorial process composed of both genetic and environmental components. Each body system, each tissue, and each cell type appears to have its own trajectory of senescence. Cellular clocks are present and operate in accordance with higher order clocks.

A great amount of data at the cellular level has accumulated from experiments conducted under well defined and controlled physiological and environmental conditions in the hopes of advancing our understanding of the aging process (182). A critical review of the theories of cellular aging was published in 1994 by Toussaint and Remacle (183). The authors demonstrated that all of the developed hypotheses can be integrated into a more general concept, *"the concept of critical threshold of error accumulation"*, taking into account the protective role of suppressor/inhibitory or defense systems in avoiding a quick increase in the level of intracellular errors.

Historically, Weissman (184, 185) was the first biologist of the evolutionary era to advance a theory of the aging process. He based his theory on Darwinian evolution and especially the concept of natural selection. He stated that the capacity of the multicellular organism to replace cells by cell division is finite and proposed: *"aging is the price somatic cells paid for their differentiation"*. Cells just wear out, due to permanent devastating actions of environmental stress. He was also the first to suggest that the failure of somatic cells to replicate indefinitely limited the life span of the individual organism. This proposal was contradicted by Carrel's tissue culture observations, in which he was able to grow chicken heart cells *in vitro* for 34 years, from 1911 (186) to 1945. This was compelling evidence that individual cell types, freed from the constraints of the whole multicellular organism, are practically immortal. Hayflick (187) reported tissue culture experiments on human amnion cells, which were altered *in vitro* to form an established cell line (WISH). He stated that to preserve diploidy, cultured human cells can divide up to 50 times (i.e. get 50 population doublings [PDs]), but this number is not related to time (188-190). The most important discoveries of Hayflick were: *a)* tumor cells or *in vitro* cell lines from normal tissue are heteroploid and have indefinite proliferative abilities; and *b)* his experimental data postulated that the clock of mammalian aging is intrinsic, existing within the cell. However, the burning question remains: Do *in vitro* PDs reflect the *in vivo* reality of cell proliferation? Kohn (191) detected 565 PDs in mouse tongue epithelium. He also reported that cell proliferation times were similar in young and older rats, and he discovered that erythrocyte precursors isolated from very old rats survived and proliferated for 13 months beyond the donor's life-span. Bell and co-investigators pointed out that *in vitro* cell life-span was monitored by experiments of forced cell divisions (192). These authors repeated the initial, basic question: Aging or Differentiation?

Developmental-genetic hypotheses consider the complex process of aging to be part of genetically programmed and regulated pre- and postnatal histogenesis and cell maturation. Although this suggestion is attractive and partly acceptable, the diverse expression of aging effects is in sharp contrast to the tightly controlled and strict ontogenetic processes of any aspects of development. An important and critical point is that genes expressed in late postnatal life probably do not play a significant role in the evolution of species (193-199).

Williams (193) clearly describes that natural selection should proceed in the direction of lengthening life which put it in opposition to aging. However, senescence might well favor vertebrate group survival by eliminating the aged individuals from competing with the younger for food and mates. Sexual drive and reproductive ability generally decrease in direct correlation with age. Thus, an evolutionary tendency may be represented by the older giving way to the young in the survival of species.

In the next several paragraphs, we wish to discuss regulation and single organ system hypotheses of aging. These are correct in many aspects, but all are only able to represent part of the complex process of aging of the whole mammalian organism.

3.1 Immunopathological Hypothesis of Aging

Moodie (200) was the first to suggest that immunity is involved in the aging of an organism. Further development and enrichment of the immunologic hypothesis was the formulation of the *"spontaneous somatic mutation theory of aging"* proposed in 1962 by Walford (201). Spontaneous cell mutations lead to great genetic diversity of cells, resulting in total immunologic disharmony and development of autoreactive cell lines and clones. He concludes: *"Aging is fundamentally concerned with histocompatibility reactions between immunologically diversifying cells"* and later he defined the aging process as *"a generalized prolonged type of autoimmune phenomena"* (202). There is no doubt that a great number of histocompatibility diseases such as arthritis, nephrosclerosis, neoplasia, etc. can be pathogenically linked to the overall aging process. Genetic instability syndromes are also associated with accelerated thymic involution, such as that reported in Down's syndrome and in ataxia telangiectasia (203, 204). Ram (205) reported secondary evidence of immune function decline with age, and this process was related to a number of autoimmune diseases. Burnet (206) combined the Hayflick limit (the allowed 50 cell divisions) and the immunological theory of aging, putting in his view the thymus at the center of mammalian aging. Burnet's conclusions can be summarized as:

1. aging is a process genetically preprogrammed and varies between species;
2. the program is mediated by a metabolic clock, manifested in the Hayflick limit;
3. the thymus dependent immune system plays the role of key system, its exhaustion being responsible for mammalian aging;
4. immune surveillance is the most significant function of the thymus dependent immune system, and such surveillance is capable to deal with the endless somatic mutations; and
5. when immune surveillance declines, neoplastic transformation and autoimmunity (autoreactivity) take over. Burnet (207) also refers to the age related increase of malignant tumor incidence as a direct correlation to alterations of age related immunoendocrine and metabolic shifts, increasing the cellular sensitivity to carcinogens. Kay (208) named the thymus *"an aging clock for the immune system"*.

The "clock" could be located in the T lymphocytic compartment of the thymus, which means a self destruction through clonal exhaustion or within the heterogeneous RE cells, where the declined production of thymic humoral factors and the changes of various surface receptors on RE cells can play an important role in immune changes. Rabin (209) also saw the thymus as a critical organ of mammalian aging, "a clock of aging", but also related T lymphocyte activity to dietary regimes, thus relating the effects of metabolism and nutrition on thymic atrophy and lowered immune defense. Certainly, when the malnutrition is severe, such as seen regularly in third world countries, thymic atrophy is great and the level of immune defense could be very low (210). New data has led to the interpretation of chronic, age dependent thymic involution as occurring due to the cytolytic depletion of the cells of stromal microenvironment (62). The usual alteration rate of single thymic self protein is in the range of 2 to 4 x 10^{-6} per gene and year. Every altered peptide is presented on the RE cell surface in context of MHC class I antigens, and only minimal alteration in a protein leads to cytolysis, and cell destruction. We cannot argue that monitoring of cell surfaces for abnormal (altered) peptides presented by MHC class I antigens represents the primary function of cytolytic, MHC Class I restricted, $CD8^{+}$ T lymphocytes. This cellular response is extremely effective, as few as 200 foreign peptides can be recognized on a target cell (211, 212). It is well established that the original T lymphocyte repertoire remains fairly constant up to senescence, demonstrating the great stability of intrathymic clonal deletion, despite the overall genetic instability (213). A great variety of peripheral self-antigens have been approved by the thymic RE cells during intrathymic self-tolerance induction and clonal deletion of self-reactive T lymphocytes. In autoimmune disorders, we can also expect disturbances in immune-endocrine regulation *via* the hypothalamo-pituitary-adrenal axis (214).

3.2 Neuroendocrine Hypotheses

Neuroendocrine regulation is well documented in cell growth, maturation and differentiation (36, 124, 215). The mice experiments of Cotzias and co-investigators (216) clearly demonstrated increase of longevity after administration of 40 mg/g body weight of the neurotransmitter L-dopamine. The rate of living (metabolism) and the neuroendocrine decay hypothesis are related in many ways, such as chronic starvation and loss of body weight leading to a significant increase in dopamine receptors (up to 50% in dietary restricted rats in direct association with a 40% increase of longevity) (217). The connection between aging and the interaction between the neuroendocrine and immune systems (i.e. the thymus) has also been discussed (218, 219).

Next, it is necessary to summarize the most acceptable, present hypotheses of aging as related to chronic thymic involution, involving the whole body.

1. As we have already mentioned above, one of the explanations of age related, chronic thymic involution is *"wear and tear"*, with progressive loss of thymic function following the deterioration of the cellular machinery of the thymus (220-223). As a criticism of this hypothesis, we support the opinion of George and Ritter (38) that this explanation is very mechanistic and does not take into account the great regenerative abilities of thymic tissue, as described in involution caused by stress, infection, pregnancy, corticosteroid and hormonal treatments (224, 225). Therefore, we think that the tissue restorative abilities of the thymus should be capable of either preventing or rebuilding the normal thymic architecture following age-related involution. As we mentioned, this chronic involution is also controlled by the neuro-endocrine-immune axis and can be slowed down significantly by procedures such as castration, which supports the argument that it is not simply a consequence of wear and tear (226, 227).
2. The *disposable soma* hypothesis was suggested by Kirkwood (228) by analogy to disposable goods that are manufactured only for a limited time. The complex process of aging is interpreted mechanistically as the result of an organism's optimization of investment in maintenance. The repair mechanisms necessary to counteract damage caused as a result of "wear and tear" require a considerable investment of energy, but a lot of energy is also needed for the sexual production of progeny to ensure further passage of genes from one generation to the next. According to Denckla's hypothesis, natural selection favors the most efficient individuals at survival and their genome will be represented in a greater extent of future generations (229-233). According to the hypothesis, the mammalian organism for the non-germline soma will spend energy only for maintenance, trying to keep a balance between the body systems in function and fitness. Maximum effort (energy) is invested in germline cells as they represent the important cells of propagation, so they show no senescence as they are passed from one generation to the next. According to this hypothesis aging is avoidable and can be detected only in somatic cells in mammalian heterosexual organisms. It is remarkable that cells in negetatively reproducing animals demonstrate no senescence. Although the thymus is known to regenerate partially in old mice after gonadectomy, a complete morphological recovery has not been observed (234).
3. Hypothesis of *direct adaptation*. Thymic tissue during its central lymphopoietic function demonstrates a very high proliferative index (128). It contains 10^{11} thymocytes of which 20-25% are produced by cell division every single day. It is accepted as very wasteful since 95% of immunologically immature thymocytes die within the thymic tissue as a result of security mechanisms designed to eliminate autoreactive T lymphocyte clones and to select an useful MHC-restricted T lymphocyte repertoire (38, 235, 236). Therefore, an immunologically efficient selection mechanism is quite energy and resource expensive. The structural reorganization of thymic structure during age related, chronic involution means that the organ might itself be at risk of neoplastic transformation. Loss of functional thymic tissue may be a protective regulatory mechanism to decrease the incidence of thymomas. Generation of immunocompetent T lymphocytes may be diminished because it is of direct benefit to the animal to supply the secondary lymphatic organs in the periphery with fewer functionally mature T lymphocytes. Certainly this suggests that the continuous intrathymic production of functional, immunocompetent T lymphocytes represents a direct disadvantage for the mammalian body, because of an increased risk of autoimmunity or a direct destructive effect of the cellular immune response on other tissues and organs in the body (237-240). An exception is the age-related enlargement of the thymus in the Buffalo rat (241). However, RE cell hyperplasia was detected only in the cortical region and was absent in the thymic medulla. Since the medullary region appears to be only

accessible to recirculating T lymphocytes, the age related immunological changes are real.

4. Theory of *genetically active factors*. Mammalian embryonal development is also regulated by genetically active chemical compounds such as retinoids (naturally occurring and synthetic analogs of vitamin A - retinol) (242-249). These compounds play an important role in normal organogenesis, but can also act as teratogens at high levels. They are also powerful inducers of differentiation in many embryonal carcinoma cell lines. Embryonal carcinoma cells are also pluripotent stem cells that in a number of aspects resemble the omnipotent stem cells of early mammalian embryogenesis (250-254). Most of the embryonal teratocarcinoma cells seem to be committed to a specific differentiation pattern regulated by their genetic programs (e.g. F9 cells differentiate into parietal or visceral endoderm) (255-258).

Zinc (Zn) is an essential trace metal for the optimal biological efficiency and function of several tissues in the mammalian organism, including the lymphatic (immune) system and the integrity of the thymic cellular microenvironment (259-262). During aging, the body's zinc pool undergoes a progressive reduction, resulting in low zinc plasma levels and the negative crude zinc balance observed in aged humans, mice and rats (210, 263-265). It has been proposed that the reduced bioavailability of zinc with advanced age may represent one of the major causes for the involution of the thymus and the ensuing immunological dysfunctions (266-270). It has also been shown that low-dose zinc supplementation leads to at least a partial restoration of nutritional and thymic status in elderly patients, without the risks associated with high-doses of zinc (271).

4. AGE-RELATED CHANGES IN THE T LYMPHOCYTE DEPENDENT IMMUNOLOGICAL SYSTEM - IMMUNOLOGICAL SENESCENCE

It is obvious that this age related, chronic thymic involution precedes the decline of immunological defense in mammals (124, 272-274). Since the thymus is the primary (central) T-lymphopoietic organ during ontogenesis in the mammalian organism, its age related involution with the already mentioned morphological alterations can be held responsible only for a decline in antigen-specific T lymphocyte immune functions (3, 127, 275-284). Adult thymectomy has been shown to result in T lymphocyte deficiency in humans (285). These results were published 22 years ago, and unfortunately every single day during open heart surgeries a great number of thymic glands are simply eliminated to ensure better access to the heart. Clinical studies of elderly humans are fraught with methodological difficulties due to the heterogeneity of the population and presence of age-related diseases of high incidence in late life including diabetes, heart disease, cancer, amyloidosis, Alzheimer's disease, atherosclerosis, vascular damages, exposure to environmental pathogens and different living conditions and stress - all defined as associated with malfunctions of cellular immunity (203, 286-290). In particular, T lymphocyte dependent immune functions seem to decline in elderly humans (see a detailed review by Steinmann [127]). The non-adaptive part of the immune response appears to remain relatively unaffected in older organism; in fact, the frequency of NK cells has been observed to be higher (127, 291). Antigen non-specific cells, such as NK cells, neutrophil granulocytes, macrophages take over in the aged organism, and their production does not require a large amount of body energy, as these cells do not undergo a self-antigen tolerance selection.

Antigen presentation and other immunological functions mediated by dendritic cells (DCs) and cells that belong to the monocyte/macrophage lineage demonstrate no serious loss of capacity with age (292). The bone marrow, the site of dendritic, B lymphocyte, and myeloid cell production also does not appear to be significantly impaired with age, although some morphological alterations have been identified, including the age related lipocyte deposits (284, 288, 293-297). Indeed, flow cytometric analyses have demonstrated that the potential of $CD34^{+}CD33^{+}CD64^{+}$ bone marrow derived hematopoietic stem cells which differentiate along myeloid cell lineages is enhanced during adult hematopoiesis in mammals, also expressing the MIC2 antigen detected using the monoclonal antibody 12E7 (298, 299). Expression of MIC2 antigen was also identified in the most immature lymphocytic (intrathymic thymocyte lineage, characterized by the antigenic profile $CD34^{weak}$ or $CD7^{++}$ surface $CD3^{-}CD1a^{weak}CD4^{weak}$ or $CD8^{weak}$) and granulocytic differentiation stages. Relatively high MIC2 antigen densities were also identified on $CD3^{-}$ $CD16^{+}CD56^{+}$ NK cells and on $CD14^{+}$ monocytes and it has been proposed that the thymic RE cellular microenvironment was the inductive site involved in the maturation of these cells from $CD34^+$ stem cells (300). In mice, myeloid-committed hematopoietic progenitor cells have also been identified by multiparameter cell sorting with the expression of antigen ER-MP58 in combination with monoclonal antibodies ER-MP12 and ER-MP20 generated against immortalized macrophage precursors (301).

Intrathymically generated immunocompetent T lymphocyte subpopulations are composed of relatively long-living cells which undergo clonal expansion in response to every single antigen stimulation during the postnatal life. Moreover, peripheral (secondary to the thymus) lymphatic organs such as the spleen and lymph nodes demonstrate no age related reduction in size, although there may be alterations in the cell proportions that populate these structures (291, 302). Once the peripheral, immunocompetent T lymphocyte pool is established, loss of the thymus will not result in life threatening loss of cellular immune responses. Extreme consequences of having a limited T lymphocyte pool were seen when novel pathogens (with virus-like life cycles) attack isolated island or elderly communities. There are several reasons why the B lymphocyte system may not undergo accelerated changes during aging. First of all, B lymphocytes need only negative selection, so their depletion is less (75%) than that of immunocompetent T lymphocytes. Second, during postnatal life, production of B lymphocytes is maintained by the bone marrow and is more important for host defense mechanisms than the production of new T lymphocytes.

In view of the several age related changes of the mammalian immune system mentioned above, it is still not fully understood why such morphologically dramatic thymic involution should take place. It appears to be a general neuroendocrine-regulated protection of the extremely effective T lymphocyte immune system and its functional integrity from the harmful consequences of the genetic instability of senescence

5. FIGURES

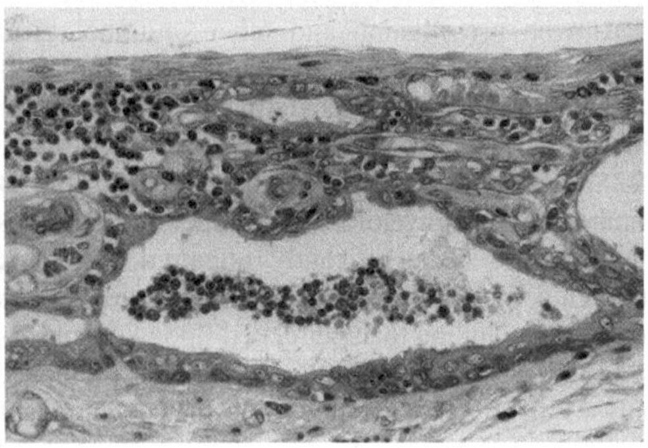

Figure 1. Postnatal dog thymus after total body irradiation (TBI) (3 x 6 Gray [Gy] dose) and autologous bone marrow transplantation. Long term (344 days) observation. Cortical thymic tissue typical of chronic involution, with a minimal number of cortical immature thymocytes and the presence of RE cells. A big cyst is present in the center, containing differentiating lymphatic cells. Schaffer tissue fixation; Methacrylate embedding; 3 µm section; Giemsa staining. Magnification: 100 x.

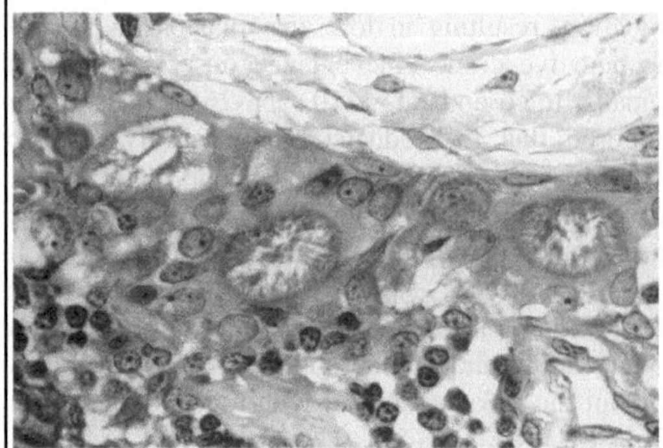

Figure 2. Postnatal dog thymus following TBI (3 x 6 Gy) and autologous bone marrow transplantation. Long term observation of the experimental beagle (sacrificed at day 344 after irradiation). This figure represents features characteristic of partially restored acute thymic involution. The thymic cortex is empty with only a small number of thymocytes. The formation of two cysts, surrounded by cortical RE cells near the interlobular connective tissue (ICT) is also evident. Schaffer tissue fixation; Methacrylate embedding; 3 µm section; Giemsa staining. Magnification: 400 x.

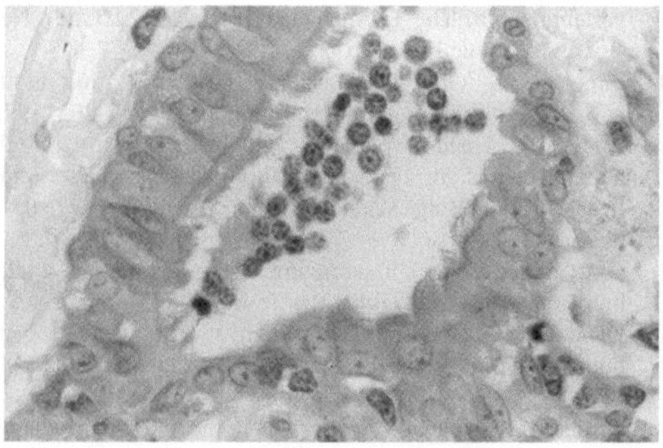

Figure 3. Postnatal dog thymus after partial, one time, upper body irradiation using a single 12 Gy dose. Middle period of the survival of the experimental animal (204 days). Cortical thymic cyst filled with developing hematopoietic cells. The cyst's wall is formed by ciliated thymic RE cells, demonstrating the great transformational capacity of these cells. Schaffer tissue fixation; Methacrylate embedding; 3 µm section; Giemsa staining. Magnification: 400 x.

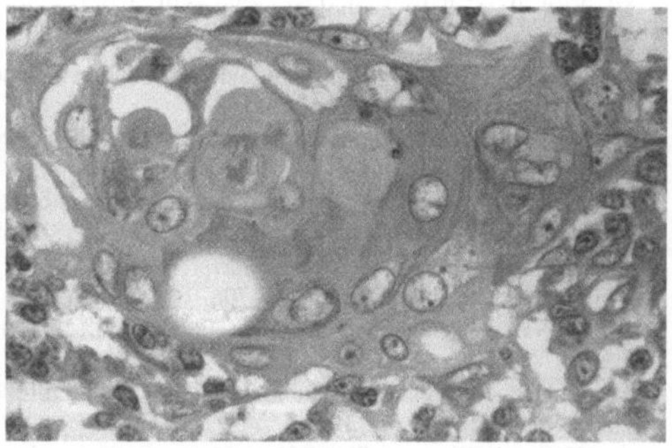

Figure 4. Postnatal dog thymus following TBI (3 x 6 Gy) and autologous bone marrow transplantation. Long term observation of the experimental animal (sacrified at day 344 after irradiation). Medullar thymic Hassall's body (HB) with a variety of histogenetic changes, as compared with HBs during prenatal ontogenesis. The commencement of cystic formation is clear, and intracystic thymic secretory (hormonal) fluid is already evident. Schaffer fixation; Methacrylate embedding; 3 µm section; Giemsa staining. Magnification: 400 x.

REFERENCES

1. Virchow R: Cellular pathology, Churchill, London, pp1-511. 1858.
2. Simon J: A Physiological Essay on the Thymus Gland. Henry Renshaw Publ, London, 1845.
3. Miller JFAP: Immunological function of the thymus. Lancet *2:* 748-749, 1961.
4. Miller JFAP: Effect of thymectomy in adult mice on immunological responsiveness. Nature *208:* 1337-1338, 1965.
5. Verheyen P: De Thymo. Louvain, 1706.
6. Bidloo GB: Exercitatio Anatomica de Thymo. London, 1706.
7. Morand S: Recherches Anatomiques sur la Structure et l'usage de Thymus. Paris, 1759.
8. von Haller A: Elementa Physiologiae Corporus Humani. vol 3, Bousquet et Sociorum M-M, Lucerne, 1766.
9. Hewson W: Experimental Inquiries: Part the Second: A Description of the Lymphatic System in the Human Subject and in Other Animals. Johnson J, London, 1774.
10. Lucae SC: Anatomische Untersuchungen der Thymus in Menschen und Tieren. vol 2, Nurnberg, 1811.
11. Cooper A: Anatomy of the Thymus Gland. Lea & Blanchard, Philadelphia, 1845.
12. Stannius H: Handbuch der Anatomie der Wirbelthiere. von Veit, Berlin, 1854.
13. Prenant A: Contribution a l'étude du développement organique et histologique du thymus, de la glande thyroide et de la glande carotidienne. La Cellule *10:* 85-184, 1894.
14. Bell ET: The development of the thymus. Am J Anat *5:* 29-62, 1905.
15. Hammar JA: Über Gewicht, Involution und Persistenz des Thymus im Postfötalleben des Menschen. Arch Anat Physiol Anat Abt [Suppl] 91-182, 1906.
16. Hammar JA: Die Menschenthymus in Gesundheit und Krankheit: Ergebnisse der numerischen Analyse von mehr als tausend menschlichen Thymusdrüsen. Teil I. Das normale Organ - zugleich eine kritische Beleuchtung der Lehre des "Status thymicus". Z Mikr Anat Forsch (Leipzig) *6:* 1-570, 1926.
17. Hammar JA: Die normal-morphologische Thymusforschung im letzten Vierteljahrhundert. Analyse und Synthese. Leipzig, Barth, pp 1-453, 1936.
18. Maximow AA: Untersuchungen über Blut und Bindegewebe. II. Über die Histogenese der Thymus bei Säugetieren. Arch mikr Anat *74:* 525-621, 1909.
19. Tamemori Y: Untersuchungen über die Thymusdrüse im Stadium der Altersinvolution. Virchows Arch pathol Anat *242:* 255-266, 1914.
20. Fenger F: On the size and composition of the thymus gland. J Biol Chem *20:* 115-118, 1915.
21. von Haberer H: Zur klinischen Bedeutung der Thymusdrüse. Arch klin Chir *109:* 193-248, 1917.
22. Gedda E: Zur Altersanatomie der Kaninchenthymus. Upsala Lakaref Forh *26:* 1-27, 1921.
23. De Sanctis S: Les enfants dysthymiques. Encephale *18:* 1, 88, 156, 1923.
24. Bratton AB: The normal weight of the human thymus. J Pathol Bacteriol *28:* 609-620, 1925.
25. Scammon RE: The prenatal growth of the human thymus. Proc Soc Exp Biol Med *24:* 906-909, 1927.
26. Young M, Turnbull HM: An analysis of the data collected by the status lymphaticus investigation committee. J Pathol Bacteriol *34:* 213-258, 1931.
27. Boyd E: The weight of the thymus gland in health and in disease. Am J Dis Child *43:* 1162-1214, 1932.
28. Carr JL: Status thymicolymphaticus. J Pediatr *27:* 1-43, 1945.
29. Manning MJ: A comparative view of the thymus in vertebrates. *In:* The Thymus Gland (Kendall MD, ed). New York, Academic Press, pp 7-20, 1981.
30. Simpson JG, Gray ES, Beck JS: Age involution in the normal adult thymus. Clin Exp Immunol *19:* 261-265, 1975.
31. Bodey B: Histomorphology and histochemistry of the human thymus during its prenatal ontogenesis. Dissertation, Inst Morphol, Bulg Acad Sci, Sofia, Bulgaria, 1977, pp 1-360.
32. von Gaudecker B: Ultrastructure of the age-involuted adult human thymus. Cell Tissue Res *186:* 507-525, 1978.
33. Kendall MD, Johnson HR, Singh J: The weight of the human thymus gland at necropsy. J Anat *131:* 483-488, 1981.
34. Oosterom R, Kater L: The thymus in the aging individual. I. Mitogen responsiveness of human thymocytes. Clin Immunol Immunopathol *18:* 187-194, 1981.
35. Oosterom R, Kater L: The thymus in the aging individual. II. Thymic epithelial function *in vitro* in aging and in thymus pathology. Clin Immunol Immunopathol *18:* 195-202, 1981.
36. Steinmann GG, Klaus B, Müller-Hermelink H-K: The involution of the ageing human thymic epithelium is independent of puberty. Scand J Immunol *22:* 563-575, 1985.
37. Simmons VP: Thymic "atrophy" at puberty is a myth. Med Hypotheses *22:* 299-301, 1987.
38. George AJT, Ritter MA: Thymic involution with ageing: obsolescence or good housekeeping? Immunol Today *17:* 267-272, 1996.
39. Strandberg A: Zur Frage des intrathymischen Bindegewebes. Anat Hefte *55:* 169-186, 1917-18.
40. Täkhä H: The weight of the thymus in children of 0-2 years of age. Acta Paediatr Scand *40:* 469-485, 1951.
41. Blau JN: Histological changes and macrophage activity in the adult guinea-pig thymus. Br J Exp Pathol *52:* 142-146, 1971.
42. von Gaudecker B: Die fortschreitende Erweiterung mesodermaler perivaskulärer Räume im Thymus des Menschen. Verh Anat Ges *71:* 783-787, 1977.
43. Kendall MD: The morphology of perivascular space in the thymus. Thymus *13:* 157-164, 1989.
44. Herrmann T: Das Auftreten des Fettgewebes im menschlichen Thymus. Anat Anz *46:* 357-359, 1914.
45. Marine D, Manley OT, Baumann EJ: The influence of thyroidectomy, gonadectomy, suprarenalectomy, and splenectomy on the thymus glands of rabbits. J Exp Med *40:* 429-443, 1924.
46. Metalnikof AE: Rôle du système nerveux et des facteurs biologiques et psychiques dans l'immunité. Masson, Paris, 1934.
47. Evans HM, Simpson ME: Reduction of the thymus by gonadotropic hormone. Anat Rec *60:* 423-435, 1934.
48. Dougherty TF: Effect of hormones on lymphatic tissue. Physiol Rev *32:* 379-401, 1952.
49. Pierpaoli W, Sorkin E: Relationship between thymus and hypophysis. Nature *215:* 834-837, 1967.
50. Trainin N: Thymic hormones and the immune response. Physiol Rev *54:* 272-315, 1974.
51. Trainin N, Kook AI, Umiel T, Albala M: The nature and mechanism of stimulation of immune responsiveness by thymus extracts. Ann NY Acad Sci *249:* 349-361, 1975.
52. Bach J-F, Carnaud C: Thymic factors. Prog Allergy *21:* 342-408, 1976.

53. Bach J-F, Dardenne M, Pleau J-M: Biochemical characterization of a serum thymic factor. Nature *266:* 55-56, 1977.
54. Grossman CJ: Regulation of the immune system by sex steroid. Endocrine Rev *5:* 435-454, 1984.
55. Berczi I: Immunoregulation by pituitary hormones. *In:* Pituitary Function and Immunity (Berczi I, ed) Boca Raton, CRC Press, pp 227-240, 1986.
56. Hiestand PC, Mekler P, Nordmann R, Grieder A, Permmongkol C: Prolactin as a modulator of lymphocyte responsiveness provides a possible mechanism of action for cyclosporin. Proc Natl Acad Sci USA *83:* 2599-2603, 1986.
57. Jankovic BD: Neuroimmunomodulation: facts and dilemmas. Immunol Lett *21:* 101-118, 1989.
58. Dardenne M, Kelly PA, Bach JF, Savino W: Identification and functional activity of prolactin receptors in thymic epithelial cells. Proc Natl Acad Sci USA *88:* 9700-9704, 1991.
59. Greenstein BD, de Bridges EF, Fitzpatrick FT: Aromatase inhibitors regenerate the thymus in aging male rats. Int J Immunopharmacol *14:* 541-553, 1992.
60. Besedovsky HO, del Rey A: Immune-neuroendocrine circuits: integrative role of cytokines. Front Neuroendocrinol *13:* 61-94, 1992.
61. Clarke AG, Kendall MD: The thymus in pregnancy: the interplay of neural, endocrine and immune influences. Immunol Today *15:* 545-551, 1994.
62. Hartwig M, Steinmann G: On a causal mechanism of chronic thymic involution in man. Mech Ageing Develop *75:* 151-156, 1994.
63. Deschaux P, Khan NA: Immunophysiology: The immune system as a multifunctional physiological unit. Cell Mol Biol Res *41:* 1-10, 1995.
64. Eidinger D, Garrett TJ: Studies of the regulatory effects of the sex hormones on antibody formation and stem cell differentiation. J Exp Med *136:* 1098-1116, 1972.
65. Pearce PT, Khalid BA, Funder JW: Progesterone receptors in rat thymus. Endocrinology *113:* 1287-1291, 1983.
66. Barr IG, Pyke KW, Pearce P, Toh B-H, Funder JW: Thymic sensitivity to sex hormones develops post-natally; an *in vivo* and *in vitro* study. J Immunol *132:* 1095-1099, 1984.
67. Luster MI, Hayes HT, Korach K, Tucker AN, Dean JH, Greenlee WF, Boorman GA: Estrogen immunosuppression is regulated through estrogenic responses in the thymus. J Immunol *133:* 110-116, 1984.
68. Carr DJJ: Neuroendocrine peptide receptors on cells of the immune system. *In:* Neuroimmunoendocrinology (Blalock JE, ed) Basel, Karger, pp 84-99, 1992.
69. Halnan ET, Marshall FHA: On the relation between the thymus and the generative organs and the influence of these organs upon growth. Proc Roy Soc s B *88:* 68-89, 1914.
70. Pappenheimer AM: The thymus gland and its possible relation to the female genital tract. Surg Gynec Obstr *25:* 276-283, 1917.
71. Paton DN: The relationship of the thymus and testes to growth. Edinborough Med J *33:* 351-356, 1926.
72. Plagge JC: The thymus gland in relation to sex hormones and reproductive processes in albino rat. J Morphol *68:* 519-545, 1941.
73. Ross MA, Korenchevsky V: The thymus of rat and sex hormones. J Pathol Bacteriol *52:* 349-360, 1941.
74. Grégoire C: Hormones sexuelles, thymus et tissu lymphoïde. Compt Rend Soc Biol *138:* 131-132, 1944.
75. Grégoire C: Sur le mécanisme de l'hypertrophie thymique déclenchée par la castration. Arch Internat Pharmacodyn Thérap *71:* 147-163, 1945.
76. Grégoire C: Failure of lactogenic hormone to maintain pregnancy involution of thymus. J Endocrinol *5:* 115-120, 1947.
77. Smith EM: Hormonal activities of lymphokines, monokines and other cytokines. Prog Allergy *43:* 121-139, 1988.
78. Smith EM: Hormonal activities of cytokines. *In:* Neuroimmunoendocrinology (Blalock JE, ed) Basel, Karger, pp 154-169, 1992.
79. Kavelaars A, Ballieux RE, Heijnen CJ: The role of interleukin in the CRF and AVP induced secretion of β endorphin by human peripheral blood mononuclear cells. J Immunol *142:* 2338-2342, 1989.
80. Besedovsky HO, del Rey A: Immune-Neuro-Endocrine Interactions: Facts and Hypotheses. Endocrine Rev *17:* 64-102, 1996.
81. Besedovsky HO, Sorkin E: Involvement of the thymus in female sexual maturation. Nature *249:* 356-358, 1974.
82. Gottschall PE, Katsuura G, Arimura A: Interleukin-1β is more potent than interleukin-1α in suppressing follicle-stimulating hormone-induced differentiation of ovarian granulosa cells. Biochem Biophys Res Commun *163:* 764-770, 1989.
83. Rivier C, Vale W: Cytokines act within the brain to inhibit luteinizing hormone secretion and ovulation in the rat. Endocrinology *127:* 849-856, 1990.
84. Kalra PS, Sahu A, Kalra SP: Interleukin-1 inhibits the ovarian steroid induced luteinizing hormone surge and release of hypothalamic luteinizing hormone-releasing hormone in rats. Endocrinology *126:* 2145-2152, 1990.
85. DiGeorge AM: Congenital absence of the thymus and its immunological consequences: Concurrence with congenital hypothyroidism. Birth Defects Orig Artic Ser *4:* 116, 1968.
86. Cardier AC, Haumont SM: Development of thymus, parathyroids, and ultimobranchial-bodies NMRI and *nude* mice. Am J Anat *157:* 227-263, 1980.
87. Lindeberg W: Ueber den Einfluss der Thymektomie auf den Gesamtorganismus und auf die Drusen mit innerer Sekretion, insbesondere die Epiphyse und Hypophyse. Folia neuropathol eston *2:* 42-108, 1924.
88. Greenstein BD, Fitzpatrick FTA, Adcock IM, Kendall MD, Wheeler MJ: Reappearance of the thymus in old rats after orchidectomy: inhibition of regeneration by testosterone. J Endocrinol *110:* 417-422, 1986.
89. Kelley KW, Davila DR, Brief S, Simon J, Arkins S: A Pituitary-Thymus Connection during Aging. Ann NY Acad Sci *521:* 88-98, 1988.
90. Charleton W: Exercitat. phys. anat. de oeconomia minimalum. Amstelod, 1659.
91. Comsa J: Hormonal interactions of the thymus. *In:* Thymic Hormones (Luckey TD, ed) Baltimore, University Park Press, 1973.
92. Smith KA, Crabtree GR, Kennedy SJ, Munck AU: Glucocorticoid receptors and glucocorticoid sensitivity of mitogen stimulated and unstimulated human lymphocytes. Nature *267:* 523-525, 1977.
93. Comsa J, Leonhardt H, Wekerle H: Hormonal Coordination of the Immune Response. Rev Physiol Biochem Pharmacol *92:* 115-191, 1982.
94. Smith LR, Brown SL, Blalock JE: Interleukin-2 induction of ACTH secretion: presence of an interleukin-2 receptor α-chain-like molecule on pituitary cells. J Neuroimmunology *21:* 249-254, 1989.
95. Cardarelli NF (ed): The Thymus in Health and Senescence, Volume 2: Aging and Endocrinology. Boca Raton, CRC Press, pp 43-78, 1989.
96. Basch K: Die Beziehung der Thymus zur Schilddruse. Ztschr exp Pathol Ther *12:* 180-206, 1912/13.

97. Fabris N, Muzzioli M, Mocchegiani E: Recovery of age-dependent immunological deterioration in Balb/C mice by short-term treatment with L-thyroxine. Mech Aging Dev *18:* 327-338, 1982.
98. Fulci F: Die Restitutionsfahigkeit des Thymus der Saugetiere nach der Schwangerschaft: vorlaufige Mitteilung. Zentralbl allgem Pathol pathol Anat *24:* 968-974, 1913.
99. Grinevich IuA, Kokhanevich EV, Iugrinova LG, Labunets IF, Patskan II, Tsip NI: Changes in the neuroendocrine and immune state of adaptational homeostasis in normal pregnancy. Fiziolog Zhurnal *39:* 61-67, 1993.
100. Pepper FJ: The effect of age, pregnancy and lactation on the thymus gland and lymph nodes of the mouse. J Endocrinol *22:* 335-348, 1961.
101. Soffer LJ, Gabrilove JL, Wolf BS: Effect of ACTH on thymic masses. J Clin Endocrinol *12:* 690-696, 1952.
102. Arrenbrecht S: Specific binding of growth hormone to thymocytes. Nature *252:* 255-257, 1974.
103. Salas MA, Evans SW, Levell MJ, Whicher JT: Interleukin-6 and ACTH act synergistically to stimulate the release of corticosterone from adrenal gland cells. Clin Exp Immunol *79:* 470-473, 1990.
104. Hussein MF, Badir N, el-Ridi R, Akef M: Differential effect on seasonal variation on lymphoid tissue of the lizard, Chalcides ocellatus. Dev Comp Immunol *2:* 297-309, 1978.
105. Hussein MF, Badir N, el-Ridi R, Akef M: Effect of seasonal variation on lymphoid tissues of the lizards, Mabuya quinquetaeniata Licht. and Uromastyx aegyptia Forsk. Dev Comp Immunol *2:* 469-478, 1978.
106. Zapata AG, Varas A, Torroba M: Seasonal variations in the immune system of lower vertebrates. Immunol Today, *142:* 142-147, 1992.
107. Browman LG, Sears HS: Cyclic variation in the mule deer thymus. Proc Soc Exp Biol Med *93:* 161-162, 1956.
108. Bargmann W: Der Thymus. *In:* Handbuch der mikroskopischen Anatomie des Menschen, Blutgefäss- und Lymphgefässapparat Innersekretorische Drüsen. Ergänzung zu Band VI/1 (von Möllendorf W, herausgeben). Berlin, Springer Verlag, pp 1-172, 1943.
109. Henry L: "Accidental" involution of the human thymus. J Pathol Bacteriol *96:* 337-343, 1968.
110. Kendall MD, Johnson HR, Singh J: The weight of the human thymus gland at necropsy. J Anat *131:* 483-497, 1980.
111. Kendall MD: Age and seasonal changes in the thymus. *In:* The thymus gland (Kendall MD, ed). London, Academic Press, pp 21-35, 1981.
112. Calvo W, Fliedner TM, Herbst EW, Hugl E, Bodey B: Degenerative changes and recovery of the thymus of lethally irradiated dogs, rescued by transfusion of cryopreserved autologous blood leukocytes. Exp Hematol *15:* 1171-1178, 1987.
113. Titova LD, Yarilin AA, Sharova NI, Oradovskaya IV: The quantitatives of T-cell subpopulation and level of α1-thymosin and autoantibodies reacting with thymic epithelium in serum of persons, who worked in 30-km zone of Chernobyl atomic electronic station. Radiatzion Biol Radioekol *36:* 601-609, 1996.
114. Metcalf D, Sparrow N, Nakamura K, Ishidate M: The behaviour of thymus grafts in high and low leukaemia strains of mice. Australian J Exp Biol Med Sci *39:* 441-453, 1961.
115. Metcalf D: The autonomous behaviour of normal thymus grafts. Australian J Exp Biol Med Sci *41:* 437-447, 1963.
116. Davies AJS, Leuchars E, Wallis V, Koller PC: The mitotic response of thymus-derived cells to antigenic stimulus. Transplantation *4:* 438-451, 1966.
117. Metcalf D, Moulds R, Pike B: Influence of the spleen and thymus on immune responses in ageing mice. Clin Exp Immunol *2:* 109-120, 1967.
118. Price GB, Makinodan T: Immunologic deficiencies in senescence. II. Characterization of extrinsic deficiencies. J Immunol *108:* 413-417, 1972.
119. Kelley KW, Brief S, Westly HJ, Novakofski J, Bechtel PJ, Simon J, Walker EB: GH_3 pituitary adenoma cells can reverse thymic aging in rats. Proc Natl Acad Sci USA *83:* 5663-5667, 1986.
120. Gottlieb M, Strober S, Kaplan HS: Allogeneic marrow transplantation after total lymphoid irradiation (TLI): effect of dose/fraction, thymic irradiation, delayed marrow infusion, and presensitization. J Immunol *123:* 379-383, 1979.
121. Blomgren H, Andersson B: Reappearance and relative importance of immunocompetent cells in the thymus, spleen and lymph nodes following lethal x-irradiation and bone marrow reconstitution in mice. J Immunol *106:* 831-834, 1971.
122. Grégoire C: Action des rayons X sur le thymus au cours de l'histogénèse. C R Assoc Anat (Paris) *27:* 328-335, 1932.
123. Tyan ML: Impaired thymic regeneration in lethally irradiated mice given bone marrow from aged donors. Proc Soc Exp Biol Med *152:* 33-35, 1976.
124. Tyan ML: Age-related decrease in mouse T cell progenitors. J Immunol *118:* 846-851, 1977.
125. Hinsull SM, Bellamy D, Franklin A: A quantitative histological assessment of cellular death, in relation to mitosis, in rat thymus during growth and age involution. Age Ageing *6:* 77-84, 1977.
126. Evans RL, Wall DW, Platsoucas CD, Siegal FP, Fikrig SM, Testa CM, Good RA: Thymus-dependent membrane antigens in man: inhibition of cell-mediated lympholysis by monoclonal antibodies to the TH_2 antigen. Proc Natl Acad Sci USA *78:* 544-548, 1981.
127. Steinmann GG: Changes in the human thymus during aging. *In:* The Human Thymus. Current Topics in Pathology, Volume 75 (Müller-Hermelink HK, ed) Berlin, Springer Verlag, pp 43-88, 1986.
128. Bodey B: Development of lymphopoiesis as a function of the thymic microenvironment. Use of $CD8^+$ cytotoxic T lymphocytes for cellular immunotherapy of human cancer. IN VIVO *8:* 915-943, 1994.
129. Tosi P, Kraft R, Luzi P, Cintorino M, Fankhauser G, Hess MW, Cottier H: Involution patterns of the human thymus. I. Size of the cortical area as a function of age. Clin Exp Immunol *47:* 497-504, 1982.
130. Dunn RJ, Luedecker CJ, Haugen HS, Clegg CH, Farr AG: Thymic overexpression of CD40 ligand disrupts normal thymic epithelial organization. J Histochem Cytochem *45:* 129-141, 1997.
131. Kaye J, Ellenberger DL: Differentiation of an immature T cell line: a model of thymic positive selection. Cell *71:* 423-435, 1992.
132. Hollander GA, Wang B, Nichogiannopoulou A, Platenburg PP, van Ewijk W, Burakoff SJ, Gutierrez-Ramos J-C, Terhorst C: Developmental control point in induction of thymic cortex regulated by a subpopulation of prothymocytes. Nature *373:* 350-353, 1995.
133. Collins C, Norris S, McEntee G, Traynor O, Bruno L, von Boehmer H, Hegarty J, O'Farrelly C: RAG1, RAG2 and pre-T cell receptor α chain expression by adult human hepatic T cells: evidence for extrathymic T cell maturation. Eur J Immunol *26:* 3114-3118, 1996.
134. Groves T, Parsons M, Miyamoto NG, Guidos CJ: TCR engagement of $CD4^+CD8^+$ thymocytes in vitro induces early

aspects of positive selection, but not apoptosis. J Immunol *158:* 65-75, 1997.
135. Kruisbeek AM, Astaldi GCB: Distinct effects of thymic epithelial culture supernatants on T cell properties of mouse thymocytes separated by the use of peanut agglutinin. J Immunol *123:* 984-991, 1979.
136. van Ewijk W, Rouse RV, Weisman IL: Distribution of H-2 microenvironments in the mouse thymus. Immunoelectron microscopic identification of I-A and H-2K bearing cells. J Histochem Cytochem *28:* 1089-1099, 1980.
137. van Vliet E, Melis M, Ewijk W: Monoclonal antibodies to stromal cell types of the mouse thymus. Eur J Immunol *14:* 524-529, 1984.
138. Kendall MD, van de Wijngaert FP, Schuurman HJ, Rademakers LH, Kater L: Heterogeneity of the human thymus epithelial microenvironment at the ultrastructural level. Adv Exp Med Biol *186:* 289-297, 1985.
139. de Maagd RA, MacKenzie WA, Schuurman HJ, Ritter MA, Price KM, Broekhuizen R, Kater L: The human thymus microenvironment: heterogeneity detected by monoclonal anti-epithelial cell antibodies. Immunology *54:* 745-754, 1985.
140. Ritter MA, Schuurman HJ, MacKenzie WA, de Maagd RA, Price KM, Broekhuizen R, Kater L: Heterogeneity of human thymus epithelial cells revealed by monoclonal anti-epithelial cell antibodies. Adv Exp Med Biol *186:* 283-288, 1985.
141. von Gaudecker B, Steinmann GG, Hansmann ML, Harpprecht J, Milicevic NM, Müller-Hermelink HK: Immunocytochemical characterization of the thymic microenvironment. A light-microscopic and ultrastructural immunocytochemical study. Cell Tissue Res *244:* 403-412, 1986.
142. van Ewijk W: T-cell differentiation is influenced by thymic microenvironments. Annu Rev Immunol *9:* 591-615, 1991.
143. Anderson G, Jenkinson EJ, Moore NC, Owen JJ: MHC class II-positive epithelium and mesenchyme cells are both required for T-cell development in the thymus. Nature *362:* 70-73, 1993.
144. Muller KP, Kyewski BA: T cell receptor targeting to thymic cortical epithelial cells *in vivo* induces survival, activation and differentiation of immature thymocytes. Eur J Immunol *23:* 1661-1670, 1993.
145. Bodey B, Kaiser HE: Development of Hassall's bodies of the thymus in humans and other vertebrates (especially mammals) under physiological and pathological conditions: Immunocytochemical, electronmicroscopic and *in vitro* observations. In Vivo *11:* 61-86, 1997.
146. Bhan AK, Reinherz EL, Poppema S, McCluskey RT, Schlossman SF: Location of T cell and major histocompatibility complex antigens in the human thymus. J Exp Med *152:* 771-782, 1980.
147. Janossy G, Thomas JA, Bollum FJ, Granger S, Pizzolo G, Bradstock KF, Wong L, McMichael A, Ganeshaguru K, Hoffbrand AV: The human thymic microenvironment: an immunohistologic study. J Immunol *125:* 202-212, 1980.
148. Wolf SS, Cohen A: Expression of cytokines and their receptors by human thymocytes and thymic stromal cells. Immunology *77:* 362-368, 1992.
149. Wekerle H, Ketelsen U-P, Ernst M: Thymic nurse cells. Lymphoepithelial complexes in murine thymuses: morphological and serological characterization. J Exp Med *151:* 925-944, 1980.
150. Vakharia DD: Demonstration of keratin filaments in thymic nurse cells (TNC) and alloreactivity of TNC-T cell population. Thymus *5:* 43-52, 1983.
151. Vakharia DD, Mitchison NA: Helper T cell activity demonstrated by thymic nurse cells (TNC-T). Immunol *51:* 269-273, 1984.
152. van Vliet E, Melis M, van Ewijk W: Immunohistology of thymic nurse cells. Cell Immunol *87:* 101-109, 1984.
153. Geenen V, Defresne MP, Robert F, Legros JJ, Franchimont P, Boniver J: The neurohormonal thymic microenvironment: immunocytochemical evidence that thymic nurse cells are neuroendocrine cells. Neuroendocrinology *47:* 365-368, 1988.
154. Penninger J: Phenotypic and functional analysis of thymic nurse cell (TNC)-lymphocytes. Wien Klin Wochenschr *103:* 45-50, 1991.
155. Carr K, Lowry T, Li LL, Tsai C, Stoolman L, Fox DA: Expression of CD60 on multiple cell lineages in inflammatory synovitis. Lab Invest *73:* 332-338, 1995.
156. Bodey B, Bodey B Jr, Kemshead JT, Siegel SE, Kaiser HE: Identification of neural crest derived cells within the cellular microenvironment of the human thymus employing a library of monoclonal antibodies raised against neuronal tissues. In Vivo *10:* 39-47, 1996.
157. Miller HC, Vito C: Action of a thymic cytokine TsIF™ in reversing the autoimmune disease state of the MRL/1pr mouse. Mol Biother *1:* 213-217, 1989.
158. Grégoire C: Recherches sur le symbiose lymphoepitheliale au niveau du thymus de mammifere. Arch Biol *46:* 717-820, 1935.
159. Grégoire C, Duchateau G: A study on lympho-epithelial symbiosis in thymus. Reactions of the lymphatic tissue to extracts and to implants of epithelial components of thymus. Arch Biol *68:* 269-296, 1956.
160. Kendall MD, van de Wijngaert FP, Schuurman H-J, Rademakers LHPM, Kater L: Heterogeneity of the human thymus epithelial microenvironment at the ultrastructural level. *In:* Microenvironment in the lymphoid system. Adv Exp Med Biol *186:* 289-297, 1985.
161. Hirokawa K, Saitoh K: Abstract # 3-3-14, 4th International Congress of Immunology, Paris, 1980.
162. Monier JC, Dardenne M, Pléau JM, Schmitt D, Deschaux P, Bach JF: Characterization of facteur thymique sérique (FTS) in the thymus. I. Fixation of anti-FTS antibodies on thymic reticuloepithelial cells. Clin Exp Immunol *42:* 470-476, 1980.
163. Schmitt D, Monier JC, Dardenne M, Pléau JM, Deschaux P, Bach JF: Cytoplasmic localization of FTS (facteur thymique sérique) in thymic epithelial cells. An immunoelectronmicroscopical study. Thymus *2:* 177-186, 1980.
164. Schmitt D, Monier JC, Dardenne M, Pléau JM, Bach JF: Location of FTS (facteur thymique sérique) in the thymus of normal and auto-immune mice. Thymus *4:* 221-231, 1982.
165. Jambon B, Montagne P, Bene MC, Brayer MP, Faure G, Duheille J: Immunohistologic localization of "facteur thymique sérique" (FTS) in human thymic epithelium. J Immunol *127:* 2055-2059, 1981.
166. Savino W, Dardenne M: Thymic hormone containing cells. VI. Immunohistologic evidence for the simultaneous presence of thymulin, thymopoietin and thymosin $\alpha 1$ in normal and pathological human thymuses. Eur J Immunol *14:* 987-992, 1984.
167. Bach JF, Dardenne M, Papiernik M, Barois A, Levasseur P, LeBrigand H: Evidence of a serum factor produced by the human thymus. Lancet *2:* 1056-1058, 1972.
168. Goldstein AL, Guha A, Zatz MM, Hardy MA, White A: Purification and biological activity of thymosin, a hormone of the thymus gland. Proc Natl Acad Sci USA *69:* 1800-1803, 1972.

169. Goldstein G: Isolation of bovine thymin: A polypeptide hormone of the thymus. Nature *247:* 11-14, 1974.
170. Hirokawa K, McClure JE, Goldstein AL: Age-related changes in localization of thymosin in the human thymus. Thymus *4:* 19-29, 1982.
171. Goldstein AL (ed): Thymic Hormones and Lymphokines. New York, Plenum Press, pp 1-600, 1984.
172. Fabris N, Mocchegiani E, Amadio L, Zannotti M, Licastro F, Franceschi C: Thymic hormone deficiency in normal aging and Down's syndrome: is there a primary failure of the thymus? Lancet *1:* 983-986, 1984.
173. Okimoto T, Kinoshita Y, Hato F, Toyokawa T, Kimura S, Kinoshita H: Effects of thymosin on soybean lectin responsiveness of thymic small lymphocyte subsets from streptozotocin-induced diabetic rats. Cell Mol Biol *34:* 465-472, 1988.
174. Kawashima I, Sakabe K, Seiki K, Fujii-Hanamoto H, Akatsuka A, Tsukamoto H: Localization of sex steroid receptor cells, with special reference to thymulin (FTS)-producing cells in female rat thymus. Thymus *18:* 79-93, 1991.
175. Goldstein AL, Hooper JA, Schulof RS, Cohen GH, Thurman GB, McDaniel MC, White A, Dardenne M: Thymosin and the immunopathology of aging. Fed Proc *33:* 2053-2056, 1974.
176. Lewis VM, Twomey JJ, Bealmear P, Goldstein G, Good RA: Age, thymic involution and circulating thymic hormone activity. J Clin Endocrinol Metab *47:* 145-150, 1978.
177. Waring AJ: The thymus gland in infancy and childhood. J Med Assoc Georgia Atlanta *13:* 295-297, 1924.
178. Andrew W: The Anatomy of Aging in Man and Animals. Grune & Stratton, New York, 1971.
179. Newberne PM: Dietary fat, immunological response, and cancer in rats. Cancer Res *41:* 3783-3785, 1981.
180. Ritter MA, Lampert IA: The thymus. *In:* Oxford Textbook of Pathology. Volume 26: Pathology of Systems (McGee JO'D, Isaacson PG, Wright NA, eds). Oxford University Press, pp 1807-1821, 1992.
181. Hirokawa K, Utsuyama M, Kasai M, Kurashima C, Ishijima S, Zeng YX: Understanding the mechanism of the age-change of thymic function to promote T cell differentiation. Immunol Lett *40:* 269-277, 1994.
182. Cristofalo VJ, Gerhard GS, Pignolo RJ: Molecular biology of aging. Surg Clinics North America *74:* 1-21, 1994.
183. Toussaint O, Remacle J: Revue critique des theories du vieillissement cellulaire. Du concept de base de Hayflick au concept de seuil critique d'accumulation d'erreurs. Pathologie Biologie *42:* 313-321, 1994.
184. Weissman A: Essays on Heredity. Clarendon Press, Oxford, England, 1889.
185. Weissman A: The duration of life. *In:* Weissman on Heredity. Edition 2 (Poulton EB, Schonland S, Shipley AE, eds) London, Oxford University Press, 1891.
186. Carrel A: On the permanent life of tissues outside of the organism. J Exp Med *15:* 516-528, 1912.
187. Hayflick L: The establishment of a line (WISH) of human amnion cells in continuous cultivation. Exp Cell Res *23:* 14-20, 1961.
188. Hayflick L, Moorhead PS: The serial cultivation of human diploid cell strains. Exp Cell Res *25:* 585-621, 1961.
189. Hayflick L: The limited *in vitro* lifetime of human diploid cell strains. Exp Cell Res *37:* 614-636, 1965.
190. Hayflick L: The cell biology of human aging. N Engl J Med *295:* 1302-1308, 1976.
191. Kohn RR: Aging and cell division. Science *188:* 203-204, 1975.
192. Bell E, Marek LF, Levinstone DS, Merrill C, Sher S, Young IT, Eden M: Loss of division potential in vitro: aging or differentiation? Science 202: 1158-1163, 1978.
193. Williams GC: Pleiotropy, natural selection and the evolution of senescence. Evolution *11:* 398-411, 1957.
194. Szilard L: On the nature of the aging process. Proc Natl Acad Sci USA *45:* 30-45, 1959.
195. Strehler BL: Time, Cells, and Aging. 2nd ed. New York, Academic Press, 1977.
196. Kohn RR: Aging of animals: Possible mechanism. *In:* Principles of Mammalian Aging. 2nd Edition. Englewood Cliffs, Prentice-Hall, 1978.
197. Medawar PB: An unsolved problem of biology. *In:* The Uniqueness of the Individual. Dover, New York, 1981.
198. Masoro EJ: Metabolism. *In:* Handbook of the Biology of Aging. 2nd Edition (Finch CE, Schneider EL, eds) New York, Van Nostrand Reinhold, p 540, 1985.
199. Rose MR: Evolutionary Biology of Aging. New York, Oxford University Press, 1991.
200. Moodie RL: General considerations of the evidence of pathological conditions found among fossil animals. *In:* Diseases in Antiquity (Brothwell D, Sandison AT, eds) Springfield, Charles C. Thomas, p 31, 1967.
201. Walford RL: Auto-immunity and aging. J Gerontol *17:* 281-285, 1962.
202. Walford RL: Further consideration towards an immunologic theory of aging. Exp Gerontol *1:* 67-76, 1964.
203. Walford RL: Immunology and aging. Philip Levine Award. Am J Clin Pathol *74:* 247-253, 1980.
204. Warner HR, Price AR: Involvement of DNA repair in cancer and aging. J Gerontol *44:* 45-54, 1989.
205. Ram JS: Aging and immunological phenomena: a review. J Gerontol *22:* 92-107, 1967.
206. Burnet FM: An immunological approach to aging. Lancet *2:* 358-360, 1970.
207. Burnet FM: Immunology, Aging and Cancer: Medical Aspects of Mutation and Selection. WH. Freeman & Co, San Francisco, CA, 1976.
208. Kay MMB: The thymus: clock for immunologic aging? J Invest Dermatol *73:* 29-38, 1979.
209. Rabin BS: The effect of diet on immune responsiveness and aging. Med Hypoth *8:* 495-503, 1982.
210. Chandra RK: Trace elements and immune response. Immunol Today, *4:* 322-325, 1983.
211. Christinck ER, Luscher MA, Barber BH, Williams DB: Peptide binding to class I MHC on living cells and quantitation of complexes required for CTL lysis. Nature *352:* 67-70, 1991.
212. McMichael A: Cytotoxic T lymphocytes and immune surveillance. Cancer Surv *13:* 5-21, 1992.
213. Segre D, Miller RA, Abraham GN, Weigle WO, Warner HR: Aging and the immune system. J Gerontol *44:* B164-B168, 1989.
214. Hu Y, Dietrich H, Herold M, Heinrich PC, Wick G: Disturbed immune-endocrine communication via the hypothalamus-pituitary-adrenal axis in autoimmune disease. Int Arch Allergy Immunol *102:* 232-241, 1993.
215. Everitt AV: The neuroendocrine system and aging. Gerontology *26:* 108-119, 1980.
216. Cotzias GC, Miller ST, Nicholson AR Jr, Maston WH, Tang LC: Prolongation of the life-span in mice adapted to large amounts of L-dopa. Proc Natl Acad Sci USA *71:* 2466-2469, 1974.
217. Levin P, Janda JK, Joseph JA, Ingram DK, Roth GS: Dietary restriction retards the age-associated loss of rat striated dopaminergic receptors. Science *214:* 561-562, 1981.

218. Goya RG, Castro MG, Saphier PW, Sosa YE, Lowry PJ: Thymus-pituitary interactions during ageing. Age Ageing *22:* S19-S25, 1993.
219. Spangelo BL: The thymic-endocrine connection. J Endocrinol *147:* 5-10, 1995.
220. Cutler RG: Evolutionary Biology of Senescence. *In:* The Biology of Aging (Behnke JA, Finch CE, Moment GB, eds). New York, Plenum Press, pp 311-360, 1978.
221. Cutler RG: Longevity is determined by specific genes; testing the hypothesis. *In:* Testing the Theories of Aging (Adelman RC, Roth GS, eds) Boca Raton, CRC Press, p 25, 1982.
222. Sacher GA: Evolution of longevity and survival characteristics in mammals. *In:* The Genetics of Aging (Schneider EA, ed) New York, Plenum Press, pp 151-167, 1978.
223. Sacher GA: 1976 Robert W. Kleemeier Award Lecture: Longevity, aging, and death: an evolutionary perspective. Gerontologist *18:* 112-119, 1978.
224. Selye H: Thymus and adrenals in the response of the organism to injuries and intoxications. Br J Exp Pathol *17:* 234-248, 1936.
225. Grégoire C: Factors involved in maintaining involution of thymus suckling. J Endocrinol *5:* 68-87, 1947.
226. Goodall A: The postnatal changes in the thymus of guinea-pigs and the effect of castration on thymus structure. J Physiol *32:* 191-198, 1905.
227. Fitzpatrick FTA, Kendall MD, Wheeler MJ, Adcock TM, Greenstein BD: Reappearance of thymus of ageing rats after orchidectomy. J Endocrinol *106:* R17- R19, 1985.
228. Kirkwood TB: Evolution of ageing. Nature *270:* 301-304, 1977.
229. Denckla WD: Systems analysis of possible mechanisms of mammalian aging. Mech Ageing Dev *6:* 143-152, 1977.
230. Holliday R, Huschtscha LI, Tarrant GM, Kirkwood TBL: Testing the commitment theory of cellular aging. Science *198:* 366-372, 1977.
231. Kirkwood TB, Holliday R: The evolution of ageing and longevity. Proc Royal Soc London Ser B Bio Sci *205:* 531-546, 1979.
232. Holliday R, Huschtscha LI, Kirkwood TBL: Cellular aging: Further evidence for the commitment theory. Science *213:* 1505-1508, 1981.
233. Kirkwood TB, Franceschi C: Is aging as complex as it would appear? New perspectives in aging research. Ann NY Acad Sci *663:* 412-417, 1992.
234. Brunelli R, Frasca D, Spano M, Zichella L, Doria G: Gonadectomy in old mice induces thymus regeneration but does not recover mitotic responsiveness. Ann NY Acad Sci *673:* 252-255, 1992.
235. von Boehmer H, Teh HS, Kisielow P: The thymus selects the useful, neglects the useless and destroys the harmful. Immunol Today *10:* 57-61, 1989.
236. Shortman K, Egerton M, Spangrude GJ, Scollay R: The generation and fate of thymocytes. Sem Immunol *2:* 3-12, 1990.
237. Aronson M: Hypothesis: involution of the thymus with aging--programmed and beneficial. Thymus *18:* 7-13, 1991.
238. Aronson M: Involution of the thymus revisited: immunological trade-offs as an adaptation to aging. Mech Ageing Dev *72:* 49-55, 1993.
239. Hartwig M: Immune control of mammalian aging: A T-cell model. Mech Ageing Dev *63:* 207-213, 1992.
240. Wick G, Hu Y, Schwarz S, Kroemer G: Immunoendocrine communication via the hypothalamus-pituitary-adrenal axis in autoimmune diseases. Endocrine Rev *14:* 539-563, 1993.
241. Hirokawa K, Utsuyama M, Kasai M, Konno A, Kurashima C, Moriizumi E: Age-related hyperplasia of the thymus and T-cell system in the Buffalo rat. Virchows Arch B Cell Pathol *59:* 38-47, 1990.
242. Jetten AM, Jetten MER: Possible role of retinoic acid binding protein in retinoid stimulation of embryonal carcinoma cell differentiation. Nature *278:* 180-182, 1979.
243. Trown PW, Palleroni AV, Bohoslawec O, Richelo BN, Halpern JM, Gizzi N, Geiger R, Lewinski C, Machlin LJ, Jetten AM, Jetten MER: Relationship between binding affinities to cellular retionic acid-binding protein and *in vivo* and *in vitro* properties for 18 retinoids. Cancer Res *40:* 212-220, 1980.
244. Jetten AM, Deluca LM: Induction of differentiation of embryonal carcinoma cells by retinol: possible mechanisms. Biochem Biophys Res Commun *114:* 593-599, 1983.
245. Eglitis MA, Sherman MI: Murine embryonal carcinoma cells differentiate *in vitro* in response to retinol. Exp Cell Res *146:* 289-296, 1983.
246. Strickland S, Breitman TR, Frickel F, Nurrenbach A, Hadicke E, Sporn MB: Structure-activity relationships of a new series of retinoidal benzoic acid derivatives as measured by induction of differentiation of murine F9 teratocarcinoma cells and human HL-60 promyelocytic leukemia cells.Cancer Res *43:* 5268-5272, 1983.
247. Jetten AM: Induction of differentiation of embryonal carcinoma cells by retinoids. *In:* Retinoids and Cell Differentiation (Sherman MI, ed). Boca Raton, CRC Press, p 105, 1986.
248. Jetten AM, Anderson K, Deas MA, Kagechika H, Lotan R, Rearick JI, Shudo K: New benzoic acid derivatives with retinoid activity: lack of direct correlation between biological activity and binding to cellular retinoic acid binding protein. Cancer Res *47:* 3523-3527, 1987.
249. Giguere V, Ong ES, Segui P, Evans RM: Identification of a receptor for the morphogen retinoic acid. Nature *330:* 624-629, 1987.
250. Pierce GB: Teratocarcinoma: model for a developmental concept of cancer. Curr Top Dev Biol *2:* 223-246, 1967.
251. Sherman MI, Solter D (eds): Teratomas and Differentiation. New York, Academic Press, 1975.
252. Martin GR: Teratocarcinomas and mammalian embryogenesis. Science *209:* 768-776, 1980.
253. Bolande RP: Models and concepts derived from human teratogenesis and oncogenesis in early life. J Histochem Cytochem *32:* 878-884, 1984.
254. Astigiano S, Sherman MI, Abarzua P: Regulation and patterns of endogenous and exogenous gene expression during differentiation of embryonal carcinoma cells. Envir Health Persp *80:* 25-38, 1989.
255. Strickland S, Mahdavi V: The induction of differentiation in teratocarcinoma stem cells by retinoic acid. Cell *15:* 393-403, 1978.
256. Strickland S, Smith KK, Marotti KR: Hormonal induction of differentiation in teratocarcinoma stem cells: generation of parietal endoderm by retinoic acid and dibutyryl cAMP. Cell *21:* 347-355, 1980.
257. Grover A, Oshima RG, Adamson ED: Epithelial layer formation in differentiating aggregates of F9 embryonal carcinoma cells. J Cell Biol *96:* 1690-1696, 1983.
258. Grover A, Adamson ED: Evidence for the existence of an early common biochemical pathway in the differentiation of F9 cells into visceral or parietal endoderm: modulation by cyclic AMP. Dev Biol *114:* 492-503, 1986.
259. Good RA, Fernandez G, West A: Nutrition, immunity, and cancer - A review. Part 1: Influence of protein or protein-calorie malnutrition and zinc deficiency on immunity. Clin Bull *9:* 3-12, 1979.

260. Iwata T, Incefy GS, Tanaka T, Fernandes G, Menendez-Botet CI, Pih K, Good RA: Circulating thymic hormone levels in zinc deficiency. Cell Immunol *47:* 100-105, 1979.
261. Prasad AS: Clinical, endocrinological and biochemical effect of zinc-deficiency. Clin Endocrinol Metab *14:* 567-589, 1985.
262. Chandra RK: Nutritional regulation of immunity and risk of infection in old age. Immunology *67:* 141-147, 1989.
263. Sugarman B, Munro HN: Altered (Zn^{65}) chloride accumulation by aged rats adipocytes in vitro. J Nutr *110:* 2317-2321, 1985.
264. Turnlund JR, Durvin N, Costa F, Margen S: Stable isotope studies of zinc absorption and retention in young and elderly men. J Nutr *116:* 1239-1247, 1986.
265. Dardenne M, Boukaiba M-C, Gagnerault F, Homo-Delarche F, Chappnis FP, Lemmonier D, Savino W: Restoration of the thymus in aging mice by *in vivo* zinc supplementation. Clin Immunol Immunopathol *66:* 127-135, 1993.
266. Fabris N, Mocchegiani E, Muzzioli M, Provinciali M: Zinc, immunity and aging. *In:* Biochemical Advances in Aging (Goldstein AL, ed) New York, Plenum Press, pp 271-281, 1990.
267. Fabris N: Neuroendocrine-immune aging: an integrative view on the role of zinc. Ann N Y Acad Sci *719:* 353-368, 1994.
268. Mocchegiani E, Santarelli L, Muzzioli M, Fabris N: Reversibility of the thymic involution and of age-related peripheral immune dysfunctions by zinc supplementation in old mice. Int J Immunopharmacol *17:* 703-718, 1995.
269. Mocchegiani E, Fabris N: Age-related thymus involution: zinc reverses *in vitro* the thymulin secretion defect. Int J Immunopharmacol *17:* 745-749, 1995.
270. Saha AR, Hadden EM, Hadden JW: Zinc induces thymulin secretion from human thymic epithelial cells *in vitro* and augments splenocyte and thymocyte responses *in vivo*. Int J Immunopharmacol *17:* 729-733, 1995.
271. Boukaiba N, Flament C, Acher S, Chappuis P, Piau A, Fusselier M, Dardenne M, Lemonnier D: A physiological amount of zinc supplementation: effects on nutritional, lipid, and thymic status in an elderly population. Am J Clin Nutr *57:* 566-572, 1993.
272. Weksler ME, Innes JB, Goldstein G: Immunologic studies of aging. IV. The contribution of thymic involution to the immune deficiencies of aging mice and reversal with thymopoietin. J Exp Med *148:* 996-1006, 1978.
273. Weksler ME: The immune system and the aging process in man. Proc Soc Exp Biol Med *165:* 200-205, 1980.
274. van der Griend RJ, Carzeno M, van Doorn R, Lempers CJM, van den Ende A, Wijermans P, Oosterhuis HJGH, Astaldi A: Changes in human T lymphocytes after thymectomy and during senescence. J Clin Immunol *2:* 289-295, 1982.
275. Good RA, Dalmasso AP, Martinez C, Archer OK, Pierce JC, Papermaster BW: The role of the thymus in development of immunologic capacity in rabbits and mice. J Exp Med *116:* 773-796, 1962.
276. Gross L: Immunologic defect in aged population and its relation to cancer. Cancer *18:* 201-204, 1965.
277. Bach MA: Lymphocyte-mediated cytotoxicity: effects of ageing, adult thymectomy and thymic factor. J Immunol *119:* 641-647, 1977.
278. Czlonkowska A, Korlak J: The immune response during aging. J Gerontol *34:* 9-14, 1979.
279. Apostoloff E, Friedel E, Apostoloff G, Jakobza D: Zelluläre Immunreaktionen *in vivo* und *in vitro* bei Menschen in höheren Lebensalter. ZFA *35:* 211-213, 1980.
280. Abe T, Morimoto C, Toguchi T, Kiyotaki M, Homma M: Evidence of aberration of T-cell subsets in aged individuals. Scand J Immunol *13:* 151-157, 1981.
281. Alder SJ, Morley AA, Seshadri RS: Reduced lymphocyte colony formation with age. Clin Exp Immunol *49:* 129-134, 1982.
282. Moore AV, Korobkin M, Olanow W, Heaston DK, Ram PC, Dunnick NR, Silverman PM: Age related changes in the thymus gland. CT-pathologic correlation. AJR *141:* 291, 1983.
283. Staiano-Coico L, Darzynkiewicz Z, Hefton JM, Dutkowski R, Darlington GJ, Weksler ME: Increased sensitivity of lymphocytes from people over 65 to cell cycle arrest and chromosomal damage. Science *219:* 1335-1337, 1983.
284. Steinmann GG: Altersveränderungen des menschlichen Thymus. Habilitationsschrift, Medizinische Fakultät, Universität Kiel, 1984.
285. Bjorkholm M, Holm G, Johansson B, Hakan M: T-lymphocyte deficiency following adult thymectomy in man. Scand J Haematol *14:* 210-215, 1975.
286. Walford RL: The Immunologic Theory of Aging. Munksgaard, Copenhagen, pp 1-248, 1969.
287. Makinodan T, Kay MMB: Age influence on the immune system. Adv Immunol *29:* 287-330, 1980.
288. Benner R, Rijnbeek AM, Bernabe RR, Martinez-Alonso C, Coutinho A: Frequencies of background immunoglobulin-secreting cells in mice as a function of organ, age, and immune status. Immunobiology *158:* 225-238, 1981.
289. Zatz MM, Goldstein AL: Thymosins, lymphokines, and the immunology of aging. Gerontology *31:* 263-277, 1985.
290. Saltzman RL, Peterson PK: Immunodeficiency of the elderly. Rev Infect Dis *9:* 1127-1139, 1987.
291. Ales-Martinez JE, Alvarez-Mon M, Merino F, Bonilla F, Martinez C, Durantez A, De la Hera A: Decreased TcR^- $CD3^+$ T cell numbers in healthy aged humans. Evidence that T cell defects are masked by a reciprocal increase of TcR^- $CD3^-$ $CD2^+$ natural killer cells. Eur J Immunol *18:* 1827-1830, 1988.
292. Brodsky FM, Lem L, Bresnahan PA: Antigen processing and presentation. Tissue Antigens *47:* 464-471, 1996.
293. Kay MMB, Makinodan T: Immunobiology of aging: evaluation of current status. Clin Immunol Immunopathol *6:* 394-414, 1976.
294. Kay MMB, Makinodan T: Relationship between aging and the immune system. Prog Allergy *29:* 134-181, 1981.
295. Bodey B, Calvo W, Prummer O, Fliedner TM, Borysenko M: Development and histogenesis of the thymus in dog. A light and electronmicroscopical study. Dev Comp Immunol *11:* 227-238, 1987.
296. Kincade PW, Lee G, Pietrangeli CE, Hayashi SI, Gimble JM: Cells and molecules that regulate B lymphopoiesis in bone marrow. Ann Rev Immunol *7:* 111-143, 1989.
297. Sharp A, Zipori D, Toledo J, Tal S, Resnitzky P, Globerson A: Age related changes in hemopoietic capacity of bone marrow cells. Mech Ageing Dev *48:* 91-99, 1989.
298. Dworzak MN, Fritsch G, Buchinger P, Fleischer C, Printz D, Zellner A, Schollhammer A, Steiner G, Ambros PF, Gadner H: Flow cytometric assessment of human MIC2 expression in bone marrow, thymus, and peripheral blood. Blood *83:* 415-425, 1994.
299. Olweus J, Lund-Johansen F, Terstappen LW: CD64/Fc γ RI is a granulo-monocytic lineage marker on $CD34^+$ hematopoietic progenitor cells. Blood *85:* 2402-2413, 1995.
300. Tjonnfjord GE, Steen R, Veiby OP, Egeland T: Thymic stromal cells support differentiation of natural killer cells

from $CD34^+$ bone marrow cells in vitro. Eur J Haematol *54:* 46-50, 1995.

301. de Bruijn MF, Ploemacher RE, Mayen AE, Voerman JS, Slieker WA, van Ewijk W, Leenen PJ: High-level expression of the ER-MP58 antigen on mouse bone marrow hematopoietic progenitor cells marks commitment to the myeloid lineage. Eur J Immunol *26:* 2850-2858, 1996.

302. Dobber R, van den Bergh P, Tielemans M, Schuitemaker J, Nagelkerken L: The sensitivity of IL-4 production for cAMP inducers is lost in $CD4^+$ T cells from aged mice. Int Immunol *5:* 1167-1176, 1993.

Section 3

THE THYMUS AND ITS SIGNIFICANCE IN ANTI-NEOPLASTIC THERAPY

Chapter 8

Thymic Hormones in Cancer Diagnosis and Treatment

Abstract: The thymus is an endocrine organ. A unified, physiological concept of humoral regulations of the immune response has emerged in the last four decades. The thymus is the major site of production of immunocompetent T lymphocytes from their hematopoietic stem cells. This complex process requires direct cell to cell, receptor based interactions, as well as *in situ* paracrine information *via* the numerous cytokines and thymic hormones produced by the cells of thymic microenvironment. Thymic hormones induce *in situ* T-cell marker differentiation, expression and functions. These polypeptide hormones have also been shown by means of immunocytochemistry to localize in the reticulo-epithelial (RE) cells of the thymic cellular microenvironment. Due to the great complexity of the intrathymic maturation sequence of T lymphocytes and the diverse immunophenotypically unique subpopulations of T lymphocytes, it is quite unlikely that a single thymic humoral factor could control all of the molecular steps and cell populations involved. It is much more likely that an extremely rich and diverse, but genetically determined, milieu is present within the thymus, and that thus the control of intrathymic T lymphocyte maturation and the functional maturation of T cells involves the orchestral interaction of various thymic-specific factors and other molecules during the differentiation process. Thymosin fraction 5 and its constituent peptides influence several properties of lymphocytes including cyclic nucleotide levels, migration inhibitory factor production, T-dependent antibody production, as well as the expression of various cell surface maturation/differentiation markers. Recently, derivatives of thymic hormones, mostly of thymosins, have been detected as products of neoplastically transformed cells and employed in the early diagnosis of neoplasms. In clinical trials, thymic hormones strengthen the effects of immunomodulators in immunodeficiencies, autoimmune diseases, and neoplastic malignancies. Combined chemo-immunotherapeutical anti-cancer treatment seems to be more efficacious than chemotherapy alone, and the significant hematopoietic toxicity associated with most chemotherapeutical clinical trials can be reduced significantly by the addition of immunotherapy.

Key words: Thymic reticulo-epithelial (RE) cells; Thymic hormones; Thymosin fraction 5 (TF5); Thymopentin (TP5); Prothymosin α1 (ProTα1); Thymosin α1 (Tα1); Thymosin α7; Thymosin β3; Thymosin β4 (Tβ4); Thymosin β10 gene; Thymosin β10; Thymosin β15; Thymic humoral factor-γ 2 (THF-γ2).

1. INTRODUCTION

During the last three centuries, detailed morphological and histological observations of the mammalian and especially the

human thymus have been conducted even before its main physiological function as the primary and central organ of the cellular immune system was revealed by Miller and others (1-47). The ontogenetical and phylogenetical development of the two arms of the immune system were well characterized:

> "A distinct dichotomy of the immune system into thymus-dependent and thymus-independent components has been demonstrated in amphibians, birds and mammals, suggesting that the B cell/T cell dichotomy of the vertebrate immune system dates back in phylogeny at least some 250 million years to the early tetrapods." (48)

Metalnik of (49), working in the Pasteur Institute (Paris), reported several experimental observations indicating a close relationship between the central nervous system (CNS) and immune response. The thymus involutes relatively early in life, while the cellular immune deficiencies of aging correspond to the decline in function of the hypothalamic-pituitary-endocrine axis (50-57). Recent studies point to important roles for the pituitary, the pineal, and the autonomic nervous system, as well as the thyroid, gonads, and adrenals in thymus integrity and function. Thymic function at the local level requires extensive and complex interactions between thymic stromal cells and developing thymocytes involving paracrine and autocrine mediators, such as interleukins (ILs) 1, 2, 6, 7, 8, colony-stimulating factors (CSFs), interferon-γ (IFN-γ), thymosin α1, zinc-thymulin, and other thymic hormones (58-73). A significant endocrine function of the thymus gland is to package zinc in the form of zinc-thymulin for more effective delivery to the periphery. The reversal of thymic involution has been attempted employing a variety of factors, including ILs, thymic hormones, growth hormone, prolactin, melatonin, zinc, and others (74). Our work to reverse thymic involution in hydrocortisone-treated, aged mice with interleukins, thymosin α1, and zinc has already been published (75). Recent efforts to successfully treat immune deficiency in aged and cancer-bearing humans will be presented in this article.

Recent work has reviewed the detailed mechanism of sex hormone actions on the thymic cells, obtained at the cellular and molecular levels (76). The *genomic* actions of sex hormones via nuclear sex hormone receptor complexes are supported by the following observations:

1. sex hormone receptors and the thymic hormone (i.e. thymulin) are co-localized in thymic epithelial cells, but not in T lymphocytes;
2. production/expression of thymic factors (thymulin, thymosin α1, etc.) are inhibited by sex hormone treatment;
3. sex hormones cause alterations in the immunomorphology of intrathymic T lymphocyte subpopulations; and
4. sex hormones, via action through their intracellular receptor, strongly influence the development of thymic neoplasms in spontaneous thymoma BUF/Mna rats.

The *non-genomic* action of sex hormones via a membrane signal transduction mechanism is based on the following:

1. the proliferation/maturation of thymic reticulo-epithelial (RE) cells is mediated through protein kinase C activity induced by sex hormones; and
2. sex hormones directly influence DNA synthesis and cdc2 kinase (cell cycle-promoting factor) activity.

Around the late 1970s, the family of polypeptides present in thymosin fraction 5 was characterized biologically and chemically (77-79). A system of nomenclature was developed, and peptides are being systematically isolated and characterized. Thymosin fraction 5 and its component parts influence a vast number of properties of lymphocytes, including cyclic nucleotide levels, migration inhibitory factor production, T-dependent antibody production, and expression of various cell surface markers. The effects of thymosin are being evaluated in clinical trials on immunodeficiency, malignant, and autoimmune diseases, and a major function of the endocrine thymus in the maintenance of immune balance and in the treatment of diseases characterized by thymic malfunction has been established. Thymosin fraction 5 contains several distinct hormone-like factors, which are effective in partially or fully inducing and maintaining immune function (77, 79). Several of the peptide components of fraction 5 have been purified, sequenced, and studied in assay systems designed to measure T lymphocyte differentiation and function. These studies indicate that several of the purified peptides act on different subpopulations of T lymphocytes. Thymosin β3 and β4 change the terminal deoxynucleotidyl transferase (TdT) expression profile of negative precursor T cells by inducing TdT positivity. Thymosin α1 induces the formation of functional helper cells and conversion of Lyt^- cells to Lyt 1^+, 2^+, 3^+ cells. Thymosin α7 induces the formation of functional suppressor T lymphocytes and also converts Lyt^- cells to Lyt 1^+, 2^+, 3^+ cells. These studies have provided further evidence that the thymus secretes a family of distinct peptides, acting at various stages of the maturation and differentiation sequences of T lymphocytes, thereby inducing and maintaining immune function. Through various functional studies it has become clear that immunological maturation under the direction of the thymic microenvironment is a process involving a number of complex steps and multiple interactions that are stage specific. Given the complexity of the maturation sequence of T lymphocytes and the increasing numbers of immunomorphologically unique T-cell subpopulations that are being identified, it would be surprising if a single thymic humoral factor could control all of the steps and populations involved. Rather, it is much more plausible that the control of the intrathymic maturation of T lymphocytes and the development of their functional capabilities involves numerous thymus-specific factors and other molecules that rigidly control the intermediary steps in lymphopoiesis.

2. THYMIC HORMONES IN CANCER DIAGNOSIS

Clonal deletion of thymocytes is a significant event in T-cell tolerance and may represent a mechanism of tumor escape (80). It has previously been shown that MHC class II-restricted, Id-specific,

$CD4^+$ T lymphocytes in T-cell receptor (TCR)-transgenic mice confer resistance against the MOPC315 plasmacytoma. The ability of monoclonal immunoglobulin produced by a plasmocytoma to induce deletion of thymocytes specific for the variable parts of Ig, i.e. the idiotype (Id) has been evaluated. Large numbers of MOPC315 tumor cells were injected s.c. in the TCR-transgenic mice to overwhelm $CD4^+$ T lymphocyte-mediated protection. When the MOPC315 plasmacytomas reached a weight of approximately 0.5g (serum myeloma protein M315 about 50μg/ml), immature, double positive $CD4^+CD8^+$ and mature $CD4^+$ transgenic thymocytes became progressively deleted. Apoptotic thymocytes were already detectable when neoplasms were 2 mm in diameter (serum M315: 5 μg/ml or 0.03 μM). The negative selection observed by the authors was Id-specific, because an Id-negative plasmacytoma failed to induce deletion. Injection of purified MOPC315-myeloma protein (M315) i.p. caused a profound reduction of Id-specific thymocytes. Enriched thymic dendritic cells (DC) from tumor-bearing animals were found to be primed with lambda2 (M315) and induced the apoptosis of the thymocytes *in vitro*. These results indicate that circulating myeloma protein is processed and presented by intrathymic antigen-presenting cells (APCs), and induces deletion of Id-specific thymocytes during their intrathymic development. Deletion of tumor-specific thymocytes may represent a tumor escape mechanism in patients with malignant neoplasms that secrete or shed neoplastic or tumor associated antigens. The very real possibility that vaccination with tumor Ig or genes coding for it may induce tolerance, instead of protection, should be taken into consideration when devising novel therapeutical approaches to the treatment of malignant disease.

The presence of prothymosin α (ProTα) in various neoplasms has been established. The proliferation index of human breast tumors can be used to identify patients at high risk for distant metastases (81, 82). In a recent observation, ProTα concentrations were measured by RIA, but an alternative nonisotopic assay that could be used in a standard clinical laboratory has been described. The main features of the ELISA method were:

1. a recombinant fusion protein glutathione S-transferase (GST)-human ProTα was used to coat the microtiter plates;
2. a polyclonal antiserum raised in rabbits against thymosin α1, the NH2-terminal fragment of ProTα was employed;
3. this method is as sensitive as the RIA; and
4. the method is faster than the RIA. ProTα concentrations in various human neoplasms (skin, esophagus, colorectal, and breast) as assessed by ELISA were comparable with, although two fold greater than, the values previously estimated by RIA.

A cDNA clone representing ProTα, was identified employing a subtraction-enhanced display technique, from rat hepatocellular carcinoma (HCC). ProTα has been reported to be involved in cell proliferation and regulated by the c-myc gene *in vitro*. In an interesting study, the gene expression pattern of ProTα was examined and its correlation with c-myc during rat hepatic carcinogenesis and liver regeneration was assessed (83). Hepatic ProTα mRNA levels, concomitant with c-myc, increased during the initial stage of hepatic carcinogenesis (6 weeks), and remained nearly 10 fold higher as the neoplasm progressed. In comparison, ProTα mRNA levels increased only slightly at the early (3-6 hr) and later stage (24-30 hr) of liver regeneration, following 70% partial hepatectomy. *In situ* hybridization revealed that overexpressed ProTα mRNA was restricted to neoplastic nodules within hepatic tissue and to tumor cells invading blood vessels.

Identification of gene products exclusively or abundantly expressed in neoplasms may yield novel tumor markers. A number of cDNA clones, including ProTα, were recently isolated from rat HCC (84, 85). ProTα is involved in cell proliferation and, as stated above, is regulated by the oncogene c-myc *in vitro*. ProTα gene expression and its correlation with c-myc were examined by the authors in various groups of patients: HCC, cirrhosis, and adenoma patients, as well as normal control patients. Hepatic ProTα mRNA levels were two- to 9.2-fold higher in neoplastically transformed tissues than in adjacent normal structures in 14 of 17 patients with HCC, regardless of coexisting cirrhosis and viral hepatitis. No marked difference in ProTα mRNA levels was detected in patients with adenoma and hepatic cirrhosis and in healthy controls. The c-myc mRNA amounts were two- to five- fold increased in 11 of 17 patients with HCC and correlated to a statistically significant degree with those of ProTα. Increased ProTα mRNA was once again localized in the neoplastic nodules of the patients with HCC.

The thymosin β10 gene is located on chromosome 2q37, as detected by fluorescence *in situ* hybridization analysis. A subtractive thyroid cDNA library was constructed from two human thyroid carcinoma cell lines originating from an anaplastic carcinoma and a papillary thyroid carcinoma. The library was employed to identify genes correlated with the progression to a highly malignant phenotype (86). The thymosin β10 gene was isolated and found to be expressed at much higher levels in the anaplastic cell line, than in the papillary cells. The thymosin β10 gene was overexpressed in five carcinoma cell lines compared with normal thyroid tissue and normal thyroid primary culture cells. The highest expression occurred in the most malignant cell lines. Thymosin β10 gene expression was also higher in surgically removed human thyroid carcinomas and was highest in the anaplastic carcinomas. Thymosin β10 gene expression correlated with the degree of the malignancy in rat thyroid cells transfected with cellular and viral oncogenes of different tumorigenicity. These results demonstrate that thymosin β10 overexpression is a general event of thyroid cell neoplastic transformation and suggest that the gene is involved in the progression of thyroid carcinogenesis. Paraffin sections from 30 human breast tissue specimens were stained with a specific antibody raised against thymosin β10, [monoclonal antibody 10 (38-43)] (87). The results revealed that thymosin β10 was detected mainly in the malignant tissue, particularly in the neoplastically transformed cells, whereas the normal cell population around the lesions showed very weak immunoreactivity. Furthermore, the intensity of staining in the neoplastic cells increased according to increasing grade of the lesion.

Thymosin β15 represents a novel 5.3 kD protein that binds actin monomers and inhibits actin polymerization, and may thus act to increase cellular motility. Thymosin β15 is upregulated at both the mRNA and protein levels in prostate carcinoma cell lines in a manner directly related to their capacity to metastasize. Due to the observations that this glycoprotein is upregulated in cells with a propensity to metastasize, its value as a prognostic marker in breast cancer was evaluated by Gold and co-workers (88). The use of an affinity-purified polyclonal antibody identified that within breast epithelium, thymosin β15 is localized diffusely throughout the cytoplasm and that thymosin β15 is upregulated in malignant (compared with benign) breast tissue. In contrast to the prostate model, thymosin β15 is upregulated in non-metastatic breast cancer and even ductal carcinoma *in situ* (compared with benign breast tissue), and, consequently, it might represent a potential marker to be used in the early diagnosis of mammary malignancy. Bao and co-workers (89) previously isolated thymosin β15 from highly metastatic Dunning rat prostatic carcinoma cells. Immunohistochemical study of human prostate cancer specimens revealed a general correlation between Gleason grade and thymosin β15 expression, with high-grade (more malignant, more dedifferentiated IP) neoplasms showing increase in intensity and cell number immunostaining compared to low-grade tumors. To determine whether thymosin β15 may be differentially expressed in cancer cells with different metastatic potential other than in the prostatic carcinoma cells, thymosin β15 mRNA levels were examined by this study in tumor cell lines from different species. The levels of thymosin β15 in human breast cancer (BC) cases were also examined. Thymosin β15 was usually upregulated in the highly metastatic mouse lung and human BC cell lines, as compared with non-metastatic lesions. Immunohistochemical staining presented clear evidence of upregulation of thymosin β15 in malignant human BCs, as compared to benign breast tumors. The expression of thymosin β15 correlated well with the metastatic potential of mouse lung carcinoma and human breast carcinoma, in addition to prostate carcinomas. Thus, thymosin β15 may represent a useful marker in the prediction of the level and progressive tendency of the metastatic potential of certain human cancers (90).

3. THYMIC HORMONES IN ANTI-CANCER THERAPY

In the last two decades, so-called non-specific immunostimulants, such as plant extracts and natural and synthetic thymic preparations have been widely employed for enhancing the reactivity and efficiency of the human defense system in chronic infections, immunodeficiency, autoimmunity, and numerous neoplastic diseases (91-97).

Thymosin fraction 5 contains a family of polypeptides with varying biological activities. Multi-laboratory efforts in the thymosin research carried out further chemical characterization of thymosin peptides and evaluation of their effectiveness in numerous clinical immunotherapeutic protocols. Twenty years of clinical studies with thymosin fraction 5 have shown therapeutic potentials for treatment of patients with primary immunodeficiency diseases and neoplasms

(98-110).

In mice bearing immunogenic tumors, adding thymic humoral factor-γ 2 (THF-γ2) immunotherapy as an adjunct to anticancer chemotherapeutic regimens not only strengthens the antitumor activity of each drug, but also repairs tumor/chemotherapy-induced damage to T lymphocyte subpopulations and functions (111). The Lewis lung carcinoma (3LL) is a weakly immunogenic, highly metastatic, aggressive neoplasm in C57BL/6 mice. To investigate whether the immunoregulatory octapeptide is also effective against a tumor that does not elicit an antitumor immune response, the effect of combination THF-γ2 immunotherapy and chemotherapy in 3LL-bearing mice was assessed. The results indicate that THF-γ2, combined with either Melphalan or 5-Fluorouracil was more effective in reducing the metastatic load than either chemotherapeutic drug alone, and was characterized by massive infiltration of cytotoxic lymphatic cells. This combined chemoimmunotherapy also prolonged survival time in all treated animals and repaired T lymphocyte defects and impaired *in vitro* cellular immune response parameters, induced either by the tumor or by chemotherapy. THF-γ2 immunotherapy reversed the decrease in the number of bone-marrow myeloid colonies (GM-CFU) caused by chemotherapy treatment of tumor-bearing mice, supporting the hypothesis that THF-γ2 directly stimulates the proliferation of myeloid stem cells. The overall results of this observation imply, that when administered as an adjunct to chemotherapy, THF-γ2 immunotherapy is equally effective against immunogenic, as well as nonimmunogenic neoplasms.

Dacarbazine (DTIC) and IL-2, as single agents, have limited anti-neoplastic activity in patients with metastatic malignant melanoma (MMM). It has been proposed that thymosin α1 (TA1) may modulate the action of IL-2. The clinical and immunological effects of these three agents employed in combination were observed by Lopez and co-workers (112). Forty-six patients with measurable MMM were treated with DTIC 850 mg i.v. on day 1, TA1 2 mg s.c. on days 4 to 7, and IL-2 18 MU/m2/d by continuous intravenous infusion on days 8 to 12, with the cycles repeated every 3 weeks. Objective responses were obtained in 15 of 42 evaluable patients. Two patients experienced complete responses, and stabilization of the neoplastic disease was observed in five. The median time to progression was 5.5 months and median survival was 11 months. Side effects were predominantly caused by IL-2. Treatment was tolerated reasonably well, and there was no overlapping toxicity or interference between chemotherapy and biotherapy. Baseline sCD4 levels have been shown to correlate well with tumor burden. Patients benefiting from treatment had lower sCD4 and higher sCD8, than did progressing patients. The combination of DTIC + TA1 + IL-2 is active in the treatment of advanced MMM, with acceptable toxicity. The interactions between thymic hormones and cytokines should be explored further, and other combinatorial approaches developed.

Cytokines, such as IL-1 and IL-6 stimulate the hypothalamic-pituitary-adrenal (HPA) axis. In addition, these proteins affect pituitary cell proliferation *in vitro*. Thymosin fraction 5 (TF5) similarly stimulates the HPA axis and the effects of this preparation on neuroendocrine tumor cell proliferation were examined in an interesting study (113). Cells of the PRL-secreting rat anterior pituitary adenoma, MMQ, were exposed to vehicle or TF5 for up to 96 hours and the proliferation of MMQ cells was monitored using the MTT assay (3-(4,5-dimethylthiazol-2-yl)-2,5-diphenyl tetrazolium bromide). TF5-mediated inhibition of cell proliferation was dependent on both TF5 concentration and the initial MMQ cell number. Minimal reductions in optical densities resulted from exposure to 100 μg/ml TF5, whereas the highest concentration of this preparation (500 μg/ml) completely blocked MMQ cell division. The concentration-dependent effects of TF5 were particularly striking at initial plating densities of 25 and 50 x 10^3 MMQ cells/well; in contrast, all concentrations of TF5 completely inhibited MMQ cell growth at 5 and 10 x 10^3 cells/well. The antiproliferative actions of TF5 on MMQ cells were demonstrable within 24 hours and were present for up to 96 hours, as determined by the MTT assay and actual cell counts. Due to the fact that the highest densities of MMQ cells were partially refractive to the antiproliferative effects of TF5, the effects of PRL (1-1000 nM) and MMQ cell conditioned medium (50%) on TF5 inhibition of MMQ adenoma proliferation was examined.. The TF5 concentration-dependent inhibition of MMQ cell growth was largely reversed by the 50% conditioned medium, whereas PRL slightly potentiated the antiproliferative actions of TF5. The proliferation of the rat C6 glioma cell line demonstrated greater sensitivity to TF5: concentrations as low as 10 μg/ml TF5 inhibited C6 cell proliferation, and near-maximal inhibition was observed at 200 μg/ml TF5. Significant reductions in MMQ and C6 cell viabilities accompanied decreases in cell number and morphological analysis indicated that these cells were undergoing apoptosis. The peptides thymosin α1, thymosin β4, MB35, and MB40 had no effect on either MMQ or C6 cell proliferation, indicating that these TF5 components are not the principle active peptides. Therefore, TF5 was further separated into 60 fractions by preparative reverse phase HPLC. HPLC fractions 17, 25, 26, and 27 significantly suppressed MMQ cell proliferation to the same extent as TF5, while other HPLC fractions had no effect. The data presented by Spangelo and co-workers demonstrate a new biological property of TF5, namely, the inhibition of cell proliferation and the induction of apoptosis in neuroendocrine tumor cells. The proliferation effects were time and concentration dependent and could be partially reversed by an activity present in the MMQ cell conditioned medium. Thus, TF5 and cytokines have opposite effects on adenoma cells, since IL-2 and IL-6 stimulate GH3 cell proliferation. Thus, circulating thymic peptides may act to prevent pituitary adenoma and glioma tumor formation, an action opposed by the growth factors secreted by these tumors and acting to promote tumor formation and progression in both an autocrine and paracrine manner.

A phase II study was performed to evaluate the clinical and immunological effects of a regimen of fluorouracil (5-FU) and folinic acid (FA) combined with thymopentin (TP-5) and interleukin-2 (IL-2) in the treatment of patients with metastatic colorectal cancer (CC) (114). Forty-five patients with CC and no prior therapy for metastatic disease were treated with 5-FU 400 mg/m2/d and FA 200 mg/m2/d i.v. on days 1-5, TP-5 50 mg s.c. on days 8-11, and IL-2 9 MU/m2 s.c. twice daily on days 12-16. Cycles were repeated at 4-week intervals, provided that toxicity was resolved. Immunological alterations as related to treatment were evaluated in 13 patients and compared with a well matched set of 13 patients, treated with the same regimen except for TP-5. Two complete responses and 17 partial responses were determined (42%). Fifteen patients (33%) were characterized by stabilization of their malignant disease. The median time to progression was 8.5 months and the median survival 13 months. Treatment was reasonably well tolerated, and there was no overlapping toxicity or interference between chemotherapy and biotherapy. Hematological and immunological changes during treatment were qualitatively similar to those expected with IL-2 ± chemotherapy. Quantitatively, significant changes (higher levels of IL-2, CD25 and IFN-γ, and lower levels of sIL-2R) were observed in patients given TP-5. The results revealed that the combination of 5-FU + FA and TP-5 + IL-2 is effective in advanced CC with acceptable toxicity. Immunological data suggest that TP-5 may modulate the action of IL-2 in the clinical setting.

Immunoregulatory effects of thymic peptides on functions of polymorphonuclear leukocytes (PMNs) have not been thoroughly investigated. The effects of ProTα1 on PMNs from patients with colorectal tumors, breast tumors and melanoma, in comparison with healthy donors, with respect to chemotaxis, cytotoxicity against HCT-116 colon tumor cells, oxidative response (chemiluminescence reaction) as well as expression of surface marker molecules were examined by Heidecke and co-authors (115). The researchers found that ProTα1 was equally effective in stimulating the chemotactic activity of PMNs from tumor patients and healthy donors (43% increase). PMNs from tumor patients, especially those derived from patients with breast tumor, were characterized by a significant enhancement of cytotoxicity against tumor target cells, as compared with healthy donors. With respect to the cytotoxity of PMNs, only about 50% of the colorectal tumor patients and healthy donors responded to ProTα1 and FMLP. As to the oxidative response of PMNs, elevated levels were found only among colorectal cancer patients. ProTα1 significantly increased the oxidative response in patients with breast and colorectal neoplasia. ProTα1 caused the expression of CD16 on PMNs of healthy donors to decrease, but had no effect on the expression of CD11a, CD11b, CD11c, CD13, CD14, CD15 and CD32.

Thymopentin (TP5) has been recently evaluated as an immunotherapeutic agent for the treatment of human neoplasms. Melanoma is a highly immunogenic malignancy, and in previous studies the treatment of metastatic melanoma with TP5 showed encouraging results. In a further study, the clinical efficacy and tolerability of high dose intravenous TP5 were evaluated in 16 patients with melanoma which had metastasized to cutaneous and subcutaneous tissue (116). All patients were given 1g intravenous TP5 every second day for 7 weeks and then assessed, with responders given a subsequent course of 2g intravenous TP5 every second day for 5 additional weeks. Six patients showed a partial response after the first course and were given the second course: one patient achieved a complete response, while the other five remained in partial response at the end of the treatment. The mean duration of response was 7.5 months. No side effects of TP5 administration were observed. Histopathological and immunohistochemical evaluation of regressing metastatic nodules showed the presence of tumor infiltrating lymphocytes (TIL), necrosis, sclerosis, intratumoral vascular proliferation and microthrombosis.

Immunophenotyping of lymphoid infiltrates demonstrated the prevalence of $CD4^+$ and $CD45RO^+$ T lymphocytes in one patient. It can be concluded that high dose intravenous TP5 three times a week may induce a clinical response in patients with cutaneous and subcutaneous metastases of melanoma without any significant side effects.

Carboplatin (CBDCA), cyclophosphamide (CTX) and etoposide (VP-16) combination chemotherapy has been shown to be quite effective in the treatment of several human malignancies. A phase I-II study designed to verify toxicity, maximum tolerated dose and activity of CBDCA, CTX and VP-16 given with granulocyte colony stimulating factor (G-CSF) and thymopentin (TP5), without bone marrow support was carried out by Recchia and co-workers (117). A group of 12 heavily pretreated patients (9 breast carcinoma, 2 small cell lung carcinoma, and 1 gastric carcinoma), received fourteen courses of the combination chemotherapy. Previous treatments were as follows: surgery in 10 patients, radiotherapy in 7 patients, chemotherapy in all patients. CBDCA, CTX, VP-16 were given over 3 days, in concentrations of 400-800 mg/m2, 1500-2500 mg/m^2, and 450-550 mg/m^2, respectively, while the G-CSF dose given was 5 mg/kg/day from day 4 to day 17. TP5 was given, on alternate days, in a dose of 1 mg/kg from day 4 to day 30. Six patients suffered from fevers >38°C for a median duration of 4 days. Four patients required platelet support, and three patients needed red blood cell support. Ten of the twelve patients had responses, while the other two experienced disease stabilization. Five of the 12 patients were still alive at the time of the report by Recchia and co-workers, and the reported median overall survival time was 13 months.

Twenty two patients with advanced non-small-cell lung cancer were randomized to receive chemotherapy (ifosfamide) or chemotherapy followed by thymosin α1, employed together with low-dose IFNα in a study by Salvati and co-workers (118). Chemo-immunotherapy induced an enhanced response rate compared with chemotherapy alone, and there was a statistically significant difference between the two groups in time to malignant disease progression. $CD4^+$ and $CD8^+$ lymphocyte, as well as natural killer (NK) cell counts were significantly depressed after two cycles of chemotherapy, while no difference in cell count were seen in chemo-immunotherapy treated patients. Hematologic toxicity was reduced by the immunotherapy, with no grade 3/4 toxicity seen, while such levels of toxicity were reported in 50% of patients treated with chemotherapy alone. This last report, as well as many of the previous data concerning clinical trials, only serves to bring to the forefront the great necessity of novel modalities of cancer therapy, as well as the improved efficacy of combined therapy, over traditional multifactorial therapy. It is our opinion that novel, radio- and chemo-immunotherapeutical approaches need to be developed, based on some of the examples presented in this article, and employed in the individualized "cocktail" therapy of human malignancies, based on individual immunophenotypical and genotypical indicators for the particular case of cancer.

4. CONCLUSIONS

1. It is well known today that the mammalian thymus is an endocrine gland, participating in local cell-to-cell regulatory interactions and it plays significant role as part of the endocrine orchestra of the body. The different thymic hormones are produced by the cells of the thymic reticulo-epithelial cellular microenvironment;
2. During physiological aging, the production of thymic hormones decreases significantly, which is related to concomitant decrease in the power of cellular and humoral immune defense;
3. During the past two decades, numerous thymic hormonal preparations were used in non-specific immmunological enhancement of the host anti-neoplastic response; and
4. Recently it has been established that in numerous neoplastic processes, the serum level of the various chemical derivatives of thymic hormones is significantly elevated because the neoplastic cells are genetically capable of producing several thymic hormone-like, chemically modified molecules. These molecules may be used as part of the early diagnostic laboratory-screening panel for the detection of mammalian neoplasms.

REFERENCES

1. Miller JFAP: Immunological function of the thymus. Lancet *2:* 748-749, 1961.
2. Miller JFAP: Effect of thymectomy in adult mice on immunological responsiveness. Nature *208:* 1337-1338, 1965.
3. Verheyen P: De Thymo. Louvain, 1706.
4. Bidloo GB: Exercitatio Anatomica de Thymo. London, 1706.
5. Morand S: Recherches Anatomiques sur la Structure et l'usage de Thymus. Paris, 1759.
6. von Haller A: Elementa Physiologiae Corporus Humani. vol 3, Bousquet et Sociorum M-M, Lucerne, 1766.
7. Hewson W: Experimental Inquiries: Part the Second: A Description of the Lymphatic System in the Human Subject and in Other Animals. Johnson J, London, 1774.
8. Lucae SC: Anatomische Untersuchungen der Thymus in Menschen und Tieren. vol 2, Nurnberg, 1811.
9. Cooper A: Anatomy of the Thymus Gland. Lea & Blanchard, Philadelphia, 1845.
10. Stannius H: Handbuch der Anatomie der Wirbelthiere. von Veit, Berlin, 1854.
11. Prenant A: Contribution a l'étude du développement organique et histologique du thymus, de la glande thyroide et de la glande carotidienne. La Cellule *10:* 85-184, 1894.
12. Bell ET: The development of the thymus. Am J Anat *5:* 29-62, 1905.
13. Hammar JA: Über Gewicht, Involution und Persistenz des Thymus im Postfötalleben des Menschen. Arch Anat Physiol Anat Abt [Suppl] 91-182, 1906.
14. Hammar JA: Die Menschenthymus in Gesundheit und Krankheit: Ergebnisse der numerischen Analyse von mehr als tausend menschlichen Thymusdrüsen. Teil I. Das normale Organ - zugleich eine kritische Beleuchtung der Lehre des "Status thymicus". Z Mikr Anat Forsch (Leipzig) *6:* 1-570, 1926.
15. Hammar JA: Die normal-morphologische Thymusforschung im letzten Vierteljahrhundert. Analyse und Synthese. Leipzig, Barth, pp 1-453, 1936.
16. Maximow AA: Untersuchungen über Blut und Bindegewebe. II. Über die Histogenese der Thymus bei Säugetieren. Arch mikr Anat *74:* 525-621, 1909.
17. Pigache R, Worms G: Le thymus considéré comme glande à secrétion interne. C R Acad Sci (Paris) *154:* 234-236, 1912.
18. Lampe AE: Die biologische Bedeutung der Thymusdrüse auf Grund neuerer Experimentalstudien. Med Klin *8:* 1117-1120, 1912.
19. Shimizu S: Beiträge zur Kenntnis der Thymusfunktion. I. Ueber das Thymolysin. Mitt Med Fak Univ Tokyo *11:* 261-291, 1913.
20. Tamemori Y: Untersuchungen über die Thymusdrüse im Stadium der Altersinvolution. Virchows Arch pathol Anat *242:* 255-266, 1914.
21. Fenger F: On the size and composition of the thymus gland. J Biol Chem *20:* 115-118, 1915.
22. von Haberer H: Zur klinischen Bedeutung der Thymusdrüse. Arch klin Chir *109:* 193-248, 1917.
23. Pappenheimer AM: The action of immunsera upon lymphocytes and small thymus cells. J Exp Med *26:* 163-179, 1917.
24. Hoskins ER: Is there a thymic hormone? Endocrinol *3:* 241-257, 1918.
25. Olkon DM: The effect of thymus gland injection on the growth and behavior of the guinea-pig. Arch Int Med *22:* 815-829, 1918.
26. Uhlenhuth E: The functions of the thymus gland. Endocrinology *3:* 285-298, 1919.
27. Takenoughi M: Studies on the reputed endocrine function of the thymus gland (albino rat). J Exp Zool *29:* 311-342, 1919.
28. Gedda E: Zur Altersanatomie der Kaninchenthymus. Upsala Lakaref Forh *26:* 1-27, 1921.
29. De Sanctis S: Les enfants dysthymiques. Encephale *18:* 1, 88, 156, 1923.
30. Bratton AB: The normal weight of the human thymus. J Pathol Bacteriol *28:* 609-620, 1925.
31. Abt JA: The present status of our knowledge concerning the thymus gland. Med J Rec *124:* 525-530, 1926.
32. Agafonow FD: Zur Physiologie der Glandula thymus. Plügers Arch *216:* 682-696, 1927.
33. Scammon RE: The prenatal growth of the human thymus. Proc Soc Exp Biol Med *24:* 906-909, 1927.
34. Fleker T: Tests with Temesvary's thymophysin. Polska Gaz Lek *8:* 537-539, 1929.
35. Nowinski VW, Beiträge zur Physiologie der Drüsen; Fortgesetzte Beiträge zur Funktion der Thymus. Die Wirkungen des Thymocrescins auf das Wachstum. Biochem Ztschr *226:* 415-428, 1930.
36. Young M, Turnbull HM: An analysis of the data collected by the status lymphaticus investigation committee. J Pathol Bacteriol *34:* 213-258, 1931.
37. Boyd E: The weight of the thymus gland in health and in disease. Am J Dis Child *43:* 1162-1214, 1932.
38. Asher D: Weitere Isolierung des wachsfördernden Thymocrescins. Biochem Z *257:* 209-212, 1933.
39. Thomas E: Physiologie des Thymus. Hirschs Handb Inn Sekr *II-1:* 423-466, 1933.
40. Asher L, Scheinfinkel N: Die Funktion der Thymus. Wien Med Woch *84:* 565-566, 1934.
41. Bomskov Ch: Das Hormon des Thymus. Arch Klin Chir *200:* 568-588, 1940.
42. Bomskov Ch, Sladovic L: Das Thymushormon und seine biologische Auswertung. Pflügers Arch *245:* 611-622, 1941.
43. Carr JL: Status thymicolymphaticus. J Pediatr *27:* 1-43, 1945.
44. Andreasen E: Function of the thymus. I. Critical remarks on the theory and experiments of Bomskov. Acta Anat *2:* 275-286, 1947.
45. Hoshino T, Takeda M, Abe K, Ito T: Early development of thymic lymphocytes in mice studied by light and electron microscopy. Anat Rec *164:* 47-65, 1969.
46. Bodey B: Histochemistry and histogenesis of the thymus in *men* during the prenatal development. D. Sc. Thesis, Inst Morphol, Bulgarian Academy of Sciences, Sofia, 1977; pp. 1-360.
47. Bodey B, Hadjioloff AI: Thymus development, structure and function. Nature (Bulgaria), *26:* 11-19, 1977.
48. Manning MJ: A comparative view of the thymus in vertebrates. *In:* The Thymus Gland (Kendall MD, ed). New York, Academic Press, pp 7-20, 1981.
49. Metalnikof AE: Rôle du système nerveux et des facteurs biologiques et psychiques dans l' immunité. Masson, Paris, 1934.
50. Ritter MA, Rozing J, Schuurman HJ: The true function of the thymus? Immunol Today *9:* 189-193, 1988.
51. Kendall MD: The thymus: new views of an old gland. Endeavour *16:* 158-163, 1992.
52. Dardenne M, Pleau JM, Savino W, Prasad AS, Bach JF: Biochemical and biological aspects of the interaction between thymulin and zinc. Prog Clin Biol Res *380:* 23-32, 1993.
53. Bodey B, Bodey B Jr, Kaiser HE: Cell culture observations of Postnatal Thymic Epithelium: An In Vitro Model for Growth and Humoral Influence on Intrathymic T

Lymphocyte Maturation. In Vivo *10:* 515-526, 1996.

54. Bodey B, Bodey B Jr, Kaiser HE: Dendritic type, accessory cells within the mammalian thymic microenvironment. Antigen presentation in the dendritic neuro-endocrine-immune cellular network. In Vivo *11:* 351-370, 1997.
55. Bodey B, Bodey B Jr, Kaiser HE: Apoptosis in the *mammalian* thymus during its normal histogenesis and under various *in vitro* and *in vivo* experimental conditions. In Vivo *12:* 123-134, 1998.
56. Bodey B, Bodey B Jr, Siegel SE, Kaiser HE: Involution of the mammalian thymus, one of the leading regulators of aging. In Vivo, *11:* 421-440, 1997.
57. Dardenne M, PapiernikM, Bach JF, Stutman O: Studies on thymus products. 3. Epithelial origin of the serum thymic factor. Immunology *27:* 299-304, 1974.
58. Low TL, Goldstein AL: Thymic hormones: an overview. Methods Enzymol *116:* 213-219, 1985.
59. Bekkum DW van, Betel I, Blankwater MJ, K
60. Ruisbeek AM, Swart AC: Biologic activities of various thymus preparations. Transplat Proc *9:* 1197-1199, 1977.
61. Oosterom R, Kater L: Effects of conditioned media of human thymus epithelial cultures on T-lymphocyte functions. Ann N Y Acad Sci *332:* 113-122, 1979.
62. Schuurman HJ, Van de Wijngaert FP, Delvoye L, Broekhuizen R, McClure JE, Goldstein AL, Kater L: Heterogeneity and age dependency of human thymus reticulo-epithelium in production of thymosin components. Thymus 7:13-23, 1985.
63. Bach J-F: Thymic hormones. Nouv Presse Med *8:* 2797-2799, 1979.
64. 63. Iwata T, Incefy GS, Tanaka T, Fernandes G, Menendez-Botet CJ, Pih K, Good RA: Circulating thymic hormone levels in zinc deficiency. Cell Immunol *47:* 100-105, 1979.
65. Comsa J, Leonhardt H, Ozminski K: Hormonal influences on the secretion of the thymus. Thymus *1:* 81-93, 1979.
66. Dardenne M, Savino W, Wade S, Kaiserlian D, Lemonnier D, Bach J-F: *In vivo* and *in vitro* studies of thymulin in marginally zinc-deficient *mice.* Eur J Immunol *14:* 454-458, 1984.
67. Dardenne M, Savino W, Berrih S, Bach J-F: A zinc-dependent epitope on the molecule of thymulin, a thymic hormone. Proc Natl Acad Sci USA *82:* 7035-7038, 1985.
68. Hall NR, McGillis JP, Spangelo BL, Healy DL, Goldstein AL: Immunomodulatory peptides and the central nervous system. Springer Semin Immunopathol *8:*153-164, 1985.
69. Deschaux P, Rouabhia M: The thymus. Key organ between endocrinologic and immunologic systems. Ann N Y Acad Sci *496:* 49-55, 1987.
70. Bach JF, Dardenne M: Thymulin, a zinc-dependent hormone. Med Oncol Tumor Pharmacother *6:* 25-29, 1989.
71. Savino W, Ban E, Villa Verde DM, Dardenne M: Modulation of thymic endocrine function, cytokeratin expression and cell proliferation, by hormones and neuropeptides. Int J Neurosci *51:* 201-204, 1990.
72. Bodey B, Calvo W, Prummer O, Fliedner TM, Borysenko M: Development and histogenesis of the thymus in dog. A light and electronmicroscopical study. Develop Comp Immunol *11:* 227-238, 1987.
73. Calvo W, Fliedner TM, Herbst EW, Hugl E, Bodey B: Degenerative changes and recovery of the thymus of lethally irradiated dogs, rescued by transfusion of cryopreserved autologous blood leukocytes. Exp Hemat *15:* 1171-1178, 1987.
74. Bodey B: Development of lymphopoiesis as a function of the thymic microenvironment. Use of $CD8^+$ cytotoxic T lymphocytes for cellular immunotherapy of human cancer. IN VIVO 8: 915-943, 1994.
75. Hadden JW: Thymic endocrinology. Ann N Y Acad Sci *840:* 352-358, 1998.
76. Bodey B, Bodey B Jr, Siegel SE, Kaiser HE: The role of zinc in pre- and postnatal mammalian thymic immunohistogenesis. In Vivo *12:* 695-722, 1998.
77. Seiki K, Sakabe K: Sex hormones and the thymus in relation to thymocyte proliferation and maturation. Arch Histol Cytol *60:* 29-38, 1997.
78. Marshall GD Jr, Thurman GB, Low TL, Goldstein AL: Thymosin: basic properties and clinical application in the treatment of immunodeficiency diseases and cancer. Recent Results Cancer Res *75:* 100-105, 1980.
79. Low TL, Thurman GB, Chincarini C, McClure JE, Marshall GD, Hu SK, Goldstein AL: Current status of thymosin research: evidence for the existence of a family of thymic factors that control T-cell maturation. Ann N Y Acad Sci *332:* 33-48, 1979.
80. Ahmed A, Wong DM, Thurman GB, Low TL, Goldstein AL, Sharkis SJ, Goldschneider I: T-lymphocyte maturation: cell surface markers and immune function induced by T-lymphocyte cell-free products and thymosin polypeptides. Ann N Y Acad Sci *332:* 81-94, 1979.
81. Lauritzsen GF, Hofgaard PO, Schenck K, Bogen B: Clonal deletion of thymocytes as a tumor escape mechanism. Int J Cancer *78:* 216-222, 1998.
82. Dominguez F, Magdalena C, Cancio E, Roson E, Paredes J, Loidi L, Zalvide J, Fraga M, Forteza J, Regueiro BJ: Tissue concentrations of prothymosin α: a novel proliferation index of primary breast cancer. Eur J Cancer *29A:* 893-897, 1993.
83. Loidi L, Vidal A, Zalvide JB, Puente JL, Reyes F, Dominguez F: Development of ELISA to estimate thymosin α1, the N terminus of prothymosin α, in human tumors. Clin Chem *43:* 59-63, 1997.
84. Wu CG, Boers W, Reitsma PR, van Deventer SJ, Chamuleau RA: Overexpression of prothymosin α, concomitant with c-myc, during rat hepatic carcinogenesis. Biochem Biophys Res Commun *232:* 817-821, 1997.
85. Wu CG, Habib NA, Mitry RR, Reitsma PH, van Deventer SJ, Chamuleau RA: Overexpression of hepatic prothymosin α, a novel marker for human hepatocellular carcinoma. Br J Cancer *76:* 1199-1204, 1997.
86. Fraga M, Garcia-Caballero T, Dominguez F, Perez-Becerra E, Beiras A, Forteza J: Immunohistochemical location of prothymosin α in regenerating human hepatocytes and hepatocellular carcinomas. Virchows Arch A Pathol Anat Histopathol *423:* 449-452, 1993.
87. Califano D, Monaco C, Santelli G, Giuliano A, Veronese ML, Berlingieri MT, de Franciscis V, Berger N, Trapasso F, Santoro M, Viglietto G, Fusco A: Thymosin β-10 gene overexpression correlated with the highly malignant neoplastic phenotype of transformed thyroid cells in vivo and in vitro. Cancer Res *58:* 823-828, 1998.
88. Verghese-Nikolakaki S, Apostolikas N, Livaniou E, Ithakissios DS, Evangelatos GP: Preliminary findings on the expression of thymosin β-10 in human breast cancer. Br J Cancer *74:*1441-1444, 1996.
89. Gold JS, Bao L, Ghoussoub RA, Zetter BR, Rimm DL: Localization and quantitation of expression of the cell motility-related protein thymosin β15 in human breast tissue. Mod Pathol *10:* 1106-1112, 1997.
90. Bao L, Loda M, Zetter BR: Thymosin β15 expression in tumor cell lines with varying metastatic potential. Clin Exp Metastasis *16:* 227-233, 1998.
91. Sasaki H, Fujii Y, Masaoka A, Yamakawa Y, Fukai I, Kiriyama M, Saito Y, Matsui H: Elevated plasma thymosin-α1 levels in lung cancer patients. Eur J Cardiothorac Surg *12:* 885-891, 1997.

92. Kater L, Oosterom R, McClure J, Goldstein AL: Presence of thymosin-like factors in human thymic epithelium conditioned medium. Int J Immunopharmacol *1:* 273-284, 1979.
93. Oosterom R, Kater L: Effects of conditioned media of human thymus epithelial cultures on T-lymphocyte functions. Ann N Y Acad Sci *332:* 113-122, 1979.
94. Schuurman HJ, Van de Wijngaert FP, Delvoye L, Broekhuizen R, McClure JE, Goldstein AL, Kater L: Heterogeneity and age dependency of human thymus reticulo-epithelium in production of thymosin components. Thymus 7:13-23, 1985.
95. Bach J-F: Thymic hormones. Nouv Presse Med *8:* 2797-2799, 1979.
96. 95. Iwata T, Incefy GS, Tanaka T, Fernandes G, Menendez-Botet CJ, Pih K, Good RA: Circulating thymic hormone levels in zinc deficiency. Cell Immunol *47:* 100-105, 1979.
97. Comsa J, Leonhardt H, Wekerle H: Hormonal coordination of the immune response. Rev Physiol Biochem Pharmacol *92:* 115-191, 1982.
98. Comsa J: Thymus hormones. Facts and problems. Med Welt *31:* 533-536, 1980.
99. Trainin N: Thymic hormones and the immune response. Physiol Rev *54:* 272-315, 1974.
100. Low TL, Goldstein AL: Thymosin β 4. Methods Enzymol *116:* 248-255, 1985.
101. Low TL, Goldstein AL: Thymosin α 1 and polypeptide β 1. Methods Enzymol *116:* 233-248, 1985.
102. Low TL, Goldstein AL: Thymosin fraction 5 and 5A. Methods Enzymol *116:* 219-233, 1985.
103. Splitter GA, Incefy G, Iwata T, McGuire TC: Evaluation of functional thymic hormones in Arabian horses with severe combined immunodeficiency. Clin Exp Immunol *38:* 37-44, 1979.
104. Kruisbeek AM: Thymic factors and T cell maturation *in vitro*: a comparison of the effects of thymic epithelial cultures with thymic extracts and thymus dependent serum factors. Thymus *1:* 163-185, 1979.
105. Kruisbeek AM: Effect of factors in thymic epithelial culture supernatants (TES) on *in vitro* maturation of thymocytes. Ann N Y Acad Sci *332:*109-112, 1979.
106. Safieh-Garabedian B, Kendall MD, Khamashta MA, Hughes GR: Thymulin and its role in immunomodulation. J Autoimmun *5:* 547-555, 1992.
107. Naylor PH, McClure JE, Spangelo BL, Low TL, Goldstein AL: Immunochemical studies on thymosin: radioimmunoassay of thymosin β 4. Immunopharmacology *7:* 9-16, 1984.
108. Savino W, Huang PC, Corrigan A, Berrih S, Dardenne M: Thymic hormone-containing cells. V. Immunohistological detection of metallothionein within the cells bearing thymulin (a zinc-containing hormone) in human and mouse thymuses. J Histochem Cytochem *32:* 942-946, 1984.
109. Evans HM, Simpson ME: Reduction of the thymus by gonadotropic hormone. Anat Rec *60:* 423-435, 1934.
110. Allen LS, McClure JE, Goldstein AL, Barkley MS, Michael SD: Estrogen and thymic hormone interactions in the female mouse. J Reprod Immunol *6:* 25-37, 1984.
111. Weusten JJ, Blankenstein MA, Gmelig-Meyling FH, Schuurman HJ, Kater L, Thijssen JH: Presence of oestrogen receptors in human blood mononuclear cells and thymocytes. Acta Endocrinol *112:* 409-414, 1986.
112. Ophir R, Pecht M, Keisari Y, Rashid G, Lourie S, Meshorer A, Ben-Efraim S, Trainin N, Burstein Y: Thymic humoral factor-γ 2 (THF-γ 2) immunotherapy reduces the metastatic load and restores immunocompetence in 3LL tumor-bearing mice receiving anticancer chemotherapy. Immunopharmacol Immunotoxicol *18:* 209-236, 1996.
113. Lopez M, Carpano S, Cavaliere R, Di Lauro L, Ameglio F, Vitelli G, Frasca AM, Vici P, Pignatti F, Rosselli M: Biochemotherapy with thymosin α 1, interleukin-2 and dacarbazine in patients with metastatic melanoma: clinical and immunological effects. Ann Oncol *5:* 741-746, 1994.
114. Spangelo BL, Farrimond DD, Thapa M, Bulathsinghala CM, Bowman KL, Sareh A, Hughes FM Jr, Goldstein AL, Badamchian M: Thymosin fraction 5 inhibits the proliferation of the rat neuroendocrine MMQ pituitary adenoma and C6 glioma cell lines in vitro. Endocrinology *139:* 2155-2162, 1998.
115. Lopez M, Di Lauro L, Paoletti G, Santini S, Gandolfo GM, Vitelli G, Frasca AM, Ameglio F, Rasi G, Garaci E: Sequential biochemotherapy for metastatic colorectal cancer using fluorouracil, folinic acid, thymopentin and interleukin-2: clinical and immunological effects. Ann Oncol *6:* 1011-1017, 1995.
116. Heidecke H, Eckert K, Schulze-Forster K, Maurer HR: Prothymosin α 1 effects in vitro on chemotaxis, cytotoxicity and oxidative response of neutrophils from melanoma, colorectal and breast tumor patients. Int J Immunopharmacol *19:* 413-420, 1997.
117. Cascinelli N, Belli F, Mascheroni L, Lenisa L, Clemente C: Evaluation of clinical efficacy and tolerability of intravenous high dose thymopentin in advanced melanoma patients. Melanoma Res *8:* 83-89, 1998.
118. Recchia F, De Filippis S, Ginaldi L, Quaglino D, Gulino A, Frati L, Rea S: Phase I-II study of dose intensified chemotherapy with filgrastim and thymopentin in patients with advanced cancer. Clin Ter *148:* 201-207, 1997.
119. Salvati F, Rasi G, Portalone L, Antilli A, Garaci E: Combined treatment with thymosin-α1 and low-dose interferon-α after ifosfamide in non-small cell lung cancer: a phase-II controlled trial. Anticancer Res *16:* 1001-1004, 1996

Chapter 9

Spontaneous Regression of Neoplasms: New Possibilities for Immunotherapy

Abstract: In mammalian cells, neoplastic transformation is directly associated with the expression of oncogenes, loss or simple inactivation of the function of tumor suppressor genes, and the production of certain growth factors. Genes for suppression of the development of the neoplastic cellular IP, as well as inhibitory growth factors have regulatory functions within the normal processes of cell division and differentiation. Telomerase (a ribonucleoprotein polymerase) activation is frequently detected in various neoplasms. Telomerase activation is regarded as essential for cell immortalization and its inhibition may result in spontaneous regression (SR) of neoplasms.

SR of neoplasms occurs when the malignant tissue mass partially or completely disappears without any treatment or as a result of a therapy considered inadequate to influence systemic neoplastic growth. This definition makes it clear that the term SR applies to neoplasms in which the overall malignant disease is not necessarily cured and to cases where the regression may not be complete or permanent.

A number of possible mechanisms of SR are reviewed, with the understanding that no single mechanism can completely account for this phenomenon. The application of the newest immunological, molecular biological and genetic insights for more individualized and adequate antineoplastic immunotherapy (alternative biotherapy) is also discussed.

Key words: Oncogenes and growth factors; Carcinogenesis; Neoplastic cell transformation; Tumor suppressor gene (TSG); Cellular immunophenotype (IP); dedifferentiation; Neoplastic progression, angiogenesis, and metastasis; Spontaneous regression (SR); Immunological mechanism of SR; Hormonal SR; Neoplastic cell redifferentiation; Monoclonal antibody (MoAB); Interleukin-2 (IL-2); Antineoplastic immunotherapy.

1. INTRODUCTION

1.1 Cellular carcinogenesis

A diverse array of mechanisms can lead to the characteristic alterations implicated in neoplastic cellular transformation. Presently, it appears that human neoplasms arise as a direct consequence of an accumulation of genetic alterations, involving two main classes of genes: proto-oncogenes and tumor suppressor genes (TSGs) (1-3). Oncogenes result from an activating mutation resulting in an enhancement of intracellular protein quantity. TSGs, on the other hand, are commonly inactivated *via* either mutation or deletion or the physiological function of the gene product is inhibited by binding of inactivating molecules. Observations of the expression of deleted in colorectal cancer (DCC) gene product, for instance, have demonstrated a significant correlation between DCC protein presence and cellular differentiation and carcinogenesis (4). Mutations resulting in an oncogene have been established as dominant genomic alterations, whereas tumor suppressor gene mutations are recessive, thus requiring loss of function at both alleles for tumor development. In all cases of neoplastic cell transformation, there are three important pre-malignant changes:

1. overexpression of a gene and its product;
2. alteration of a gene product; and
3. inactivation of an encoded protein (5-9). Neoplastic cells with a highly malignant immunophenotype (IP) are not stable; their genetic alterations can be rapid and dramatic, resulting in cell dedifferentiation and regional tumor heterogeneity (10-13).

2. THE PROGRESSION OF NEOPLASTICALLY TRANSFORMED CELLS

Neoplastically transformed cells progress towards an increasingly dedifferentiated (resembling an embryonic), more and more malignant IP and eventually reach a metastatic IP in a stepwise series of genetic and epigenetic changes (14-16). It seems that neoplastic cells are constantly undergoing rapid cell biological changes, a phenomenon termed *"dynamic heterogeneity"* (17). Malignant cells continually change their IP from non-metastatic to metastatic and back again and the progression of a neoplasm towards a more malignant IP always remains reversible. In addition, the metastatic propensity of a given malignancy can also be modulated.

Initially, a great number of the neoplastic cells tend to penetrate adjacent tissues, so that they may later metastasize to particular preferred tissues, and that they may eventually conquer as many organs as possible. Changes in metastatic properties also include increased adhesion of such cells to capillary endothelium, increased invasiveness through extracellular matrix and basement membranes, and increased cell motility in response to autocrine and paracrine chemotactic factors. During neoplastic recurrence (reactivation of neoplastic cell proliferation and tumor progression), the particular genetical, biochemical and immunocytochemical markers listed above once again become detectable.

The pathophysiological process of neoplastic tissue destruction or the complex mechanism of neoplastic tissue elimination have not been observed in mammalian oncology with the same thoroughness as the IP markers, and cell biological changes during the progression of a neoplastically transformed cell. In Table I, I listed some immunomorphological and cell biological characteristics of neoplastic cells.

Table 1. Markers of Neoplastic Cell Screening [modified after Parsons (18)]

Histological grading
Nuclear cell grade
Invasion of adjacent tissues
Grade of neo-angiogenesis
Cell Proliferation
S-phase antibodies against cell cycle cyclins ($p34^{cdc2}/p58^{cyclin\ A}$; etc.), proliferating cell nuclear antigen (PCNA), Ki-67
Mitotic index
Biochemical Markers
Proteases: cathepsins B and D, plasminogen activator, activated protein kinases
Matrix degrading metalloproteinases (MMPs); Collagenase IV; etc.
Glutathione S-transferases (GSTs)
Tyrosine-kinases (hormone receptors; epidermal growth factor receptor [EGFR] and other growth factor receptors; transforming growth factor receptor family [endoglin, morphogenetic protein 6]; etc.)
Immunocytochemical Markers
Oncogenes (c-myc; c-ras; c-kit; c-fos; c-jun; c-MET or hepatocyte growth factor receptor; int-2 oncoprotein or protein of recurrent malignancy; bcl-2 oncoprotein or inhibitor of apoptosis; etc.)
Oncofetal antigens (carcinoembryonic antigen family [CEA], etc.)
Tumor specific genes (Wilms' tumor gene; deleted in colorectal cancer gene [DCC])
Cell surface antigens: (multidrug resistance P-glycoprotein; tumor-associated antigens [TAAs]; cancer/testis antigens etc.)
Changes in the quantity of cell adhesion molecules (CAM; talin, tenascin, etc.)
Tumor suppressor genes (p53 family and related proteins, DCC, etc.)
Anti-metastasis gene (nm-23)
Tissue "specific" antigens (prostate specific antigen [PSA]; prostatic inhibin peptide [PIP]; epithelial membrane antigen [EMA]; epithelial specific antigen [ESA]; melan A, a melanocyte differentiation marker expressed on melanoma cells; renal cell carcinoma marker [RCC]; ovarian cancer antigen [CA125]; plasma cell marker, rhabdomyosarcoma markers [Myf-3, Myf-4, MyoD1])
Heat-shock/stress proteins (HSPs 27 and 70)

3. HISTORICAL OVERVIEW OF SPONTANEOUS REGRESSION (SR) OF NEOPLASMS

Papac (19) recently published a very well written review article concerning the SR of neoplasms and, therefore it is not necessary for us to repeat all of her interpretations here. Boyd (20) in his monograph suggested that neoplasms associated with SR should be termed "Saint Peregrine tumors". This term is derived from a legend from the 13th century. Saint Peregrine was a young priest who developed a large osteosarcoma. On the night preceding the amputation of his leg, he prayed for quite some time, and, it is alleged, he awoke without any trace of the bone tumor. It is noted that he died in 1345, at the age of 80, without any recurrence of the bone tumor. In more recent reports, Meares (21, 22) reported SR of osteogenic sarcoma metastases and other neoplasms following intense meditation.

The first case of SR of malignant melanoma (MM) was published in 1899 (23). According to Papac (19), only 70 of a total of 302 tumor cases collected by Rohdenberg (24) demonstrated acceptable temporary or permanent SR. It is important to realize that incomplete surgical removal or some acute infection with high fever preceded the majority of cases of SR. The first case of SR of neuroblastoma by recognized cell differentiation into benign ganglioneuroma (termed sympathicoblastoma by the authors) was observed and described by Cushing and Wollbach (25). The next year, the medical literature was enriched with a report concerning SR of lung metastases of hypernephroma following surgical removal of the neoplastically transformed kidney tissue (26).

In the Everson and Cole (27) review of SR cases from 1900 to 1964, only 176 met some important diagnostic criteria:

1. observed and properly documented histological regression of biopsy proven metastases;
2. radiological documentation of presumptive malignant neoplastic disease; and
3. advanced neoplastic disease (i.e. with metastases) regression following a therapeutical approach generally deemed ineffective.

Surprisingly, neoplastic cases of squamous cell carcinoma, lymphoma and leukemia were excluded due to their different natural history. The authors demonstrated that the majority of SR cases occurred in renal cell carcinoma, neuroblastoma, melanoma, and choriocarcinoma. Boyd (20) described SR in 61 tumor cases, excluding lymphomas and leukemias. In this work, most attention was focused on two common neoplasm types: retinoblastoma and breast cancer. No cases of renal cell cancer or choriocarcinoma were discussed.

In 1990, Challis and Stam (28) reviewed cases of SR between 1966 and 1987 to update the monographs of Everson and Cole, and Boyd. Practically the entire history of SR of neoplasms was once again described and the authors made an attempt to specify the possible cause of SR. The possible mechanisms mentioned by the authors included immunological and endocrine alterations, concurrent infection, surgical trauma, and intratumoral cell necrosis. Interestingly, an extremely controversial mechanism based on psychological factors was also taken into consideration.

O'Regan and Hirschberg (29) published a bibliography of more than 3,500 SR cases, which encompassed both benign, as well as malignant neoplasms. In this compilation neoplastic remission was clearly distinguished from the more effective process of neoplastic regression. Attention was assigned to the SR cases that were reported with complete biopsy or histological confirmation of the neoplasms and adequate disease follow-up to verify the SR. According to this review, the five most common types of neoplasm undergoing SR were renal cell carcinoma, lymphomas and leukemias, neuroblastoma, breast carcinoma, and melanoma. The authors proposed the immune mechanism of SR.

4. SPONTANEOUS REGRESSION (SR) IN VARIOUS HUMAN NEOPLASMS

The true existence of SR of neoplasms is often questioned and some simply do not accept this pathophysiological phenomenon. On the other hand, there are physicians who think that SR is the most fascinating event in all of medicine. Osler (30) characterized SR as "one of the most remarkable phenomena observed in medicine". The significance of the acceptance of SR as a realistic possibility demonstrates a physician's position concerning the endogenous control of neoplastic cell transformation and progression. The original definition of SR is clear: when the malignant tumor mass partially or completely disappears without any treatment or as a result of therapy considered inadequate to influence systemic neoplastic disease (24, 27, 28, 31, 32). This definition makes it clear that the term SR applies to cases of neoplasms in which the malignant disease is not necessarily cured, and also in those where regression may be neither complete nor permanent (32). It is important to notice that most neoplasms patients relapse after SR. A difficulty in determining whether SR has occurred is the diversity of opinions on what is considered adequate anticancer treatment. In addition, Everson and Cole (27) excluded leukemias, lymphomas, as well as Hodgkin's disease from their review because of the natural changes in growth rates, which these entities undergo. In any case, SR of neoplasms is viewed by many clinical oncologists and researchers as a relatively rare event, only 10 to 20 cases are reported each year (28, 33). During the present century, a great majority of authors who reported SR of mammalian neoplasms failed to speculate or specify possible causes for the regression. Those who did, postulated that factors such as the host's immunological and endocrine response, additional infections (bacterial or viral), apoptotic and necrotic intratumoral changes, surgical or other trauma, etc. may have caused the SR. In cases of SR, patients remain free of symptoms and the entire laboratory, histological and immunocytochemical parameters revert to a normal range primarily as a result of regression in the actual tumor mass. Far-reaching species-specific aspects of SR also present the opportunity for intensive comparison of neoplastic tissue progression and regression in various taxonomic units (34-38).

Since 1999, numerous publications have interpreted cases of SR of breast cancer (BC) (39), of carcinoma of the stomach (40), a germinoma in the pineal body after placement of a ventriculoperitoneal shunt (41), SR of hepatocellular carcinomas (HC) (42) and of hepatocellular carcinoma (HC) after portal vein tumor thrombi occurred (43), of pilocytic astrocytoma in a patient without neurofibromatosis type 1 (44), regression of a gastric T cell lymphoma (45) and gastric lymphoma of mucosa-associated lymphatic tissue after eradication of Helicobacter pylori in a kidney graft recipient (46). SR was also observed in congenital leukemia (CL), which is a rare disorder with extramedullary infiltrates and a myeloid phenotype. CL can progress rapidly without adequate treatment, but, can inconsistently remit spontaneously. Because of the significant toxicity of chemotherapy to newborns, it is important to identify those newborns that may not require treatment. The authors reported an infant who was diagnosed with congenital myeloid leukemia at one week of age. Cytogenetic analysis revealed a t(8; 16)(q11; p13) translocation and the infant's leukemia underwent SR (47). Recently, SR of a cerebral oligodendroglioma following infarction in the territory of its feeding artery was also reported (48). Parker and co-workers (49) reported a case of advanced stage IV clear cell carcinoma of the endometrium in a 73 year old, which underwent SR on the possible contribution of the patient's thrombocytosis. In the last year, SR was also reported in extremely malignant melanoma (50), with the author detailing a deeper understanding into the immunology of mammalian neoplasms.

5. POSSIBLE MECHANISMS OF SR

5.1 The immunological mechanism

Since the early years of our century, the cytolytic effect of fresh human (particularly pregnant women's) serum on certain neoplastic cells was well known (51-53). The cytolytic activity of serum derived from pregnant animals was first observed on murine cancer cells and was demonstrated to be the result of natural IgM antibody that activates and amplifies the lytic action of the complement system after binding to the surface of neoplastic cells (54, 55). Both classical and alternative pathways of complement were involved in this cytolytic process. Serum activity was in direct correlation with levels of IgM antibody and complement. It seems that gangliosidic oncofetal antigen (G_{D2}) may be involved in this cell surface complement interaction (56, 57). Later the same immunological pathway was detected in the cytolytic form of SR of neuroblastoma, and additionally the G_{D2} ganglioside was detected in over 98% of neuroblastoma cases (58-62). Melanoma associated antigens, such as G_{D2}, G_{D3}, and glycoproteins 75 and 110 were shown to be expressed at higher levels in regressing lesions, as detected by immunocytochemistry (63).

The increased incidence of formation of neoplastically transformed cell groups in immunosuppressed individuals (e.g. following organ transplantation) and SR following reduction of immunosuppressive therapy suggests that the host's immune system plays a critical role in both the initial process of neoplastic transformation and the subsequent progression of the tumor (64, 65). Immunological factors must thus also be considered crucial in certain aspects of SR of mammalian neoplasms. It has been observed that renal cell carcinoma and melanomas undergo SR after the patients received plasma infusion from tumor-bearing patients whose neoplasms were in the process of regression, suggesting that humoral factors may play a role (32, 66, 67).

Mammalian tumor infiltrating leukocytes (TILs), including the specialized T lymphocyte clone of highly active, $CD8^+$, MHC class I restricted, cytotoxic T lymphocytes (CTLs), are drawn to the sites of tumor cell targets spontaneously and possess powerful cytolytic abilities and great target specificity (68-70) due to the fact that they are tumor-associated antigen (TAA) or tumor-specific antigen (TSA) directed. The transfer of such "killer" autologous T lymphocytes with specific antitumor activity to the tumor-bearing host was termed "adoptive cellular immunotherapy" (70-73). CTLs have been shown to be 50-150 times as potent as lymphokine activated killer (LAK) cells in mediating tumor and micrometastasis regression (70, 74). Heterogeneous populations of TILs have now been isolated from all kinds of solid human and mammalian neoplasms. Use of cytokines and lymphokines in the adoptive immunotherapy of cancer has been furthered by observations that the differentiation of T cells within the TILs to CTLs can be accomplished in the presence of recombinant Interleukin-2 (rIL-2) and/or autologous tumor cells *in vitro*, thus providing cells with great target specificity for the tumor mass (75). Identification of the target specificity and the evolution of the cell surface IP of these CTL involved in *in situ*, intratumoral immune responses to malignant neoplastic transformations are important in the field of human immunobiology and was described by us previously (76, 77). As mentioned above, the presence of autologous TILs, including cells directed specifically against TAAs and/or TSAs is extremely significant. In many malignancies, these infiltrates have been shown to possess a great number of CTLs, as well as many $CD4^+$ helper T cells with increased expression of IL-2 receptor (63, 78). The composition of the heterogeneous poly- and mononuclear cell infiltrate present within solid tumors includes leukocytes, monocyte/macrophages, natural killer (NK) cells, LAK cells, various subpopulations of T lymphocytes and in some cases such rare cell populations as plasma cells and mast cells (79). The infiltration may vary from florid to none at all, and the phenomenon rarely follows a consistent or predictable pattern. CTLs and NK cells kill target cells using two different mechanisms: the granzyme system and the Fas/APO-1 to Fas-Ligand interaction inducing apoptosis (programmed cell death) (80-82). It seems that perforin mediated membrane damage caused by these cells is followed by the apoptotic death of the target neoplastically transformed cells. NK cells also produce tumor necrosis factor-α (TNF-α) and the immune modulating interferon-γ (IFN-γ), and it has been shown that binding of TNF-α to its receptor is capable of triggering programmed cell death (83, 84). In malignant melanoma (MM) patients who had regional lymph node metastases, suppressor T cell activity was detected against the CTL response at the lymph node level (85, 86). A significant migration of $CD4^+$, helper T lymphocytes from the metastatic lymph nodes to the regressing melanoma lesions was suggested. As noted previously, lysis of malignant cells is accomplished mostly by the CTLs. Tumors effectively evade this antigen-specific immune response by down-regulating or losing their cell-surface MHC class I molecules (87, 88). Other neoplasms have been found to be MHC class I-deficient and thus effectively evade a specific immune response and lysis by CTLs (89). Transfection of MHC class I molecules into MHC class I-negative tumors led to tumor rejection by $CD8^+$ T lymphocytes *in vivo*, further substantiating the importance of the major histocompatibility complex in this cellular immune response (90, 91). NK cells have been recognized as the most effective immune cells in elimination of blood-borne metastases (92, 93).

During the last few years additional physiological regulatory functions of IL-2 (initially termed T lymphocyte growth factor) have been identified. Dendritic cells (DCs) play an important role as professional antigen presenting cells (APCs) *in vivo* (94, 95). It has been determined that DCs accumulate in the lung in response to parenteral injections of IFN-γ. In this publication the authors reported that rat DCs express the interleukin-2 receptor (IL-2R) α chain (OX-39) and obtain enhanced motility in response to rIL-2 in Boyden chamber and video time-lapse microscopy assays. This enhanced motility was inhibited by pretreating DCs with anti-OX-39 (an anti-IL-2R MoAB), but not with anti-OX-22 (a MoAB which reacts with CD45RC). The intratracheal administration of rIL-2 increased the number of $OX\text{-}6^+$ intrapulmonary DCs, located around pulmonary venules and in the interstitium. When rIL-2 was injected into the rats' footpad, DC numbers increased around dermal venules 24 hours after the injection. This effect was blocked by the parenteral injection of MoAB anti-OX-39. It is clear that exogenous rIL-2 is a potent enhancer of DC motility and may cooperate with other Th-1 proinflammatory cytokines in attracting

DCs to sites of tissue inflammation.

Our laboratory previously conducted systematic, immunocytochemical, cell-surface antigen expression profile of 76 primary childhood brain tumors [34 medulloblastomas (MED)/primitive neuroectodermal tumors (PNETs) and 42, various types of astrocytomas (ASTR)], using immunocytochemistry (76). The library of MoABs that we employed identified the expression of various leukocyte-associated, lymphocyte cell-line differentiation, cell-surface antigens. A strong, well-developed immune response was detected in the brain tumors observed. Leu-2/a$^+$ cells comprise the most significant CD8$^+$, CTL population of the heterogeneous infiltrate and were identified in 58/76 (76.32%) childhood brain tumors. The CTL population usually represented 1-10% of total cells counted, but in some cases 30-44% of the cells were CD8$^+$. CD4$^+$ (MHC class II restricted helper lymphocytes) were detected using MoAB anti-Leu-3/a, and were present in 65/76 (85.53%) brain tumors, representing 1-10% of the total cell population. Macrophages (Leu-M5 antigen positive cells) were detected in 74/76 (97.37%) brain tumors. Their number also represented 1-10% of all observed cells. All 76 (100%) brain tumors contained many cells that reacted positively with MoABs anti-HLA-A,-B,-C and anti-HLA-DR, confirming the expression of MHC class I and II molecules on the surface of the neoplastically transformed cells, although the reaction with the anti-HLA-A,-B,-C was much stronger. Leukocyte common antigen (LCA) expression was demonstrated by the positive reaction of cellular antigens with MoAB anti-HLe-1, in all 76 (100%) brain tumors studied. MoAB UJ 308 detected the presence of premyelocytes and mature granulocytes in 60/76 (78.95%) brain tumors. They were localized perivascularly, within the neoplastic tissue, or close to necrotic regions. Natural killer (NK) cells were not identified in the childhood brain tumors observed in our immunocytochemical studies.

The above described observations on intratumoral host's immune effector cells lends four conclusions:

1. various *effector cells* of the host cellular immunologic defense were present in over 75% of the childhood brain tumors studied;
2. the tumor cell population expressed both MHC class I and II molecules, although class I-restriction was predominant;
3. granulocytes and premyelocytes were among the TIL population, with an as yet unclear function; and
4. the infiltration by tumor infiltrating poly- and mononuclear cells occurs due to inflammatory "signals" that cause a nonspecific immune response to occur initially and is therefore, at least in the very first stages, independent of the cell-surface IP of the tumor cell population.

In screening MM infiltrating mononuclear and polynuclear immunological effector cells we utilized ten, newly developed and well-characterized MoABs, directed against leukocyte differentiation antigens (77). Employing immunocytochemistry we were able to characterize the IP of the heterogeneous TILs of human primary (n=30) and metastatic (n=10) malignant melanomas (MM). Our study was the first to employ formalin fixed, paraffin embedded tissue sections in seeking to determine the *ex vivo* presence of tumor infiltrating leukocytes, including T lymphocytes, in human melanomas. We established the presence of some type of melanoma infiltrating host's immunological effector cells in all 40 observed melanoma cases. Such a presence of infiltrating mononuclear cells is always indicative of a favorable response (76, 77, 96-99). More specifically, we found NK cells, macrophages and granulocytes in 30 out of 30 PMs and 10 out of 10 MMs. These effector cells represented the vast majority (>80%) of the melanoma infiltrating immunocompetent cells. T lymphocytes were observed in 20 out of 30 PMs and 6 out of 10 MMs, but their numbers represented only between 5% to 10% of the heterogeneous leukocytic infiltrate. B cells were found in 22 out of 30 PMs and 8 out 10 MMs, their numbers representing less than 5%. Presence of cells of the dendritic reticulum, involved in antigen presentation was not determined in any of the observed PMs and MMs. There were more than enough macrophages and neutrophils, cells that are also involved in antigen presentation and, as previously mentioned, we found these cells within the melanomas. Thus defective antigen presentation cannot be the reason for the small proportion of T lymphocytes. The most probable explanation is the increased dedifferentiation of the melanoma IP to a degree that MHC class I molecules and the antigens complexed with them are lost and therefore there is nothing left to be recognized by the T lymphocytes. In view of these results it is our opinion that following the initial non-specific immune reaction, which may not be tumor specific, the specific immune response of effector T cells takes over. After significant IP changes by the majority of melanoma cells, including the internalization or shedding of MHC class I antigens, these highly specialized "killer cells" are no longer effective and the immune system responds by causing the reappearance of NK cells, macrophages and granulocytes (100). Thus, the relationship between melanoma and the host's immune system is evolutionarily dynamic: either side when necessitated readily makes changes.

6. THE SIGNIFICANCE OF ONCOGENES AND GROWTH FACTORS

Normal cell survival requires the expression of certain genes and the action of various trophic growth factors. The irregular activation of cellular proto-oncogenes, normally activated during processes of development and differentiation, is a key moment in the development of neoplasms (101-104). Inactivation, mutation, or loss of tumor suppressor genes, as well as inactivation of inhibitory growth factors are also regarded as important events in neoplastic cell transformation (105). Three members of the *ras* gene family (*H-*, *K-*, and *N-ras*) are frequently activated oncogenes in both human neoplasms and animal model systems (106, 107).

During our earlier research we observed the expression (overexpression) of the *c-erbB-2* and *c-erbB-3* oncoproteins in 30 primary cutaneous malignant melanomas MMs (CMMs), 10 already metastasized malignant melanomas (MMMs) and 15 lymph-node negative breast carcinomas (BCs) (108). Both oncoproteins were expressed as a result of either oncogene amplification or post-translational stabilization. c-erbB-2 alone is unable to bind neuregulins, but it is able to act as a pan c-erbB receptor subunit. Heterodimerization between c-erbB-2 and c-erbB-3 is required to initiate neuregulin directed signal transduction. Presence of c-erbB-2 oncoprotein was detected in 12/30 CMMs, 8/10 MMMs and 6/15 BCs, while c-erbB-3 was identified in 14/30 CMMs, 7/10 MMMs and 6/15 BCs. The intensity of the cell membrane localized immunoreactivity was observed to be greater when the c-erbB-2 oncoprotein was targeted (A, AB and B). The c-erbB-3 oncoprotein was also detected in the cytoplasm with a medium intensity (B, BC and C). Unfortunately, little is known concerning the range of oncoprotein overexpression after formalin fixation and paraffin embedding. We demonstrated overexpression localized to several cell clones within the oncoprotein positive population of neoplastically transformed cells. The immunocytochemically defined extent of expression of both oncoproteins was between 10-40% (+ to ++) of the total cell population in the malignant melanomas and 20-35% (++) of the total cell population in the BCs.

These results lead to the following conclusions:

1. there is a direct correlation between the expression (overexpression) of both oncoproteins and the aggressive clinical behavior and poor prognosis of the observed human malignancies; and
2. oncogene receptor directed immunotherapy, as part of a more individualized anticancer treatment can also be recommended.

An indirect effect of growth factors on SR was reported in a case report of SR of advanced lung cancer following myxoedema coma (109). Thyroid hormone (T3) is a potent enhancer of growth factors, notably epidermal growth factor (EGF) which is frequently present in lung malignancies, nerve growth factor (NGF) and insulin-like growth factor (IGF). We have found that transforming growth factor β (TGF-β), an endogenous inhibitor, plays a significant role in the ontogenetic development of thymic lymphopoiesis and in normal hematopoiesis (110). The diverse actions and target cells of growth regulatory factors has established the fact that dysregulation of these molecules is a crucial step in carcinogenesis. Observation of chronic myeloid leukemia (CLM) cells in tissue culture indicates an alteration in the balance between stimulatory and inhibitory factors, as compared with normal hematopoietic cells (111), further substantiating the role of growth regulatory factors in the etiology and maintenance of neoplastic growth and progression.

6.1 Hormonal mechanisms

Several cases of pregnancy associated complete SR of MMs have been reported (27, 112). The remission of acute myelogenous leukemia and rare cases of SR of sarcoma metastases following termination of pregnancy have also been described (27, 29, 113, 114).

In SR of BCs, complex hormonal mechanisms have been proposed as the main mediators (115-117). The development of ovarian metastases, which induce an endogenous oophorectomy, may also lead to SR of the BC. Estrogen receptors have been detected on the cell surface of a number of human malignancies, including BCs, MMs, myeloid leukemia cell lines, ovarian cancer, and prostate cancer. Five cases of SR of ovarian carcinoma following oophorectomy were described by Everson and Cole (27). Estrogen along with colony stimulating factor (CSF) has been reported to have a stimulatory effect on the growth of a number of myeloid leukemia cell lines (118). Estrogen receptor related hormonal therapy of MMs has generally produced disappointing results (119-121).

6.2 Induction of intratumoral cell differentiation

Mammalian embryonal development is partly regulated by genetically active chemical compounds such as retinoids (naturally occurring and synthetic analogs of vitamin A [retinol]) (122-129). These compounds play an important role in normal organogenesis, but can also act as teratogens at high levels. They are also powerful inducers of differentiation in many embryonal carcinoma cell lines. Embryonal carcinoma cells are also pluripotent stem cells that, in a number of aspects, resemble the omnipotent stem cells of early mammalian embryogenesis (130-134). Most of the embryonal teratocarcinoma cells seem to be committed to a specific differentiation pattern regulated by their genetic programs (e.g. F9 cells differentiate into parietal or visceral endoderm) (135-138).

Embryonal cell carcinoma of the testis, leukemia, retinoblastoma, neuroblastoma, and choriocarcinoma are neoplasms in which cell redifferentiation, a process of redevelopment of cellular non-malignant IP is an accepted phenomenon, and can be a major factor in SR (62, 139-143). *In vitro* experiments with neuroblastoma cells demonstrated that cell redifferentiation can be induced employing nerve growth factor (NGF), cyclic nucleotides, retinoic acid, among other compounds (61, 144).

6.3 The effects of stress

The vital processes of cellular growth, tissue remodeling, and morphogenesis are partially regulated by stress (145). Solid stress imposed by the extracellular matrix, in addition to the stress generated by neoplastically transformed cell growth and tissue reforming may also be important in the progression of neoplasms. A number of experimental observations have hypothesized that solid stress may affect the neoplasm's rate of growth (146-148), growth patterns *in vivo* and *in vitro* (149, 150), intratumoral cellular interactions and blood flow (151-153), and metastasis of neoplastically transformed cells, expressing a unique IP (154, 155). At the level of a single blood capillary, the local stress caused by a growing neoplastic spheroid mass has been shown to generate a significant 45 to 120 mm Hg intra-arterial pressure. It may be responsible for the collapse of the vessel wall and the related chronic vascular or lymphatic occlusion, regardless of mammalian species, tissue of origin, or stage of cellular dedifferentiation. This kind of blood or lymphatic vessel collapse can explain the poor blood flow and inhibited lymphatic drainage in solid tumors, which is also capable of inducing a sub-optimal delivery of anticancer agents to the site of the malignancy. The proliferation rate of tumors under such stresses did not demonstrate any significant alteration, but apoptosis among these neoplastically-transformed cells was downregulated. The experimental growth of multicellular neoplastic spheroids *in vitro* mimics the complex process of *in vivo* tumor progression in the mammalian host, and may serve as another experimental model (156).

7. SIGNIFICANCE OF SR IN ANTINEOPLASTIC THERAPY

As was mentioned above, the continuous process of neoplastic transformation consists of thousands of discrete steps beginning with the initial expression of oncogenes and neoplastic transformation of the cellular IP and culminating in uncontrollable cell proliferation and the development of metastatic potential. Theoretically, carcinogenesis could be inhibited at each one of these steps. It is made clear in the medical literature that the rate of occurrence of SR in uncommon malignancies is much higher. This observation may also be true since it is also possible that these neoplasms are considered uncommon because SR is much more frequent in such cases. In addition, SR cases are hard to come by since it is hard to accept that at the end of the twentieth century and at the beginning of the twenty-first century, some cases of malignant human neoplasms would still be allowed to grow without any therapeutical intervention or are not properly treated. Indeed a thorough literature search reveals a number of reports of SR in which the employed therapy may have played a regression modulating action. However, even today some patients are refusing anticancer surgery, total or partial body irradiation, and chemotherapeutic clinical trials.

During the last decade, a number of novel discoveries in basic research were successfully applied in the immunotherapy (biotherapy) of various human neoplasms in specially designed, genetically altered and more individualized clinical trials (157, 158). Neoplasm associated antigens recognized by cellular or humoral effectors of the immune system are potential targets for antigen specific anticancer immunotherapy (159). Different categories of neoplasm antigens have been identified as inducing cytotoxic T lymphocyte (CTL) responses *in vitro* and *in vivo*, namely:

1. "cancer testis" (CT) antigens, expressed in different tumors and normal testis,
2. melanocyte differentiation antigens,
3. point mutations of normal genes,
4. self antigens that are overexpressed in malignant tissues, and
5. viral antigens.

7.1 Newly developed hormone replacement strategies

During the last 20 years it has become clear that estrogen replacement therapy dramatically increases a woman's chance of developing uterine and breast cancer. Progestins have been added to the estrogen administration since the late 1980's, first for 7 days every month, and later for 10 days. Pike and co-investigators (160) observed 833 women who had developed endometrial neoplasia and 791 women who had not. 10 day or continuous combined (estrogen and progestin) therapeutical regimens completely eliminated the increased carcinogenic risk on the endometrium. These findings should be helpful in the devisement of effective hormone replacement strategies that are not associated with increased risk of developing BC. It is important to mention that complete regression of hepatocellular adenoma was reported after withdrawal of oral contraceptives (161).

7.2 Adoptive cellular immunotherapy employing autologous CTLs

Adoptive cellular immunotherapy can be defined as a representative type of fourth modality anticancer therapy involving the active transfer of autologous immunological effector cells, expanded *in vitro* with antitumor activity to the tumor-bearing host. The *in vitro* expansion and long-term maintenance of spontaneously generated TAA- and/or TSA-specific and directed, MHC class I restricted, $CD8^+$ CTLs, along with the killer cells of natural immunity represents the ultimate goal of adoptive cellular anticancer immunotherapy. As we have already stated, the intratumoral poly- and mononuclear cell infiltrate represents a heterogeneous cell population, both in their surface IP and functional profile (79, 162). Single color flow-cytometry (FACScan) analysis has revealed a predominance of T lymphocytes among tumor infiltrating immunological effector cells, and a significant reduction of natural killer (NK) cells or "large granular lymphocytes" (LGL) compared with the controls in most solid malignancies. Two color analysis demonstrated that the number of single positive, mature $CD4^+CD8^-$ (helper T lymphocytes) and $CD8^+11b^-$ (cytotoxic T lymphocytes) were significantly increased. This increase of both helper and cytotoxic T lymphocytes, which are indispensable components of the cellular arm of the immune system, strongly implies that tumor infiltrating T lymphocytes mediate an antigen-specific cellular immune response against neoplasms (162). Due to the heterogeneous nature of the leukocytic infiltrate, it is a reasonable hypothesis that TILs are first attracted to the tumor site as part of a nonspecific immune response to an inflammatory "signal." It is following necrotic changes in the neoplastic cell mass that monocytes among other APCs, acting in their antigen presenting cell role, consume the necrotic cells and present previously "hidden" TAAs and/or TSAs to the other effector cells among the TILs thus accomplishing the immunization and subsequent differentiation of the T lymphocytes into CTLs, with tumor-specific lytic capabilities *in situ*. These CTLs are responsible for the specific, immunological process of antibody-dependent cell-mediated cytotoxicity (ADCC), and represent the dominant cell population among the TILs, following the initial, non-specific immunological response (163). It has also been determined that NK cells also participate in cell-mediated cytotoxicity (without MHC class I restriction), and are able to induce apoptosis in a variety of neoplastic cell targets resistant to secretory (perforin/granzyme mediated) cytolysis (92, 93). The authors demonstrated that NK cells are unique immunological effector cells endowed with multiple mechanisms of destroying neoplastically transformed cells. Only NK cells have been observed to possess an inherent ability to rapidly induce programmed cell death in solid tissue derived neoplastic cells displaying a malignant IP. Furthermore, a subpopulation of IL-2 activated NK cells, the so-called A-NK cells, is capable of extravasation, movement within solid tissues, localization to the sites of neoplastic metastases, and destruction of malignant cells *in situ*. Therefore systemic adoptive immunotherapy employing A-NK cells is a tolerable and highly recommended form of anticancer therapy for solid tissue neoplasms and hematologic malignancies (92, 93).

TILs from renal cell carcinomas, lung adenocarcinomas and other malignancies have been expanded *in vitro* and been shown to mediate significant lysis of autologous tumor cell preparations, as detected by a ^{51}Cr release cytotoxicity assay (164, 165, Bodey unpublished results). When the specificity of TIL antitumor cytotoxicity was examined, the renal anticancer effectors lyzed autologous and allogeneic renal and non-renal tumor targets equally well (166). The lytic capacity of these TILs was dependent on CTLs expanded *in vitro* in the presence of autologous tumor cells and exogenous IL-2. If the culture medium was enriched only with rIL-2 during the *in vitro* cultivation, several "killer" clones were also developed, therefore the mixture of these cells also lyzed allogeneic tumor cell targets. It has recently been established that the *in vivo* administration of rIL-2 during the immune response does not enhance, but rather eliminates the antitumor activity of T cells and thus the addition of irradiated tumor cells along with the T lymphocytes has been proposed as an effective way of modulating the immune response (167).

Rosenberg reported that from 50% of patients with MM, TILs with specific cytolytic antitumor reactivity can be isolated (168-174). Rosenberg and co-investigators also isolated TILs from 15 of 20 surgical specimens of non-renal, uro-genital malignancies (transitional cell carcinoma of the urinary bladder, prostate cancer, testicular cancer, Wilms' tumor and adrenal cancer) (175). Expansion was carried out in four different tissue culture media enriched with 1000 U/ml rIL-2:

1. RPMI-1640 medium supplemented with 20% (by volume) of LAK-culture supernatant;
2. RPMI-1640 medium without the supernatant supplement;
3. AIM V medium supplemented with 20% (by volume) of LAK-culture supernatant;
4. AIM V medium without any supplement.

A higher proliferation rate was detected in AIM V medium than in RPMI-1640. Addition of LAK-culture supernatant significantly enhanced the growth conditions in RPMI-1640 medium, but did not improve the ability of AIM V medium to culture TILs. The TILs that were maintained for the longest period of time were $CD3^+$. The ratio of $CD4^+/CD8^+$ T cells varied with time in culture and culture conditions (employed medium, various supplements, etc.), but most cultures in time became mostly $CD4^+$. Mononuclear cells bearing NK, B, and macrophage surface markers disappeared from the culture quite early. Overall 14/15 TIL samples were cytolytic against one of the autologous and allogeneic tumor targets tested. The results suggest that the TIL phenomenon may reflect a combination of immunological cellular responses, including the numerous non-specific LAK effectors, as well as the highly specific CTL clone (176). Cellco Inc. (distributed by Spectrum Inc., Laguna Hills, CA, USA) has developed an artificial capillary cell culture system called Cellmax®, designed to emulate the human capillary network. Cellmax® provides superb *in vitro* conditions for T lymphocyte proliferation and expansion, within its semi-permeable artificial capillaries. Perfused with cell culture medium, the capillaries supply the cells with oxygen and the necessary nutrients, and also provide an outlet for metabolic waste and other growth inhibitory substances produced in culture. Cellmax® represents an advanced, more natural microenvironment for the cultured TILs, resulting in exceptionally dense cell growth. These conditions allow better cell to cell communication and rapid accumulation of autocrine growth factors for enhanced cellular proliferation (permits culture of up to 5×10^9 TILs) and viability, and also permits the use of a reduced quantity of serum or even serum-free culture conditions (177-180).

Neoplastic cell IP heterogeneity has been proposed as a possible basis for immunotherapeutic failure when specific expanded TILs are employed for cancer treatment (181). TILs from a melanoma patient (#397) were co-cultured with the autologous melanoma cell line in form of bulk culture. Four cycles of immunoselection produced tumor 397-R4, completely resistant to 397 TILs, but not to allogeneic LAK cell lysis, as examined in 4 hour ^{51}Cr release cytotoxicity assays. Cell surface alteration was at least one of the mechanisms causing resistance to TILs. Treatment of 397-R4 cells with IFN-α or IFN-γ with or without TNF-α 72 hours before cytolytic assays enhanced lysis be TILs, possibly due to the enhanced cell surface expression of MHC class I and II molecules and the cell adhesion molecule ICAM-1. The up-regulation of ICAM-1 allows for the more effective lysis of low susceptibility tumors by CTLs. A TIL subpopulation grown independently from tumor #397 overcame the resistance of the tumor cells.

Unfortunately, the full potential of adoptive cellular immunotherapy has yet to be explored since clinical trials (Phase II and III) with TILs are limited to the treatment of advanced, metastatic, usually stage IV neoplasms in humans. The sheer size of these tumor masses poses a great barrier to the efficacy of adoptive cellular immunotherapy. The much greater cellular proliferation rate of the neoplasm, coupled with the heterogeneous nature of the tumor cell population simply overpowers the ADCC immunological response of the TILs against the malignancy.

It is well known that IL-2-dependent activated, cytotoxic cells undergo apoptotic death when IL-2 is withdrawn either *in vitro* or after *in vivo* cell transfer. To attempt to sustain their survival after IL-2 withdrawal, melanoma-reactive human T lymphocytes were retrovirally transduced with an exogenous human IL-2 gene (182). Transduced PBMC and cloned CD8($^+$) T lymphocytes produced IL-2 and maintained viability after IL-2 withdrawal. Upon restimulation, IL-2 transductants proliferated in the absence of exogenous IL-2 and were able to be actively grown. Their survival could be maintained without adding any IL-2 for over 8 weeks. However, PBMCs similarly transduced with a control vector did not produce IL-2 and failed to proliferate in the absence of IL-2. A CD8($^+$) T cell clone, when transduced with an IL-2 gene, manifested the same phenotypes as PBMCs in the absence of exogenous IL-2. Furthermore, an antibody reactive with the α-chain of IL-2R complex reduced the viability mediated by IL-2 secretion of the IL-2 transductants. Moreover, transduction of an IL-2 gene did not affect the high degree of recognition and specificity of transductants against melanoma targets. These tumor-reactive IL-2 transductants were valuable for *in vitro* studies and for improved adoptive transfer therapies for patients with metastatic melanoma.

7.3 Retargeting of CTLs using bi-specific antibodies

Cytotoxic immune effector cells express special cell surface receptors by which they are able to distinguish altered autologous or foreign cells from the host's cells. Recently a new method has been developed by which this natural recognition system of cytotoxic cells can be manipulated and directed to give rise of cytotoxic cells with any desired antigenic epitope specificity, including specificity against altered autologous, neoplastically transformed cells. This method, termed "retargeting of cytotoxic cells" (or "antigen-directed retargeting of cytotoxic cells" would probably be more accurate) employs heterocross-linked, bi-specific antibodies in which one antibody is directed against the cytotoxic cell receptor (CCR or in the case of CTLs part of the Ti-CD3 complex) involved in the lytic response, while the second antibody is directed against a target cell structure (in the case of tumor, this would be represented by the TAAs and/or TSAs). By linking the CCR directly to the target cell, the bi-specific MoABs promote the formation of effector-to-target conjugates and signal cytotoxic cells to deliver a lethal hit (183-185).

Several types of ADCC effector cells (NK/K, T lymphocytes, macrophages, neutrophilic granulocytes) can be specifically retargeted, using heteroconjugates containing antibodies against Fc-γ receptors (186). Antigen directed, retargeted human peripheral blood lymphocytes with a bi-specific antibody targeting c-erbB-2 and CD16 caused the eradication of xenografted human ovarian carcinoma in 70% of murine models (187). In a more recent study, the immunomodulatory function of an anti-CD3/CA19-9 bi-specific MoAB in combination with a costimulatory anti-CD28 antibody was examined (188). Such a combination may result in not only antigen-specific, T lymphocyte mediated tumor cell lysis, but also in the recruitment of other cellular effector functions. Indeed, the authors found that specific activation of resting T cells with CD3-associated bi-specific MoABs in combination with an anti-CD28 antibody lead to a Th1 differentiation pathway (characterized by secretion of IL-2 and IFN-γ) and is accompanied by recruitment of MHC-independent LAK and NK cell cytotoxicity. Such methods may hold great promise in the treatment of malignancies, which are always characterized by a great deal of cellular heterogeneity. Many other combinations have also been tested with the hopes of developing a more specifically targeted and effective immune response against neoplastically transformed cells (189-197).

7.4 Adoptive cellular immunotherapy employing genetically modified cells

The introduction of genes of cytokines such as TNF, IFN-α. IFN-γ, IL-2, and IL-4 into malignant tumor cells to improve their immunogenicity is being studied extensively (71, 168, 198-203). A preclinical study in mice indirectly raised a host CTL-mediated antitumor response using autologous renal cell carcinoma cells engineered by gene transfection to secrete the cytokines IL-2 or IL-4 (so-called Renca-IL-2 and Renca-IL-4 cells). Renca cells became immunogenic and a CD8$^+$, MHC class I restricted, TAA and/or TSA directed cellular immune response ensued (204, 205). This *in situ* secretion of lymphokines critical for CTL activation, proliferation, and function appears to bypass a deficient helper T cell arm of the immune system in tumor-bearing host. Small numbers of Renca cells (1 X 10^3) were not rejected by the BALB/c mouse. When injected under the renal capsule, Renca cells displayed a metastatic pattern of growth similar to human renal cell carcinoma. The genetically engineered autologous tumor cells were altered to secrete IL-4, the helper cell lymphokine that participates in the regulation of growth and differentiation of B cells and T cells in the generation of CTLs. In fact, the administration of a tumor "vaccine" consisting of non-immunogenic BF16-F10 murine melanoma cells genetically engineered to express granulocyte-macrophage colony-stimulating factor (GM-CSF), but not IL-2, IL-4, or IFN-γ resulted in the induction of resistance to rechallenge (206). In another recent study, tumor cell clones derived from the MT-901 chemically induced murine mammary adenocarcinoma were generated to either express the T cell costimulatory molecule B7-1 or secreting GM-CSF (207). Upon reintroduction into Balb/c mice, these genetically modified tumor cells exhibited reduced tumorigenicity, but not complete growth inhibition. *In vivo* subcutaneous inoculation of a transgenic cell clone secreting GM-CSF as a tumor "vaccine" resulted in enhanced T-cell reactivity of tumor-draining lymph node (TDLN) cells as compared with wild-type TDLN cells. The additional transfection of the IL-12 gene into these cells was shown to result in complete abrogation of growth of the malignancy. A plethora of studies concerning the use of genetically modified tumor cells as "vaccines" in inducing enhanced T lymphocyte responses and even immunological memory have been published during the last few years (for representative articles see 208-221).

A novel and possibly even more effective method for activation of the host's immune system lies in the injection of tumor-extract pulsed antigen presenting cells (APCs) instead of genetically engineered tumor cells. In a recent study, the efficacy a vaccine comprised of APCs (both DCs and macrophages) that had been exposed to unfractionated tumor material *in vitro* was investigated. The malignancies observed included two poorly immunogenic murine models: murine bladder tumor (MBT-2) and B16 melanoma lung metastasis (222). The efficacy of APCs (especially DCs) pulsed with unfractionated extracts from these tumors in eliciting tumor-specific CTLs was significant. A measurable CTL response could be observed following just a single immunization with tumor extract-pulsed DC.

The viral and viral vaccine therapy of human neoplasms is also being investigated and involves two possibilities:

1. the appearance of a virus-induced peptide on the surface of the neoplastic cell targets that allows an enhanced CTL response against neoplastically transformed cells, and
2. the development of virally-modified vaccines using neoplastic cell membranes that can provoke a "rejection strength" immunological reaction against the malignancy (223).

The genetic manipulation of TILs employing various immune response modifiers such as IL-2, IL-4, IL-7, TNF, and others is also being investigated so that the effectiveness of TILs against neoplastically transformed target cells would be enhanced (71, 168, 198-201, 224-226). This genetically induced secretion of various cytokines may even provoke secondary cytokine secretion *in situ* and thus further amplify the cellular immune response (227). The genetic modification of the T lymphocyte receptor to form chimeric molecules expressed on the surface of CTLs and targeted against TAAs or TSAs represents another elegant possibility in the immunotherapy of cancer (228).

8. CONCLUSIONS

SR is defined as an occurrence when the malignant neoplastic mass partially or completely disappears without any treatment or as a result of therapy considered inadequate to influence systemic neoplastic disease. This definition is highly accepted, but it should be noted that the true existence of SR of neoplasms is often questioned and some simply do not accept this pathophysiological phenomenon. On the other hand, there are physicians who think that SR is one of the most fascinating events in all of medicine and should be further explored. Even the term SR, in fact, reflects our quite incomplete knowledge concerning possible neoplastic regression without the utilization of the three classical antineoplastic modalities of treatment. A number of authors postulated that factors such as the diseased host's immunological and endocrine response, additional infections (bacterial or viral), apoptotic and necrotic intratumoral changes, surgical or other trauma, etc. may have caused the SR. There a number of possible mechanisms of SR, but no single mechanism can completely account for this phenomenon. It becomes obvious that much more research is needed to properly explore this phenomenon, with knowledge of it possibly opening a new door for treatment.

We strongly recommend that TILs should be isolated, expanded, genetically engineered and primed against the existing tumor mass through autologous neoplastic cells obtained through biopsy *in vitro*, and should be employed as soon as following the initial surgical resection of the bulk of the primary neoplasm. These highly activated immune effector cells should be employed in the *in situ* targeting of residual neoplastically transformed cells and consequently preventing metastasis as a result of the dissemination of cancer cells during surgical resection. Employed together with a mixture of appropriate cytokines and MoABs, the adoptive cellular immunotherapy could be a successful antineoplastic therapeutic modality.

REFERENCES

1. Marshall C: Tumor suppressor genes. Cell *64:* 313-326, 1991.
2. Bodey B, Kaiser HE, Goldfarb RH: Immunophenotypically varied cell subpopulations in primary and metastatic human melanomas. Monoclonal antibodies for diagnosis, detection of neoplastic progression and receptor directed immunotherapy. Anticancer Res *16:* 517-531, 1996.
3. Bodey B, Groger AM, Bodey B Jr, Siegel SE, Kaiser HE: Immunocytochemical detection of p53 protein overexpression in primary human osteosarcomas. Anticancer Res *17:* 493-498, 1997.
4. Hedrick L, Cho KR, Fearon ER, Wu T-C, Kinzler KW, Vogelstein B: The DCC gene product in cellular differentiation and colorectal tumorigenesis. Genes Dev *8:* 1174-1183, 1994.
5. Perry ME, Levine AJ: P53 and mdm-2: interactions between tumor suppressor gene and oncogene products. Mount Sinai J Med *61:* 291-299, 1994.
6. Talib VH, Pandey J, Dhupia JS: Molecular markers in cancer diagnosis. Ind J Pathol Microbiol 38: 1-3, 1995.
7. Jacobson DR, Fishman CL, Mills NE: Molecular genetic tumor markers in the early diagnosis and screening of non-small-cell lung cancer. Annals Oncol *6:* S3-S8, 1995.
8. Bugert P, Kovacs G: Molecular differential diagnosis of renal cell carcinomas by microsatellite analysis. Am J Pathol *149:* 2081-2088, 1996.
9. Dietzmann K, von Bossanyi P, Sallaba J, Kirches E, Synowitz HJ, Warich-Kirches M: Immunohistochemically detectable p53 and mdm-2 oncoprotein expression in astrocytic gliomas and their correlation to cell proliferation. General Diagn Pathol *141:* 339-344, 1996.
10. Volpe JP: Genetic instability of cancer: Why a metastatic tumor is unstable and a benign tumor is stable. Cancer Genet Cytogenet 34: 124-134, 1988.
11. Bodey B, Zeltzer PM, Saldivar V, Kemshead J: Immunophenotyping of childhood astrocytomas with a library of monoclonal antibodies. Int J Cancer *45:* 1079-1087, 1990.
12. Bodey B, Bodey B Jr, Groger AM, Siegel SE, Kaiser HE: Clinical and prognostic significance of Ki-67 and proliferating cell nuclear antigen expression in childhood primitive neuroectodermal brain tumors. Anticancer Res *17:* 189-196, 1997.
13. Bodey B, Bodey B Jr, Groger AM, Siegel SE, Kaiser HE: Nm23/nucleoside diphosphate (NDP) kinase expression in human malignant melanomas. Significance and implications in tumor biology. Anticancer Res *17:* 505-512, 1997.
14. Vogelstein B, Fearon E, Hamilton SR, Kern SE, Preisinger AC, Leppert M, Nakamura Y, White R, Smits AM, Bos JL: Genetic alterations during colorectal-tumor development. N Engl J Med *319:* 525-532, 1988.
15. Sirica AE: Multistage carcinogenesis in cellular and molecular pathogenesis. Philadelphia, Lippincott-Raven, pp 283-320, 1996.
16. Pollard M, Luckert PH: Phenobarbital promotes multistage pulmonary carcinogenesis in MNU-inoculated L-W rats. In Vivo *11:* 55-60, 1997.
17. Nicolson GL: Growth mechanism and cancer progression. Hosp Prac *28:* 43-53, 1993.
18. Parsons DF: Some tumor cell protein kinases activated by receptors as markers, including elastin receptors. Cancer Invest *13:* 629-636, 1995.
19. Papac RJ: Spontaneous regression of cancer. Cancer Treat Rev *22:* 395-423, 1996.
20. Boyd W: The Spontaneous Regression of Cancer. Springfield, Charles C Thomas Publ, 1966.
21. Meares M: Regression of cancer after intensive meditation. Med J Austr *5:* 184, 1976.
22. Meares M: Regression of osteogenic sarcoma metastases associated with intensive meditation. Med J Austr *2:* 433, 1978.
23. Bennett WH: Some peculiarities in the behavior of certain malignant and innocent growths. Lancet *1:* 3-7, 1899.
24. Rhodenberg GL: Fluctuations in the growth rate of malignant tumors in man, with special reference to spontaneous regression. J Cancer Res *3:* 193-221, 1918.
25. Cushing H, Wollbach SB: The transformation of a malignant paravertebral sympathicoblastoma into a benign ganglioneuroma. Am J Pathol *3:* 203-216, 1927.
26. Bumpus HC Jr: The apparent disappearance of pulmonary metastasis in a case of hypernephroma following nephrectomy. J Urol *20:* 185-191, 1928.
27. Everson TC, Cole WH: Spontaneous Regression of Cancer. Philadelphia, WB Saunders, pp 1-1187, 1966.
28. Challis GB, Stam HJ: The spontaneous regression of cancer. A review of cases from 1900 to 1987. Acta Oncol *29:* 545-550, 1990.
29. O'Regan B, Hirschberg C: Spontaneous Remission. An Annotated Bibliography. Sausalito (CA, USA), Institute of Noetic Sciences, 1993.
30. Osler W: The medical aspects of carcinoma of the breast, with a note on the spontaneous disappearance of secondary growths. Am Med *17-19:* 63-66, 1901.
31. Rae MV: Spontaneous regression of hypernephroma. Am J Cancer *24:* 839-841, 1935.
32. Cole WH: Spontaneous regression of cancer. The metabolic triumph of the host? Ann NY Acad Sci *230:* 111-141, 1974.
33. Papac RJ: Spontaneous regression of cancer. Conn Med *54:* 179-182, 1990.
34. Kaiser HE: A biological approach to an understanding of regression. In: Cancer Growth and Progression (series ed Kaiser HE), vol 4: Influence of Tumor Development of the Host (volume ed Herberman RB), Kluwer Academic Publishers, Dordrecht, The Netherlands, p 42, 1989.
35. Kaiser HE: Biological viewpoints of neoplastic regression. In Vivo *8:* 155-165, 1994.
36. Cooper KR: Regression of neoplasms in invertebrates with special emphasis on mollusca. In: Cancer Growth and Progression (Kaiser HE, series ed). Vol 4 "Influence of Tumor Development of the Host" (Herberman RB, vol ed), Dordrecht, Netherlands, Kluwer Academic Publishers, 1989, pp 30-36.
37. Meins F Jr: Tumor reversal and tumor suppression in plants. In: Cancer Growth and Progression (Kaiser HE, series ed). Vol 4 "Influence of Tumor Development of the Host" (Herberman RB, vol ed), Dordrecht, Netherlands, Kluwer Academic Publishers, 1989, pp 37-41.
38. Xylander WER: Immune defense in arthropods - How chelicerates, crustaceans, insects and myriapods protect themselves against disease. Spiegel der Forschung *11:* 27-30, 1994.
39. Kruse J: Spontaneous remission of breast cancer. Ugeskr Laeger *161:* 4001-4004, 1999.
40. Sharma DN, Mohanti BK, Shukla NK, Rath GK: Spontaneous regression of carcinoma of the stomach. Clin Oncol (R Coll Radiol) *12:* 335-336, 2000.
41. Murai Y, Kobayashi S, Mizunari T, Ohaki Y, Adachi K, Teramoto A: Spontaneous regression of a germinoma in the pineal body after placement of a ventriculoperitoneal shunt. J Neurosurg *93:* 884-886, 2000.
42. Okada S, Abo T: Spontaneous regression of hepatocellular carcinoma. J Gastroenterol Hepatol *15:* 965-966, 2000. Comment on: J Gastroenterol Hepatol *15:* 1079-1086, 2000.

43. Uenishi T, Hirohashi K, Tanaka H, Ikebe T, Kinoshita H: Spontaneous regression of a large hepatocellular carcinoma with portal vein tumor thrombi: report of a case. SurgToday *30:* 82-85, 2000.
44. Gallucci M, Catalucci A, Scheithauer BW, Forbes GS: Spontaneous involution of pilocytic astrocytoma in a patient without neurofibromatosis type 1: case report. Radiology *214:* 223-226, 2000.
45. Bariol C, Field A, Vickers CR, Ward R: Regression of gastric T cell lymphoma with eradication of Helicobacter pylori. Gut *48:* 269-271, 2001.
46. Le Meur Y, Pontoizeau-Potelune N, Jaccard A, Paraf F, Leroux-Robert C: Regression of a gastric lymphoma of mucosa-associated lymphoid tissue after eradication of Helicobacter pylori in a kidney graft recipient. Am J Med *107:* 530, 1999.
47. Weintraub M, Kaplinsky C, Amariglio N, Rosner E, Brok-Simoni F, Rechavi G: Spontaneous regression of congenital leukaemia with an 8;16 translocation. Br J Haematol *111:* 641-643, 2000.
48. Choudhari KA, McLorinan GC, Byrnes DP: Spontaneous regression of a primary cerebral tumour following postoperative middle cerebral artery territory infarction. Br J Neurosurg *15:* 177-179; discussion 179-180, 2001.
49. Parker R, Lanvin D, Gilks B, Miller D: Spontaneous regression of stage IV clear cell carcinoma of the endometrium in a patient with essential thrombocytosis. Gynecol Oncol *82:* 395-399, 2001.
50. Printz C. Spontaneous regression of melanoma may offer insight into cancer immunology. J Natl Cancer Inst *93:* 1047-1048, 2001.
51. Wells H: Chemical Pathology. Philadelphia, WB Saunders, 1925.
52. Glumac G, Mates A, Eidinger D: The heterocytotoxicity of human serum. III. Studies of the serum levels and distribution of activity in human populations. Clin Exp Immunol *26:* 601-608, 1976.
53. Hellstrom KE, Hellstrom I: Spontaneous tumor regression: Possible relationships to in vitro parameters of tumor immunity. Natl Cancer Inst Monogr *44:* 131-134, 1976.
54. Bolande RP, Todd EW: The cytotoxic action of normal human serum on certain cells propagated in vitro. AMA Arch Pathol *66:* 720-732, 1958.
55. Bolande RP: Spontaneous regression of neuroblastoma: an experimental approach. Ped Pathol *10:* 195-206, 1990.
56. Gross J, Simon E, Szwarc-Bilotynski L, Durst A, Pfefferman R, Polishuk Z: Complement-dependent lysis of Ehrlich ascites cells by human serum (ascitolysin) is lowered in cancer patients and raised in pregnant women. Eur J Cancer Clin Oncol *22:* 13-19, 1986.
57. Bolande RP, Arnold JL, Mayer DC: Natural cytotoxicity of human serum. A natural IgM "antibody" sensitizes transformed murine cells to the lytic action of complement. Pathol Immunopathol Res *8:* 46-60, 1989.
58. Evans AE, Gerson J, Schnaufer L: Spontaneous remission of neuroblastoma. Natl Cancer Inst Monogr *44:* 49-54, 1976.
59. Cheung NK, Saarinen UM, Neely JE, Landmeier B, Donovan D, Coccia PF: Monoclonal antibodies to a glycolipid antigen of human neuroblastoma cells. Cancer Res *45:* 2642-2649, 1985.
60. Saarinen UM, Coccia PF, Gerson SL, Peeley R, Cheung NK: Eradication of neuroblastoma cells in vitro by monoclonal antibodies and complement: Method for purging autologous bone marrow. Cancer Res *45:* 5969-5975, 1985.
61. Carlsen NLT: How frequent is spontaneous remission of neuroblastomas? Implications for screening. Br J Cancer *61:* 441-446, 1990.
62. Brodeur GM, Nakagawara A: Molecular basis of clinical heterogeneity in neuroblastoma. Am J Pediatr Hem Oncol *14:* 111-116, 1992.
63. Tefany FJ, Barnetson RS, Halliday GM, McCarthy SW, McCarthy H: Immunocytochemical analysis of the cellular infiltrate in primary regressing and non-regressing malignant melanoma. J Invest Dermatol *97:* 197-202, 1991.
64. Baker HW: Biologic control of cancer. Arch Surg *121:* 1237-1241, 1986.
65. Penn I: Cancers complicating organ transplantation. N Engl J Med *323:* 1767-1769, 1990.
66. Horn L, Horn H: An immunologic approach to the treatment of cancer? Lancet *2:* 466-469, 1971.
67. Cole WH: Efforts to explain spontaneous regression of cancer. J Surg Oncol *17:* 201-209, 1981.
68. Rosenberg SA: Adoptive immunotherapy of cancer using lymphokine activated killer cells and recombinant interleukin-2. Important Adv Oncol pp: 55-91, 1986.
69. Sinkovics JG: Oncogenes and growth factors. CRC Critical Reviews in Immunol *8:* 217-298, 1988.
70. Culliton BJ: Fighting cancer with designer cells. Science 244: 1430-1433, 1989.
71. Hersey P: Cellular therapy. Curr Opinion Oncol *5:* 1049-1054, 1993.
72. Rosenberg SA, Terry W: Passive immunotherapy of cancer in animals and man. Adv Cancer Res *25:* 323-388, 1977.
73. Rosenberg SA: Adoptive immunotherapy of cancer: accomplishments and prospects. Cancer Treat Rep *68:* 233-255, 1984.
74. Sondel PM, Hank JA, Kohler PC, Chen BP, Sosman J: "Lymphokine activated killing" as treatment for human cancer: clinical extrapolations from laboratory studies with interleukin-2 expanded leukocytes. Prog Clin Biol Res *288:* 151-160, 1989.
75. Katano M, Nagumo F, Kubota E, Yamamoto H, Matsuo T, Hisatsagu T, Tadano J: Autologous tumour-specific cytotoxic T cell clone established from tumour infiltrating lymphocytes (TIL) of malignant ascites in the absence of recombinant interleukin 2 (rIL-2): activation by autologous tumour cell alone. Biotherapy *6:* 25-32, 1993.
76. Bodey B, Bodey B Jr, Siegel SE: Immunophenotypic characterization of infiltrating poly- and mononuclear cells in childhood brain tumors. Mod Pathol *8:* 333-338, 1995.
77. Bodey B, Bodey B Jr, Siegel SE, Luck JV, Kaiser HE: Immunophenotypic characterization of human primary and metastatic melanoma infiltrating autologous immunological effector cells. Anticancer Res *16:* 3439-3446, 1996.
78. Hawkins MJ: Interleukin-2 antitumor and effector cell response. Semin Oncol *20:* 52-59, 1993.
79. Maccali C, Mortarini R, Parmiani G, Anichini A: Multiple sub-sets of $CD4^+$ and $CD8^+$ cytotoxic T-cell clones directed to autologous human melanoma identified by cytokine profiles. Inter J Cancer *57:* 56-62, 1994.
80. Shi L, Kam CM, Powers JC, Aebersold R, Greenberg AH: Purification of three cytotoxic lymphocyte granule serine proteases that induce apoptosis through distinct substrate and target cell interactions. J Exp Med *176:* 1521-1529, 1992.
81. Nagata S, Golstein P: The Fas death factor. Science *267:* 1449-1456, 1995.
82. Khar A, Parshasaradhi BVV, Varalakshmi Ch, Ali MA, Kumari AL: Natural killer cells as the effector which mediates in vivo apoptosis in AK-5 tumor cells. Cell Immunol *177:* 86-92, 1997.
83. Trinchieri G: Biology of natural killer cells. Adv Immunol *47:* 187-376, 1989.
84. Fesus L, Davies PJA, Piacentini M: Apoptosis: molecular

mechanism in programmed cell death. Eur J Cell Biol *56:* 170-177, 1991.
85. Hoon DB, Bowker RJ, Cochran AJ: Suppressor cell activity in melanoma-draining lymph nodes. Cancer Res *47:* 1529-1533, 1987.
86. Shaw HM, McCarthy SW, McCarthy WH, Thompson JF, Milton GW: Thin regressing malignant melanoma: significance of concurrent regional lymph node metastases. Histopathology *15:* 257-265, 1989.
87. Restifo NP, Kawakami Y, Marincola F, Shamamian P, Taggarse A, Esquivel F, Rosenberg SA: Molecular mechanisms used by tumors to escape immune recognition: immunogenetherapy and the cell biology of major histocompatibility complex class I. J Immunother *14:* 182-190, 1993.
88. Schrier PI, Bernards R, Vaessen RTMJ, Houweling A, van der Eb AJ: Expression of class I major histocompatibility antigens switched off by highly oncogenic adenovirus 12 in transformed rat cells. Nature *305:* 771-775, 1983.
89. Rooney CM, Rowe M, Wallace LE, Rickinson AB: Epstein-Barr virus positive Burkitt's lymphoma cells not recognized by virus specific T-cell surveillance. Nature *317:* 629-631, 1985.
90. Weber JS, Kay G, Tanaka K, Rosenberg SA: Immunotherapy of a murine tumor with interleukin-2: increased sensitivity after MHC class I gene transfection. J Exp Med *166:* 1716-1722, 1987.
91. Tanaka K, Yoshioka T, Bieberich C, Jay G: Role of the major histocompatibility complex class I antigens in tumor growth and metastasis. Annu Rev Immunol *6:* 359-380, 1988.
92. Vujanovic NL, Nagashima S, Herberman RB, Whiteside TL: Nonsecretory apoptotic killing by human NK cells. J Immunol *157:* 1117-1126, 1996.
93. Vujanovic NL, Basse P, Herberman RB, Whiteside TL: Antitumor functions of natural killer cells and control of metastases. Methods *9:* 394-408, 1996.
94. Kradin RL, Xia W, Pike M, Byers HR, Pinto C: Interleukin-2 promotes the motility of dendritic cells and their accumulation in lung and skin. Pathobiology *64:* 180-186, 1996.
95. Bodey B, Bodey B Jr, Kaiser HE: Dendritic type, accessory cells within the mammalian thymic microenvironment. Antigen presentation in the dendritic neuro-endocrine-immune cellular network. In Vivo *11:* 351-370, 1997.
96. Underwood JC: Lymphoreticular infiltration in human tumours: prognostic and biological implications: a review. Br J Cancer *30:* 538-548, 1974.
97. Ioachim HL: The stromal reaction of tumors: an expression of immune surveillance. J Natl Cancer Inst *57:* 465-475, 1976.
98. Ruiter DJ, Bhan AK, Harrist TJ, Sober AJ, Mihm MC Jr: Major histocompatibility antigens and mononuclear inflammatory infiltrate in benign nevomelanocytic proliferations and malignant melanoma. J Immunol *129:* 2808-2815, 1982.
99. Mikhail GR, Gorsulowsky DC: Spontaneous regression of metastatic malignant melanoma. Dermatol Surg Oncol *12:* 497-500, 1986.
100. Herberman RB, Holden HT: Natural cell-mediated immunity. Adv Cancer Res *27:* 305-337, 1978.
101. Nowell P: The clonal evolution of tumor cell populations. Science *194:* 23-28, 1976.
102. Nicolson GL: Tumor cell instability, diversification and progression to the metastatic phenotype: from oncogene to oncofetal expression. Cancer Res *47:* 1473-1487, 1987.
103. Corominas M, Leon J, Kamino H, Cruz-Alvarez M, Novick SC, Pellicer A: Oncogene involvement in tumor regression: H-ras activation in the rabbit keratoacanthoma model. Oncogene *6:* 645-651, 1991.
104. Alexandrow MG, Moses HL: Transforming growth factor β and cell cycle regulation. Cancer Res *55:* 1452-1457, 1995.
105. Klein G: The approaching era of the tumor suppressor genes. Science *238:* 1539-1545, 1987.
106. Barbacid M: ras genes. Annu Rev Biochem *56:* 779-827, 1987.
107. Bos JL: The ras gene family and human carcinogenesis. Mutat Res *195:* 255-271, 1988.
108. Bodey B, Bodey B Jr, Groger AM, Luck JV, Siegel SE, Taylor CR, Kaiser HE: Clinical and prognostic significance of the expression of the c-erbB-2 and c-erbB-3 oncoproteins in primary and metastatic malignant melanomas and breast carcinomas. Anticancer Res *17:* 1319-1330, 1997.
109. Hercbergs A, Leith JT: Spontaneous remission of metastatic lung cancer following myxedema coma - an apoptosis-related phenomenon? J Natl Cancer Inst *85:* 1342-1343, 1993.
110. Bodey B, Bodey B Jr, Hall FL: Expression of Transforming Growth Factor-β type II receptors in the cells of the human thymic microenvironment during ontogenesis. In: Molecular Biology of Hematopoiesis, Volume 5, Chapter 78 (Abraham NG, Asano S, Brittinger G, Maestroni GJM, Shadduck R, eds) New York, Plenum Press, pp 645-658, 1996.
111. Eaves C, Eaves A: Differential manipulation of normal and chronic myeloid leukemia stem cell proliferation in vitro. Blood Cells *20:* 83-95, 1994.
112. Nathason L: Spontaneous regression of malignant melanoma: A review of the literature on incidence, clinical features, and possible mechanisms. Natl Cancer Inst Monogr *44:* 67-76, 1976.
113. Birge RF, Jenks AL, Davis SK: Spontaneous remission in acute leukemia: Report of a case complicated with eclampsia. JAMA *140:* 589-592, 1949.
114. Cantini E, James B: Acute myelogenous leukemia in pregnancy. South Med J *77:* 1050-1052, 1984.
115. Smithers DW: Cancer of the breast and menopause. Clin Radiol *4:* 89-96, 1952.
116. Lewison EF: Spontaneous regression of breast cancer. Natl Cancer Inst Monogr *44:* 23-26, 1976.
117. Ross MB, Buzdar AU, Hortobagyi GN, Lukeman JM: Spontaneous regression of breast carcinoma: follow-up report and literature review. J Surg Oncol *19:* 22-24, 1982.
118. Taetle R, Guittarad JP: Modulation of leukaemia blast colony growth by steroid hormones. Br J Haemat *50:* 247-255, 1982.
119. Fisher RI, Neifeld JP, Lippman ME: Oestrogen receptors in human malignant melanoma. Lancet *2:* 337-339, 1976.
120. Papac RJ, Kirkwood J: High dose tamoxifen in malignant melanoma. Cancer Treat Rep *67:* 1049, 1983.
121. Walker MJ: Role of hormones and growth factors in melanomas. Semin Oncol *15:* 512-523, 1988.
122. Jetten AM, Jetten MER: Possible role of retinoic acid binding protein in retinoid stimulation of embryonal carcinoma cell differentiation. Nature *278:* 180-182, 1979.
123. Trown PW, Palleroni AV, Bohoslawec O, Richelo BN, Halpern JM, Gizzi N, Geiger R, Lewinski C, Machlin LJ, Jetten A, Jetten ME: Relationship between binding affinities to cellular retinoic acid-binding protein and in vivo and in vitro properties for 18 retinoids. Cancer Res *40:* 212-220, 1980
124. Jetten AM, Deluca LM: Induction of differentiation of embryonal carcinoma cells by retinol: possible mechanisms. Biochem Biophys Res Commun *114:* 593-599, 1983.
125. Eglitis MA, Sherman MI: Murine embryonal carcinoma cells differentiate in vitro in response to retinol. Exp Cell Res *146:* 289-296, 1983.

126. Strickland S, Breitman TR, Frickel F, Nurrenbach A, Hadicke E, Sporn MB: Structure-activity relationships of a new series of retinoidal benzoic acid derivatives as measured by induction of differentiation of murine F9 teratocarcinoma cells and human HL-60 promyelocytic leukemia cells. Cancer Res *43:* 5268-5272, 1983.
127. Jetten AM: Induction of differentiation of embryonal carcinoma cells by retinoids. In: Retinoids and Cell Differentiation (Sherman MI, ed). Boca Raton, CRC Press, p 105, 1986.
128. Jetten AM, Anderson K, Deas MA, Kagechika H, Lotan R, Rearick JI, Shudo K: New benzoic acid derivatives with retinoid activity: lack of direct correlation between biological activity and binding to cellular retinoic acid binding protein. Cancer Res *47:* 3523-3527, 1987.
129. Giguere V, Ong ES, Segui P, Evans RM: Identification of a receptor for the morphogen retinoic acid. Nature *330:* 624-629, 1987.
130. Pierce GB: Teratocarcinoma: model for a developmental concept of cancer. Curr Top Dev Biol *2:* 223-246, 1967.
131. Sherman MI, Solter D (eds): Teratomas and Differentiation. New York, Academic Press, 1975.
132. Martin GR: Teratocarcinomas and mammalian embryogenesis. Science *209:* 768-776, 1980.
133. Bolande RP: Models and concepts derived from human teratogenesis and oncogenesis in early life. J Histochem Cytochem *32:* 878-884, 1984.
134. Astigiano S, Sherman MI, Abarzua P: Regulation and patterns of endogenous and exogenous gene expression during differentiation of embryonal carcinoma cells. Envir Health Persp *80:* 25-38, 1989.
135. Strickland S, Mahdavi V: The induction of differentiation in teratocarcinoma stem cells by retinoic acid. Cell *15:* 393-403, 1978.
136. Strickland S, Smith KK, Marotti KR: Hormonal induction of differentiation in teratocarcinoma stem cells: generation of parietal endoderm by retinoic acid and dibutyryl cAMP. Cell *21:* 347-355, 1980.
137. Grover A, Oshima RG, Adamson ED: Epithelial layer formation in differentiating aggregates of F9 embryonal carcinoma cells. J Cell Biol *96:* 1690-1696, 1983.
138. Grover A, Adamson ED: Evidence for the existence of an early common biochemical pathway in the differentiation of F9 cells into visceral or parietal endoderm: modulation by cyclic AMP. Dev Biol *114:* 492-503, 1986.
139. Hong WK, Wittes RE, Hajdu ST, Cvitkovic E, Whitmore WF, Golbey RB: The evolution of mature teratoma from malignant testicular tumors. Cancer *40:* 2987-2992, 1977.
140. Friedman SJ, Skehan P: Morphological differentiation of human choriocarcinoma cells induced by methotrexate. Cancer Res *39:* 1960-1967, 1979.
141. Haas D, Ablin AR, Miller C, Zoger S, Matthay KK: Complete pathologic maturation and regression of stage IVS neuroblastoma without treatment. Cancer *62:* 818-825, 1988.
142. Castaine S, Chomienne C, Daniel MT, Ballerini P, Berger R, Fenaux P, Degos L: All trans retinoic acid as differentiation therapy for acute promyelocytic leukemia. I. Clinical results. Blood *76:* 1704-1709, 1990.
143. Kurie JM, Bosl GJ, Dmitrovsky ED: The genetic and biologic aspects of treatment response and resistance in male germ cell cancer. Semin Oncol *19:* 197-205, 1992.
144. Stoll BA: Spontaneous regression of cancer: new insights. Biotherapy *4:* 23-30, 1992.
145. Helmlinger G, Netti PA, Lichtenbeld HC, Melder RJ, Jain RK: Solid stress inhibits the growth of multicellular tumor spheroids. Nature Biotech *15:* 778-783, 1997.
146. Fung YC: Biomechanical aspects of growth and tissue engineering. In: Biomechanics, motion, flow, stress, and growth. (Fung YC, ed) New York, Springer-Verlag, pp 499-546, 1990.
147. Vaage J: Fibrosis in immune control of mammary-tumor growth. Int J Cancer *51:* 325-328, 1992.
148. Gartner MFRM, Fearns C, Wilson EL, Campbell JAH, Dowdle EB: Unusual growth characteristics of human melanoma xenografts in the nude mouse: a model for desmoplasia, dormancy and progression. Br J Cancer *65:* 487-490, 1992.
149. Young JS: The invasive growth of malignant tumours: an experimental interpretation based on elastic-jelly models. J Pathol Bact *77:* 321-326, 1959.
150. Eaves G: The invasive growth of malignant tumours as a purely mechanical process. J Pathol *109:* 233-237, 1973.
151. Falk P: Patterns of vasculature in two pairs of related fibrosarcomas in the rat and their relation to tumour responses to single large doses of radiation. Eur J Cancer *14:* 237-250, 1978.
152. Falk P: The vascular pattern of the spontaneous C3H mouse mammary carcinoma and its significance in radiation response and in hyperthermia. Eur J Cancer *16:* 203-217, 1980.
153. Tozer GM, Lewis S, Michalowski A, Aber V: The relationship between regional variations in blood flow and histology in a transplanted rat fibrosarcoma. Br J Cancer *61:* 250-257, 1990.
154. Jain RK: Determinants of tumor blood flow: a review. Cancer Res *48:* 2641-2658, 1988.
155. Li L, Price JE, Fan D, Zhang RD, Bucana CD, Fidler IJ: Correlation of growth capacity of human tumor cells in hard agarose with their in vivo proliferative capacity at specific metastatic sites. J Natl Cancer Inst *81:* 1406-1412, 1989.
156. Sutherland RM: Cell and environment interactions in tumor microregions: the multicell spheroid model. Science *240:* 177-184, 1988.
157. Oldham RK, Lewko WM, Good RW, Sharp E: Cancer biotherapy with interferon, interleukin-2 and tumor-derived activated cells (TDAC). IN VIVO *8:* 653-663, 1994.
158. Rosenberg SA, Blaese RM, Brenner MK, Deisseroth AB, Ledley FD, Lotze MT, Wilson JM, Nabel GJ, Cornetta K, Economou JS, Freeman SM, Riddell SR, Brenner M, Oldfield E, Gansbacher B, Dunbar C, Walker RE, Schuening FG, Roth JA, Crystal RG, Welsh MJ, Culver K, Heslop HE, Simons J, Wilmott RW, Boucher RC, Siegler HF, Barranger JA, Karlsson S, Kohn D, Galpin JE, Raffel C, Hesdorffer C, Ilan J, Cassileth P, O'Shaughnessy J, Kun LE, Das TK, Wong-Staal F, Sobol RE, Haubrich R, Sznol M, Rubin J, Sorcher EJ, Rosenblatt J, Walker R, Brigham K, Vogelzang N, Hersh E, Eck SL: Human gene marker/therapy clinical protocols. Hum Gene Ther *11:* 919-979, 2000.
159. Jager D, Jager E, Knuth A: Immune responses to tumour antigens: implications for antigen specific immunotherapy of cancer. J Clin Pathol 54: 669-674, 2001; Comment in: J Clin Pathol 54: 675-676, 2001.
160. Pike M, Peters R, Cozen W, Probst-Hensch NM, Felix JC, Wan PC, Mack TM: Estrogen-Progestin replacement therapy and endometrial cancer. J Natl Cancer Inst *89:* 1110-1116, 1997.
161. Steinbrechner UP, Lisbona R, Huang SN, Mishkin S: Complete regression of hepatocellular adenoma after withdrawal of oral contraceptives. Dig Dis Sci *26:* 1045-1050, 1981.
162. Tahara H, Shiozaki H, Kobayashi K, Yano T, Yano H, Tamura S, Oku K, Miyata M, Wakasa K, Sakurai M: Phenotypic characteristics of tumour-infiltrating-lymphocytes in human oesophageal cancer tissues defined

by quantitative two colour analysis with flow-cytometry. Virchows Arch *416:* 329-334, 1990.
163. Belldegrun A, Kasid A, Uppenkamp M, Topalian SL, Rosenberg SA: Human tumor infiltrating lymphocytes. Analysis of lymphokine mRNA expression and relevance to cancer immunotherapy. J Immunol *142:* 4520-4526, 1989.
164. Muul LM, Spiess PJ, Director EP, Rosenberg SA: Identification of specific cytolytic immune responses against autologous tumor in humans bearing malignant melanoma. J Immunol *101:* 171-181, 1987.
165. Somers SS, Guillou PJ: Isolation and expansion of lymphocytes from gastrointestinal tumour tissue. Surg Oncol *2:* 283-291, 1993.
166. Belldegrun A, Muul LM, Rosenberg SA: Interleukin-2 expanded tumor-infiltrating lymphocytes in human renal cell cancer: Isolation, characterization, and antitumor activity. Cancer Res *48:* 206-214, 1988.
167. Yoshida K, Tachibana T: Prevention of lymph node metastases by adoptive transfer of $CD4^+$ T lymphocytes admixed with irradiated tumor cells. Cancer Immunol Immunother *36:* 323-330, 1993.
168. Rosenberg SA: The development of new immunotherapies for the treatment of cancer using interleukin-2. Am Surg *208:* 121-135, 1988.
169. Panelli MC, Bettinotti MP, Lally K, Ohnmacht GA, Li Y, Robbins P, Riker A, Rosenberg SA, Marincola FM: A tumor-infiltrating lymphocyte from a melanoma metastasis with decreased expression of melanoma differentiation antigens recognizes MAGE-12. J Immunol *164:* 4382-4392, 2000.
170. Rosenberg SA: Progress in human tumour immunology and immunotherapy. Nature *411:* 380-384, 2001.
171. Dudley ME, Wunderlich J, Nishimura MI, Yu D, Yang JC, Topalian SL, Schwartzentruber DJ, Hwu P, Marincola FM, Sherry R, Leitman SF, Rosenberg SA: Adoptive transfer of cloned melanoma-reactive T lymphocytes for the treatment of patients with metastatic melanoma. J Immunother *24:* 363-373, 2001.
172. Harada M, Li YF, El-Gamil M, Ohnmacht GA, Rosenberg SA, Robbins PF: Melanoma-Reactive $CD8^+$ T cells recognize a novel tumor antigen expressed in a wide variety of tumor types. J Immunother *24:* 323-333, 2001.
173. Riley JP, Rosenberg SA, Parkhurst MR: Identification of a New Shared HLA-A2.1 Restricted Epitope From the Melanoma Antigen Tyrosinase. J Immunother *24:* 212-220, 2001.
174. Riley JP, Rosenberg SA, Parkhurst MR: Identification of a new shared HLA-A2.1 restricted epitope from the melanoma antigen tyrosinase. J Immunother *24:* 212-220, 2001.
175. Haas GP, Solomon D, Rosenberg SA: Tumor infiltrating lymphocytes from non-renal urological malignancies. Cancer Immunol Immunother *30:* 342-350, 1990.
176. Miescher S, Whiteside TL, Moretta L, von Fliedner V: Clonal and frequency analyses of tumor-infiltrating T lymphocytes from human solid tumors. J Immunol *138:* 4004-4011, 1987.
177. Schiltz PM, Beutel LD, Nayak SK, Dillman RO: Characterization of tumor-infiltrating lymphocytes derived from human tumors for use as adoptive immunotherapy of cancer. J Immunother *20:* 377-386, 1997.
178. Freedman RS, Tomasovic B, Templin S, Atkinson EN, Kudelka, Edwards CL, Platsoucas CD: Large-scale expansion in interleukin-2 of tumor infiltrating lymphocytes from patients with ovarian carcinoma for adoptive immunotherapy. J Immunol Methods *167:* 145-160, 1994.
179. Freedman RS, Edwards CL, Kavanagh JJ, Kudelka AP, Katz RL, Carrasco CH, Atkinson EN, Scott W, Tomasovic B, Templin S: Intraperitoneal adoptive immunotherapy of ovarian carcinoma with tumor-infiltrating lymphocytes and low-dose recombinant interleukin-2: a pilot trial. J Immunother *16:* 198-210, 1994.
180. Lewko WM, Good RW, Bowman D, Smith TL, Oldham RK: Growth of tumor derived activated T-cells for the treatment of cancer. Cancer Biotherapy *9:* 211-224, 1994.
181. Topalian SL, Kasid A, Rosenberg SA: Immunoselection of a human melanoma resistant to specific lysis by autologous tumor-infiltrating lymphocytes. Possible mechanisms for immunotherapeutic failures. J Immunol *144:* 4487-4495, 1990.
182. Liu K, Rosenberg SA: Transduction of an IL-2 Gene into Human Melanoma-Reactive Lymphocytes Results in Their Continued Growth in the Absence of Exogenous IL-2 and Maintenance of Specific Antitumor Activity. J Immunol 167: 6356-6365, 2001.
183. Beun GD, van de Velde CJ, Feuren GJ: T-cell based cancer immunotherapy: direct or redirected tumor-cell recognition? Immunol Today *15:* 11-15, 1994.
184. Demanet C, Brissinck J, Leo O, Moser M, Thielemans K: Role of T-cell subsets in the bispecific antibody (anti-idiotype x anti-CD3) treatment of the BCL1 lymphoma. Cancer Res *54:* 2973-2978, 1994.
185. Thibault C, Nelson H, Chapovl AI: Tumor-infiltrating lymphocytes can be activated in situ by using in vivo activants plus $F(ab')_2$ bispecific antibodies. Int J Cancer *67:* 232-237, 1996.
186. Ferrini S, Cambiaggi A, Sforzini S, Canevari S, Mezzanzanica D, Colnaghi MI, Moretta L: Use of anti-CD3 and anti-CD16 bispecific monoclonal antibodies for the targeting of T and NK cells against tumor cells. Cancer Detect Prevent *17:* 295-300, 1993.
187. Weiner LM, Holmes M, Adams GP, LaCreta F, Watts P, Garcia de Palazzo I: A human tumor xenograft model of therapy with a bispecific monoclonal antibody targeting c-erbB-2 and CD16. Cancer Res *53:* 94-100, 1993.
188. Hombach A, Tillmann T, Jensen M, Heuser C, Sircar R, Diehl V, Kruis W, Pohl C: Specific activation of resting T cells against tumour cells by bispecific antibodies and CD28-mediated costimulation is accompanied by Th1 differentiation and recruitment of MHC-independent cytotoxicity. Clin Exp Immunol *108:* 352-357, 1997.
189. Chapoval AI, Nelson H, Thibault C, Penna C, Dean P: Bifunctional antibody retargeting in vivo-activated T lymphocytes: simplifying clinical application. J Hematother *4:* 571-577, 1995.
190. Csoka M, Strauss G, Debatin KM, Moldenhauer G: Activation of T cell cytotoxicity against autologous common acute lymphoblastic leukemia (cALL) blasts by CD3xCD19 bispecific antibody. Leukemia *10:* 1765-1772, 1996.
191. Demanet C, Brissinck J, De Jonge J, Thielemans K: Bispecific antibody-mediated immunotherapy of the BCL1 lymphoma: increased efficacy with multiple injections and CD28-induced costimulation. Blood *87:* 4390-4398, 1996.
192. Haas C, Schirrmacher V: Immunogenicity increase of autologous tumor cell vaccines by virus infection and attachment of bispecific antibodies. Cancer Immunol Immunother *43:* 190-194, 1996.
193. Holliger P, Brissinck J, Williams RL, Thielemans K, Winter G: Specific killing of lymphoma cells by cytotoxic T-cells mediated by a bispecific diabody. Protein Eng *9:* 299-305, 1996.
194. Kroesen BJ, Wellenberg GJ, Bakker A, Helfrich W, The TH, de Leij L: The role of apoptosis in bispecific antibody-

mediated T-cell cytotoxicity. Br J Cancer *73:* 721-727, 1996.
195. Kuwahara M, Kuroki M, Arakawa F, Senba T, Matsuoka Y, Hideshima T, Yamashita Y, Kanda H: A mouse/human-chimeric bispecific antibody reactive with human carcinoembryonic antigen-expressing cells and human T-lymphocytes. Anticancer Res *16:* 2661-2667, 1996.
196. Mack M, Gruber R, Schmidt S, Riethmuller G, Kufer P: Biologic properties of a bispecific single-chain antibody directed against 17-1A (EpCAM) and CD3: tumor cell-dependent T cell stimulation and cytotoxic activity. J Immunol *158:* 3965-3970, 1997.
197. Renner C, Held G, Ohnesorge S, Bauer S, Gerlach K, Pfitzenmeier JP, Pfreundschuh M: Role of perforin, granzymes and the proliferative state of the target cells in apoptosis and necrosis mediated by bispecific-antibody-activated cytotoxic T cells. Cancer Immunol Immunother *44:* 70-76, 1997.
198. Blaese RM, Mullen CA, Ramsey WJ: Strategies for gene therapy. Pathologie Biologie *41:* 672-676, 1993.
199. Hwu P, Rosenberg SA: The Genetic Modification of T Cells for Cancer Therapy: An Overview of Laboratory and Clinical Trials. Cancer Detect Prevent *18:* 43-50, 1994.
200. Kawakami Y, Rosenberg A, Lotze MT: Interleukin-4 promotes the growth of tumor infiltrating lymphocytes cytotoxic for human autologous melanoma. J Exp Med *168:* 2183-2191, 1988.
201. Kim TS, Cohen EP: MHC antigen expression by melanomas recovered from mice treated with allogeneic mouse fibroblasts genetically modified for interleukin-2 secretion and the expression of melanoma-associated antigens. Cancer Immunol Immunother *38:* 185-193, 1994.
202. Abdel-Wahab Z, Weltz C, Hester D, Pickett N, Vervaert C, Barber JR, Jolly D, Seigler HF: Phase I clinical trial of immunotherapy with interferon-γ gene-modified autologous melanoma cells: monitoring the humoral immune response. Cancer *80:* 401-412, 1997.
203. Jaffee EM, Pardoll DM: Considerations for the clinical development of cytokine gene-transduced tumor cell vaccines. Methods *12:* 143-153, 1997.
204. Fearon ER, Pardoll DM, Itaya T, Golumbek P, Levitsky HI, Simons JW, Karasuyama H, Vogelstein B, Frost P: IL-2 production by tumor cells bypasses T helper function in the generation of an anti-tumor response. Cell *60:* 387-403, 1990.
205. Golumbek PT, Lazenby AJ, Levitsky HI, Jaffee LM, Karasuyama H, Baker M, Pardoll DM: Treatment of established renal cancer by tumor cells engineered to secrete interleukin-4. Science *254:* 713-716, 1991.
206. Lipshy KA, Kostuchenko PJ, Hamad GG, Bland CE, Barrett SK, Bear HD: Sensitizing T-lymphocytes for adoptive immunotherapy by vaccination with wild-type or cytokine gene-transduced melanoma. Ann Surg Oncol *4:* 334-341, 1997.
207. Aruga E, Aruga A, Arca MJ, Lee WM, Yang NS, Sminth JW 2nd, Chang AE: Immune responsiveness to a murine mammary carcinoma modified to express B7-1, interleukin 12, or GM-CSF. Cancer Gene Ther *4:* 157-166, 1997.
208. Arienti F, Sule-Suso J, Belli F, Mascheroni L, Rivoltini L, Melani C, Maio M, Cascinelli N, Colombo MP, Parmiani G: Limited antitumor T cell response in melanoma patients vaccinated with interleukin-2 gene-transduced allogeneic melanoma cells. Hum Gene Ther *7:* 1955-1963, 1996.
209. Baskar S: Gene-modified tumor cells as cellular vaccine. Cancer Immunol Immunother *43:* 165-173, 1996.
210. Dar MM, Abdel-Wahab Z, Vervaert CE, Darrow T, Barber J, Seigler HF: Immunological memory induced by genetically transduced tumor cells. Ann Surg Oncol *3:* 247-254, 1996.
211. De Wit D, Flemming CL, Harris JD, Palmer KJ, Moore JS, Gore ME, Collins NS MK: IL-12 stimulation but not B7 expression increases melanoma killing by patient cytotoxic T lymphocytes (CTL). Clin Exp Immunol *105:* 353-359, 1996.
212. Dunussi-Joannopoulos K, Weinstein HJ, Nickerson PW, Strom TB, Burakoff SJ, Croop JM, Arceci RJ: Irradiated B7-1 transduced primary acute myelogenous leukemia (AML) cells can be used as therapeutic vaccines in murine AML. Blood *87:* 2938-2946, 1996.
213. Knight BC, Souberbielle BE, Rizzardi GP, Ball SE, Dalgleish AG: Allogeneic murine melanoma cell vaccine: a model for the development of human allogeneic cancer vaccine. Melanoma Res *6:* 299-306, 1996.
214. Levitsky HI, Montgomery J, Ahmadzadeh M, Staveley-O'Carroll K, Guarnieri F, Longo DL, Kwak LW: Immunization with granulocyte-macrophage colony-stimulating factor-transduced, but not B7-1-transduced, lymphoma cells primes idiotype-specific T cells and generates potent systemic antitumor immunity. J Immunol *156:* 3858-3865, 1996.
215. Mule JJ, Custer M, Averbook B, Yang JC, Weber JS, Goeddel DV, Rosenberg SA, Schall TJ: RANTES secretion by gene-modified tumor cells results in loss of tumorigenicity in vivo: role of immune cell subpopulations. Hum Gene Ther *7:* 1545-1553, 1996.
216. Ohno K, Yoshizawa H, Tsukada H, Takeda T, Yamaguchi Y, Ichikawa K, Maruyama Y, Suzuki Y, Suzuki E, Arakawa M: Adoptive immunotherapy with tumor-specific T lymphocytes generated from cytokine gene-modified tumor-primed lymph node cells. J Immunol *156:* 3875-3881, 1996.
217. Tung C, Federoff HJ, Brownlee M, Karpoff H, Weigel T, Brennan MF, Fong Y: Rapid production of interleukin-2-secreting tumor cells by herpes simplex virus-mediated gene transfer: implications for autologous vaccine production. Hum Gene Ther *7:* 2217-2224, 1996.
218. Wakimoto H, Abe J, Tsunoda R, Aoyagi M, Hirakawa K, Hamada H: Intensified antitumor immunity by a cancer vaccine that produces granulocyte-macrophage colony-stimulating factor plus interleukin 4. Cancer Res *56:* 1828-1833, 1996.
219. Gaken JA, Hollingsworth SJ, Hirst WJ, Buggins AG, Galea-Lauri J, Peakman M, Kuiper M, Patel P, Towner P, Patel PM, Collins MK, Mufti GJ, Farzaneh F, Darling DC: Irradiated NC adenocarcinoma cells transduced with both B7.1 and interleukin-2 induce $CD4^+$-mediated rejection of established tumors. Hum Gene Ther *8:* 477-488, 1997.
220. Lee CT, Wu S, Ciernik IF, Chen H, Nadaf-Rahrov S, Gabrilovich D, Carbone DP: Genetic immunotherapy of established tumors with adenovirus-murine granulocyte-macrophage colony-stimulating factor. Hum Gene Ther *8:* 187-193, 1997.
221. Schmidt W, Maass G, Buschle M, Schweighoffer T, Berger M, Herbst E, Schilcher F, Birnstiel ML: Generation of effective cancer vaccines genetically engineered to secrete cytokines using adenovirus-enhanced transferrinfection (AVET). Gene *190:* 211-216, 1997.
222. Nair SK, Snyder D, Rouse BT, Gilboa E: Regression of tumors in mice vaccinated with professional antigen-presenting cells pulsed with tumor extracts. Int J Cancer *70:* 706-715, 1997.
223. Sinkovics JG, Horvath J: New developments in the virus therapy of cancer: a historical review. Intervirology *36:* 193-214, 1993.
224. Kasid A, Morecki S, Aebersold P, Cornetta K, Culver K, Freeman S, Director E, Lotze MT, Blaese RM, Anderson WF: Human gene transfer: Characterization of human

tumor-infiltrating lymphocytes as vehicles for retroviral-mediated gene transfer in man. Proc Natl Acad Sci USA *87:* 473-477, 1990.
225. Abe J, Wakimoto H, Tsunoda R, Okabe S, Yoshida Y, Aoyagi M, Hirakawa K, Hamada H: In vivo antitumor effect of cytotoxic T lymphocytes engineered to produce interferon-γ by adenovirus-mediated genetic transduction. Biochem Biophys Res Commun *218:* 164-170, 1996.
226. Nordon RE, Schindhelm K: Ex vivo manipulation of cell subsets for cell therapies. Artif Organs *20:* 396-402, 1996.
227. Tada M, Sawamura Y, Sakuma S, Suzuki K, Ohta H, Aida T, Abe H: Cellular and cytokine responses of the human central nervous system to intracranial administration of tumor necrosis factor α for the treatment of malignant gliomas. Cancer Immunol Immunother *36:* 251-259, 1993.
228. Moritz D, Groner B: A spacer region between the single chain antibody- and the CD3 zeta-chain domain of chimeric T cell receptor components is required for efficient ligand binding and signaling activity. Gene Ther *2:* 539-546, 1995.

APPENDIX

1. MATERIALS AND METHODS

1.1 Experimental Subjects

During our thymic studies, 212 fetal and 25 postnatal human thymuses were observed immunomorphologically employing light histological, histochemical, immunocytochemical and electron microscopical (transmission and scanning) methods. Thymic indices were calculated for each dissected organ separately as the percentage that thymic weight represented of the total body weight. Postnatal thymus samples were obtained fresh during open-heart surgery on patients aged 0 (i.e. newborn) to 21 years. The patients suffered from various congenital cardiovascular diseases without any other systemic or immunological disorders. Several thymic tissues were used for *in vitro*, tissue culture investigations, followed by detailed histological, immunofluorescent (FACS) and immunocytochemical observation. Detailed descriptions of the histological, histochemical, and immunocytochemical techniques employed were published earlier in our other reports on the thymus (160, 339, 462-471).

Twenty-six adult male and female dogs (beagle) were used for radiation and stem cell transplantation experiments. The animals were dewormed and immunized against distemper, leptospirosis, and hepatitis contagiosa several weeks before the commencement of the experiments. Twenty of them were irradiated and six served as the control group. The dogs were kept in the animal facilities of our laboratory, in compliance with the recommendations of the Institute of Laboratory Resources, National Research Council (USA) for the care and use of laboratory animals (166).

Male Balb/c mice were housed in plastic cages and fed with pelleted food and water *ad libitum*. The life span of our experimental mice ranged between 24 and 28 months. We considered a mouse as being "young" at 2 months of age and "old" at 22 months.

1.2 Conditioning of Recipients for Hematopoietic Stem Cell Grafting

A Siemens Stabilipan x-ray therapy machine was employed as a radiation source. The following factors were used: 300 kV, 12 mA, half value layer 4 mm Cu, and a source-to-target distance of 1.2 m. The total dose (12 Gy) and the dose rate (6.5 cGy/min) were measured at the midline of a phantom. The dogs were anesthetized with Combalen (Bayer-Leverkusen, Germany) and exposed to half the dose from each side. No part of the body received less than 80% of the dose measured at the midline (166, 449).

1.3 Collection of Cells for Grafting

Leukocytes were obtained from the blood by continuous flow centrifugation, using the NIH-IBM experimental blood cell separator described by Buckner et al. (450). The blood was collected from the carotid artery and returned to the jugular vein. During a 4-6 hour leukapharesis, about 10,000-15,000 ml of blood went through the machine at a flow rate of 40 ml/min, separating 16.1-24.8 x 10^9 cells (mean 22.5 ± 5.3) leukocytes from the blood, as calculated by Herbst and co-workers (451).

1.4 Tissue Cultures of Thymic RE cells on Polycarbonate Filters

Prenatal thymic lobules from 14 day old mice which had just begun forming were placed on 0.4 mm thick polycarbonate filters floating on 6-7 ml RPMI 1640 culture medium supplemented with 10%

heat-inactivated fetal calf serum, at 24°C for 7 days to produce a thymocyte free RE cell matrix. After 7 days, the cultured cells were warmed to 37°C, then harvested or collected for morphological observation or used in co-culturing experiments with bone marrow derived hematopoietic cells (170).

1.5 Cryopreservation and Transplantation Procedure

The separated blood leukocytes were frozen in a cell freezer (type BV4, Cryoson, Medden Beemster, The Netherlands) and stored for several weeks in liquid nitrogen. The frozen cells were rapidly thawed, suspended in Hanks' balanced salts solution (HBSS), and then injected i.v. about 2 hours after irradiation. The dogs received 0.17-2.44 x 10^9 blood mononuclear cells (BMNC) per kg of body weight. Each animal received its own cells, which had been cryopreserved prior to irradiation. The number of mononuclear cells in a leukocyte suspension was minimally influenced by the freezing and thawing procedures. For further information about the procedure, see Bruch and co-investigators (452).

1.6 Assay for the Presence of Granulopoietic Colonies

The number of colony-forming units in agar culture (CFU-c) present in the transfusate was calculated by an assay of samples of the cell suspension taken after thawing. The dogs for the short-term study received 2,000-176,000 CFU-c per kg of body weight. Those for the long-term study received 19,000-138,000 CFU-c per kg of body weight. The method was performed according to Bradley and Metcalf (453), modified by Kovacs *et al.* (454).

1.7 Determination of Mouse T Lymphocyte Subsets

Mouse T lymphocyte subpopulations were analyzed in spleen and thymic cell suspensions by double cytofluorimetric assay (Epics V-Coulter) and employing a library of monoclonal antibodies (Becton-Dickinson Inc).

1.8 Cell Proliferation Assay and Analysis of IL-2 Secretion

Splenocytes (2 to 4 x 10^5) from 2 month old mice were stimulated with the mitogens ConA (2.5 mg/ml), PMA (5 ng/ml) plus A23187 (100 ng/ml), cross linked for two days with MoAB anti-CD3ε, and pulsed with 37 kBq [^{3}H] thymidine (Amersham Corp., Arlington Heights, IL) and kept *in vitro* for 8 to 16 hours and harvested. [^{3}H] thymidine uptake was measured employing a Microβ™ liquid scintillation counter (Pharmacia, Uppsala, Sweden). Production of interleukin-2 (IL-2) was measured by the proliferation rate of the IL-2 dependent cell line CTLL-2 (6 x 10^3 cells). The cells were co-cultured using supernatants from the proliferation assay after stimulation for 2 days. The CTLL-2 cells were pulsed with [^{3}H] thymidine and harvested in the manner described above.

1.9 Histochemical Stainings

1.10 Immunocytochemistry

1.10.1 Libraries of Antibodies

Our detailed immunohistochemical studies were performed on several human and other mammalian prenatal and postnatal thymuses, employing fresh frozen and formalin fixed, paraffin-wax embedded tissue sections with well selected libraries of poly- and monoclonal antibodies (MoABs) to observe the maturation level and pathway of thymocytes and to identify the immunophenotypes (IPs) of developing and functionally mature stromal and non-stromal, accessory cells of the thymic microenvironment (455). During the detailed immunocytological studies, we were able to characterize the intrathymic professional antigen presenting cell types and functional changes in IP during maturation of thymocytes into immunocompetent, functional subsets of T lymphocytes and the differentiation of the stromal elements of the thymic microenvironment. The following antibody (mouse, anti-human, monoclonal and rabbit, anti-human, polyclonal) library was employed in our immunocytochemical observations on human and other mammalian thymuses:

1. *Anti-endocrine cell and anti-RE cell:* A_2B_5, Thy-1, and anti-epithelial membrane antigen (EMA);
2. *Anti-neuroectodermal and anti-epithelial*: UJ 13/A, UJ 127.11, UJ 167.11, UJ 181.4, UJ

223.8, UJ 308, J1153 and PI 153/3 (produced and provided to us by the laboratory of Dr. Kemshead in England); 215, 275, and 282.1;

3. *Anti-hematopoietic and anti-T lymphocyte differentiation markers*: CLA (anti-HLe-1 directed against a 200-220 kD cell surface receptor, the CD45 LCA is a family of five or more high mol. weight glycoproteins present on the surface of the majority of human leukocytes but absent from erythrocytes and platelets. The CD45R subfamily is divided into three isoforms - CD45RA, CD45RB and CD45RO (DAKO); Leu-2/a (against $CD8^+$, cytotoxic/suppressor T-lymphocytes), Leu-3/a,3/b (for identification of $CD4^+$, helper lymphocytes), Leu-4 (CD3), Leu-5b (CD2), Leu-6 (CD1), Leu-7 (HNK1), Leu-8, Leu-9 (CD7), Leu-11/b (CD16), anti-CD57 for detection of natural killer cells, and neuroectodermal cells. $CD57^+CD8^+$ lymphocytes are identified as suppressor/cytotoxic T lymphocyte subset; Myelomonocytic antigens: Leu-7, anti-CD11a or LFA, Leu-14, Leu-19, Leu-M5 or anti-CD11b (for monocytes/macrophages); Activated immune cells: Interleukin-2 receptor (CD25 or anti-Tac or IL-2R) and anti-transferrin receptor (Neomarkers Inc., Fremont, CA); Intracellular Adhesion Molecule (ICAM-I or CD54), LFA-3 (gp55-70 or CD58) and Leu-M5 (for monocytes/macrophages); *mouse T lymphocyte subsets* (mouse lymphocytes): anti-Lyt2 (for $CD8^+$), L3T4 ($CD4^+$), and Thy1.2 ($CD3^+$) (all monoclonal antibodies from Becton-Dickinson Inc.,);
4. *Anti-Major Histocompatibility Complex (MHC) molecules*: HLA-A,B,C (Class I molecules); and HLA-D (DP, DQ, DR), HLA-DR (CD74), HLA-DQ and HLA-DP (Class II molecules) (Neomarkers Inc., Fremont, CA);
5. *Anti-Thymic Microenvironment Stromal Cells*: TE-3, TE-4, TE-7, TE-8, TE 15, TE 16, and TE 19 (developed in the laboratory of Dr. Haynes and provided to us);
6. *Anti-Neuronal*: Neurophysin and Synaptophysin;
7. *Anti-Intermediate Filament proteins*: Neurofilament-L (molecular weight 68 kD), Neurofilament-M (molecular weight 160 kD), Neurofilament-H (molecular weight 200 kD), Glial fibrillary acidic protein (GFAP), Vimentin, MAP-1 (microtubules), MAP-2, MAP-5, cytokeratins 13 and 18, desmosomal cytokeratin, AE1, AE2, AE3, AE5, and AE8;
8. *Anti-Tumor markers*: B18.7.7 and D14 [carcinoembryonic antigen (CEA)];
9. *Anti-Oncogenes and related protein products*: p53 (also tumor suppressor gene related), c-erbB-2, c-erbB-3, c-myc, and c-ras;
10. *Cell cycle related cyclins, proliferation markers, and DNA repair related proteins:* anti-p53, Ki-67, anti-$p34^{cdc2}$, anti-cyclin A, and anti-cyclin D; and
11. *Anti-growth factors*: anti-TGF-β type II receptor (TGF-βIIR).
12. *Anti-Estrogen Receptor*: MoAB specific to ERa and a polyclonal antibody specific for ERb (Affinity BioReagents, Inc., Golden, CO);
13. *Anti-Progesterone* Receptor *(PgR)*: regulated by estrogen; two MoABs against A and B forms of PR (Novocastra Lab., US Distributor: Vector Lab.,);
14. *Anti-Androgen Receptor*: Androgen receptor is a member of the superfamily of ligand responsive transcription regulators - two MoAB, named NCL-AR-2F12 and PG-21 and two polyclonals specific to AR;
15. *Anti-Thyroid Hormone Receptor*: five polyclonal and two MoAB against the various TR isoforms;
16. *Anti-Prolactin Receptor*: MoAB (MAI-610) detects rat, human, rabbit, and pig prolactin receptors. The MoAB reacted with the extracellular portion of the receptor, but not with the ligand-binding domain. This antibody does not inhibit the interaction between prolactin and its receptor.
17. *Anti-Glucocorticoid Receptor*: MoAB BuGR2, and others reacting with GRa and GRb.
18. *anti-Retinoic Acid Receptor*: three MoABs against the α, β, and γ isoforms of RAR;
19. *anti-Homeobox Gene Products:* polyclonal antibodies directed against the HOX-B3, -B4 and -C6 gene products.

1.11 Immunoalkaline Phosphatase Antigen Detection Technique

We used the following immunoalkaline phosphatase cytochemical method, modified by us for antigen detection in formalin fixed, paraffin wax embedded thymus tissues (65-68). The technique is a highly sensitive, indirect, four to six step immunocytochemical method, which combines the biotin-streptavidin based ABC-method with enzyme-linked (alkaline phosphatase) immunohistochemistry. Briefly, following deparaffinization in three changes of Xylene substitute (Shandon-Lipshaw, Pittsburgh, PA, USA) for 20 to 30 minutes, rehydration was carried out employing descending dilutions of alcohol (100% to 50%) to TBS. An initial blocking step using 1% glacial acetic acid mixed with the working buffer for 10 minutes was necessary to eliminate the endogenous AP activity from the tissues. Use of levamisole solution is also described in our earlier observations. As we explained earlier (68), GAA inhibition was preferred because of the possible presence of levamisole-resistant AP iso-enzyme (69). The second blocking step was conducted with a purified mixture of proteins (Shandon-Lipshaw) from various species for 5-10 minutes to block cross-reactive antigenic epitopes. Excess serum was removed from the area surrounding the sections. The tissue sections were then incubated for 60-120 minutes with the particular primary antibody. Next, incubation with the secondary antibody, which was either a biotinylated, whole goat anti-rabbit or goat anti-mouse IgG molecule (IgG molecule diluted by ICN Biomedicals, Inc., Aurora, OH, USA) was carried out for 20 minutes. Streptavidin conjugation was accomplished by incubation with AP conjugated streptavidin for 20 minutes (70-73). Color visualization of the primary antigen-antibody (Ag-Ab) reaction was accomplished with an alkaline phosphatase (AP) kit I (Vector Laboratories, Burlingame, CA, USA) which contains AS-TR with Tris-HCl buffer at pH 8.2, added for 28-40 minutes to allow formation of a stable red precipitate. Sections were counterstained with a diluted solution of Gill's hematoxylin (Richard-Allan, Kalamazoo, MI, USA). The tissue slides were then dehydrated in ascending concentrations of alcohol (60% to 100%) to xylene substitute (Shandon-Lipshaw), in which they were kept overnight to ensure complete morphological clearing. The stained tissue sections were mounted using a solution specially designed for use following morphological clearing in xylene substitute (Shandon-Lipshaw).

1.12 Immunoperoxidase Antigen Detection Technique

The following ABC method (74) was employed for the immunocytochemical detection of antigens in thymus specimens. Briefly, following deparaffinization in three fresh changes of xylene substitute (Shandon-Lipshaw), tissue rehydration employing descending grades of alcohols (100% to 40%) to PBS was performed. The tissue sections were never allowed to dry-out before being moved to the next solution. After 30 minutes incubation with 0.6% H_2O_2 in methanol to block the endogenous H_2O_2, a good rinse with running tap water and PBS for 2-8 minutes was performed. The second blocking step required 5 minutes incubation in non-specific protein mixture solution (Shandon-Lipshaw). The incubation time with the primary MoABs, as with any primary antibody, depended on the developers instructions, but it was usually between 60 to 120 minutes at room temperature. Next, we applied the secondary antibody for 20 minutes (on paraffin-wax sections, the whole anti-rabbit or anti-mouse IgG antibody was used as secondary antibody directed against the IgG of the original species of the primary antibody), followed by incubation for 30-40 minutes in 8 to 10 drops of the streptavidin/biotinylated horseradish peroxidase H complex (ABC). The binding of the biotinylated antibody to streptavidin/peroxidase complexes occurs with an extremely high affinity (10^{-19}M) (BioWhitaker, Inc., Walkersville, MD, USA). Color visualization of the primary Ag-Ab reaction was accomplished using diamino-benzidine (DAB) or amino-ethyl-carbazol (AEC). The brown color was enhanced with copper sulfate and diluted 1:100 Gill's hematoxylin (Richard-Allan) was used for gentle nuclear counterstaining. After the employment of DAB solution and the appropriate counterstain, the tissue slides were dehydrated, cleared, and mounted in a manner identical to that described above.

1.13 Evaluation of the Immunoreactivity

Qualitative and quantitative evaluation of the percent of antigen positive cells and the intensity of immunostaining were conducted using a light microscope (Olympus America, Inc., Melville, NY) counting 100-300 cells from each of six to ten distinct areas in non-necrotic thymic and positive and negative control tissues. Artifacts were avoided, while, on the other hand, morphologically characteristic areas were sought out.

Quantitative evaluation (171, 175, 339): **(++++)** over 90% of the total cell number are positive; **(+++)** 50% to 90% of the total cell number are positive; **(++)** 10% to 50% of the total cell number are positive; **(+)** 1% to 10% of the total cell number are positive; **(±)** under 1% of the total cell number are positive; **(-)** negative.

Qualitative evaluation (171,175,339): **(A)** very intense red/brown staining; **(B)** strong red/brown staining; **(C)** light red/brown staining; **(D)** negative staining.

2. TRANSMISSION ELECTRON MICROSCOPY (TEM) OF CULTURED THYMIC MEDULLARY CELLS (RE, DC, LC, & IDC) AND MACROPHAGES

The cultured DCs, RE cells and macrophages following a short time *in vitro* stay attached to the culture dish surface, forming a monolayer. They were removed using gentle digestion with collagenase IV (Millipore Corp., Bedford, MA) for 5 minutes at 37 °C, followed by pelleting at 600-800 rpm for an additional 5 minutes. The cell pellets were fixed in 3% glutaraldehyde (Fischer Scientific), diluted in 0.1 M Sorenson's buffer at pH 7.2 for one hour at room temperature (20-23 °C), followed by three washes in phosphate buffer. Post-fixation was performed with 1% solution of osmium tetroxide for 30 minutes, followed by washing in 0.9% NaCl for another 30 minutes. Staining *"en bloc"* with 0.5% uranyl magnesium acetate, diluted in 0.9% NaCl for one hour in the dark, at 4 °C, was followed by washing in saline for an additional 30 minutes. After dehydration in ascending concentrations of ethanol, the tissue microcultures were embedded in Epox (E.F. Fullam, Schenectady, NY). Ultrathin sections were stained secondarily with uranyl acetate and lead citrate and were examined and photographed under the Siemens Elmiskop IA and Philips EM 300 transmission electronmicroscopes at 80 kV. The cell types were identified by their TEM nuclear and cytoplasmic characteristics.

2.1 Scanning Electron Microscopic (SEM) Procedure for Tissue Samples

Small portions (1-2 mm^2) of thymic tissue were fixed for 2-4 hours in 2.5-4% glutaraldehyde diluted in 0.1M phosphate buffer at room temperature, and at 4°C in a regular refrigerator. Post-fixation was carried out employing 1% OsO_4 diluted in phosphate buffer (kept in dark). After three washes in phosphate buffer, dehydration was carried out in ascending dilutions of ethanol. After processing in the critical point dryer (CPD), sputter coating was performed. Scope and photography of the thymic tissues was carried out with JEOL-JSM-35C Scanning Electron microscope (SEM) at 380 to 60,000 times magnifications.

2.2 Immuno-Electron Microscopy

This experimental technique was employed on various postnatal thymic cell suspensions. After 3 to 4 rinses in Hanks' medium, the cells were fixed in 4% glutaraldehyde in 0.07 M cacodylate buffer (pH 7.35) at room temperature. Thereafter we used 30-60 minutes incubation with the primary antibody. The primary antigen-antibody reaction was revealed by the immunoperoxidase technique using the ABC method and 3' diaminobenzidine (DAB) as the chromogen. After complementary fixation in glutaraldehyde, post-fixation was accomplished in 2% osmium tetroxide. Thereafter the pellet was dehydrated in ascending concentrations of ethanol and propylene oxide before being embedded in Epon. Controls were set up by omitting the primary antibody and using a mixture of proteins to detect non-specific protein staining.

Index

A

B

C

D

E

F

G

H

I

L

M

N

O

P

R

S

T

MIX
Papier aus verantwortungsvollen Quellen
Paper from responsible sources
FSC® C105338

If you have any concerns about our products,
you can contact us on
ProductSafety@springernature.com

In case Publisher is established outside the EU,
the EU authorized representative is:
Springer Nature Customer Service Center GmbH
Europaplatz 3, 69115 Heidelberg, Germany

Printed by Libri Plureos GmbH
in Hamburg, Germany